# Advances in Experimental Medicine and Biology
Volume 1413

*Advances in Experimental Medicine and Biology* provides a platform for scientific contributions in the main disciplines of the biomedicine and the life sciences. This series publishes thematic volumes on contemporary research in the areas of microbiology, immunology, neurosciences, biochemistry, biomedical engineering, genetics, physiology, and cancer research. Covering emerging topics and techniques in basic and clinical science, it brings together clinicians and researchers from various fields.

*Advances in Experimental Medicine and Biology* has been publishing exceptional works in the field for over 40 years, and is indexed in SCOPUS, Medline (PubMed), EMBASE, BIOSIS, Reaxys, EMBiology, the Chemical Abstracts Service (CAS), and Pathway Studio.

2021 Impact Factor: 3.650 (no longer indexed in SCIE as of 2022)

Chelsea M. Magin
Editor

# Engineering Translational Models of Lung Homeostasis and Disease

Springer

*Editor*
Chelsea M. Magin
University of Colorado, Denver | Anschutz Medical Campus
Aurora, CO, USA

ISSN 0065-2598 ISSN 2214-8019 (electronic)
Advances in Experimental Medicine and Biology
ISBN 978-3-031-26624-9 ISBN 978-3-031-26625-6 (eBook)
https://doi.org/10.1007/978-3-031-26625-6



This Springer imprint is published by the registered company Springer Nature Switzerland AG
The registered company address is: Gewerbestrasse 11, 6330 Cham, Switzerland

# Foreword

Lung is a vital organ that works in concert with the heart and vasculature to provide gas exchange, by delivering oxygen to all cells in the body and removing carbon dioxide released by the cells. The respiratory system has 24 generations of branches, from large airways all the way to the alveolar sacs, generating an astonishing number of 2^24 airway units. This hierarchical network is interfaced with the equally hierarchical vasculature, to form 140 $m^2$ of gas exchange surface (just about 70% of the size of a tennis court). Additional complexities arise because over 50 types of cells comprise various compartments of the lung (epithelial, vascular, mesenchymal, immune) which collectively support its function.

A healthy lung can effectively maintain itself, but has limited ability to regenerate following injury or disease. Lung injury, whether acute or chronic, can lead to end-stage lung disease, the third leading cause of death worldwide, accounting for 400,000 deaths per year in the United States alone. The burden of lung disease has further increased during the COVID-19 pandemic, and therapies for rebuilding damaged lung tissue are urgently needed.

The last decade brought major advances in our understanding of lung development, modeling, and treating lung injury and disease. While the key morphogenic steps such as cell lineage specification, branching morphogenesis, and formation of alveoli were already identified, we are now learning that these steps are less discrete than previously thought and are overlapping in space and time during the lung development. Just recently, through the use of single cell-RNA sequencing, Krasnow and colleagues have provided us with the most complete cellular and genetic map of the lung.

In parallel, the field has become increasingly aware of the limitations of cell culture and animal models in recapitulating human lung development and disease pathology. The murine lung differs substantially from its human counterpart, including the 6000-fold smaller tidal volume, 8000 times smaller surface area, and about a half of the airway generations. Similarly, cell cultures are unable to emulate the compositional, architectural, and functional features of the lung. The critical need for human tissue models of lung function resulted in major strides in lung bioengineering over the last 10 years, which are enabling lung research and its translation.

The emergence of organs-on-a-chip, micro-sized human tissues designed to recapitulate organ-level functions, was propelled by the development of lung-on-a-chip models of the epithelial-endothelial interface that also incorporated key mechanical forces. This micro-scale model has been used to recapitulate a number of lung diseases and to evaluate therapeutic regimens. On the macro-scale, we have witnessed the emergence of whole-lung models allowing multi-day ex vivo studies of lung injury and regeneration (to provide more lungs for transplant), and disease (to model lung diseases and evaluate therapeutic modalities).

This interesting and timely book covers the engineering models of human lung, in the context of both homeostasis and disease. To this end, the book discusses modeling of lung development, large airway, and the mesenchymal, parenchymal, and vascular compartments. The approach is highly interdisciplinary, reflecting the need for biologically inspired engineering models that combine biological relevance with the control of environmental and regulatory factors. The chapters discuss the experimental and computational methodologies across multiple scales – from lung-on-a-chip to the whole human lung. The book will be helpful to a lung biologist, bioengineer, and clinician, in line with the highly interdisciplinary nature of the current research in lung bioengineering. We live in an era of collaboration and can overcome the challenges involved in translating scientific insights into benefits to patients by working together and training the next generation of investigators and practitioners.

Columbia University
New York, NY, USA
December 2022

Gordana Vunjak-Novakovic

# Preface

In the fall of 2016, as the Director of Product Development at Sharklet Technologies, Inc, I never imagined that I would soon move to the adjacent building to become the Principal Investigator of the Bio-inspired Pulmonary Engineering Laboratory at the University of Colorado, Denver | Anschutz Medical Campus. Less than a year later, however, I started my first faculty position and met a small but incredible group of scientists, physicians, and engineers working together to understand lung disease and repair at the Stem Cells, Cell Therapies, and Bioengineering in Lung Biology and Disease Conference in Vermont. Coming from a background where we used sophisticated biomaterial to address challenges ranging from algae accumulation on ocean-going ships to improving healing times for full-thickness skin wounds, I was surprised that so few engineers were working in the pulmonary healthcare space. I was also shocked to learn that chronic respiratory diseases are the third leading cause of death globally even before the COVID-19 pandemic started.

The idea for this book originated over tapas with Dr. Darcy Wagner during the Society for Biomaterials Annual Meeting. Our goal was to write a book that would introduce engineers to lung biology and biologists to engineering models for studying lung disease. I would like to extend my immense gratitude to Darcy and all the contributing authors in this text for creating a book that not only describes translational models of lung homeostasis and disease but also overviews the relevant biology required to understand each topic: (1) Developing Lung, (2) Large Airways, (3) Mesenchyme and Parenchyma, (4) Vasculature, and (5) Interfaces Between Medical Devices and the Lung. I must also thank my friends and family for their support during the process of putting this text together, especially my husband, David Wilson, for proof-reading everything I write and my son, Brennan Wilson, for making me smile every day. I envision that collaborations among engineers, scientists, and physicians will enable us to treat and cure the most devastating lung diseases, and I hope this text inspires you to seek out these connections and take on a new challenge!

Aurora, CO, USA

Chelsea M. Magin

# Contents

# Chapter 1
# An Introduction to Engineering and Modeling the Lung

**Alicia E. Tanneberger, Daniel J. Weiss, and Chelsea M. Magin**

## 1.1 Introduction

Over the last decade, the field of lung biology has evolved considerably due to many advancements, including the advent of single-cell RNA (scRNA) sequencing, induced pluripotent stem cell (iPSC) reprogramming, and 3D cell and tissue culture. Despite rigorous research and tireless efforts, chronic pulmonary diseases remain the third leading cause of death globally, with transplantation being the only option for treating end-stage disease. This chapter will introduce the broader impacts of understanding lung biology in health and disease, provide an overview of lung physiology and pathophysiology, and summarize the key takeaways from each chapter describing engineering translational models of lung homeostasis and disease. This book is divided into broad topic areas containing chapters covering basic biology, engineering approaches, and clinical perspectives related to (1) the

---

A. E. Tanneberger
Department of Bioengineering, University of Colorado, Denver | Anschutz Medical Campus, Aurora, CO, USA

D. J. Weiss
Department of Medicine, University of Vermont, Larner College of Medicine, Burlington, VT, USA

C. M. Magin (✉)
Department of Bioengineering, University of Colorado, Denver | Anschutz Medical Campus, Aurora, CO, USA

Division of Pulmonary Sciences and Critical Care Medicine, Department of Medicine, University of Colorado, Anschutz Medical Campus, Aurora, CO, USA

Department of Pediatrics, University of Colorado, Anschutz Medical Campus, Aurora, CO, USA
e-mail: Chelsea.Magin@cuanschutz.edu



C. M. Magin (ed.), *Engineering Translational Models of Lung Homeostasis and Disease*, Advances in Experimental Medicine and Biology 1413,
https://doi.org/10.1007/978-3-031-26625-6_1

developing lung, (2) the large airways, (3) the mesenchyme and parenchyma, (4) the pulmonary vasculature, and (5) the interface between lungs and medical devices. Each section highlights the underlying premise that engineering strategies, when applied in collaboration with cell biologists and pulmonary physicians, will address critical challenges in pulmonary health care.

## 1.2 Broader Impacts of Understanding Lung Biology in Health and Disease

Chronic pulmonary diseases were ranked as the third leading cause of death globally in 2019 [1, 2]. Chronic obstructive pulmonary disease (COPD) and asthma, two of the leading chronic respiratory diseases, together affect nearly 400 million people worldwide [1–3]. The emergence of the infectious coronavirus disease (COVID-19) caused by the SARS-CoV-2 virus quickly and negatively impacted global respiratory health. There have been over 550 million documented cases worldwide, an astonishing statistic expected to continue rising [4]. Emerging studies suggest that chronic pulmonary fibrosis may follow acute respiratory distress syndrome (ARDS) caused by severe SARS-CoV-2 infection [5, 6]. Pulmonary fibrosis, an interstitial lung disease which causes scarring of gas exchanges surfaces, makes up a worrisome and growing proportion of patients experiencing respiratory distress. More specifically, idiopathic pulmonary fibrosis (IPF) is an uncurable and progressive disease that has a current median survival rate of only 2–3 years after diagnosis [7]. As the long-term effects of COVID-19 are still largely unknown, the burden of acute and chronic pulmonary diseases is expected to overwhelm clinical settings for years to come [4].

Overall, the number of diseases that still have limited or no therapeutic options is surprising and troublesome [8, 9]. Approval for new drugs is expensive, time intensive, and must undergo many rigorous drug development stages. As such, the only cure for many end-stage chronic pulmonary diseases remains lung transplantation. In the United States, the number of individuals requiring an organ transplant far outweighs the number of organs available [10]. Currently, there are over 100,000 patients listed on the national transplant list, yet only about 40% of individuals on the list will receive a transplant [11]. As such, there remains a critical need for both better in vitro models of lung tissue that advance the drug development timeline and technologies that improve quality of life until transplant.

Mechanistic in vitro modeling makes up a large percentage of current pulmonary research efforts [12]. While each approach has its own strengths and weaknesses, adopting multidimensional and systematic engineering methods will enhance studies of complex pulmonary diseases. The emergence of engineering approaches to model lung homeostasis and disease has created opportunities for cutting-edge research and interdisciplinary collaboration. Therefore, we envision that engineering strategies will revolutionize current approaches to modeling both healthy and diseased lung tissues, accelerate the development and validation of personalized therapeutics that are so desperately needed, and provide a bridge-to-transplant for patients able to receive this treatment. The next section will briefly introduce differences between healthy and diseased lung anatomy and physiology to provide background information on how deviations due to injuries or disease affect the lung cells and tissues.

## 1.3 Lung Physiology in Homeostasis and Disease

The chapters of this book contain information on five broad topic areas: (1) the developing lung, (2) the large airways, (3) the mesenchyme and parenchyma, (4) the pulmonary vasculature, and (5) the interface between lungs and medical devices. As such, this introductory section will briefly introduce these topics. First, the phenomenon of human lung organogenesis defines how hollow tube-like structures divide and branch repetitively during early fetal lung development and then evolve into mature lung structures with age. Mechanical forces are extremely important during this phase as they provide physical cues that give the tissue its structure [13, 14]. Furthermore, both the coordinated crosstalk between mesenchymal and epithelial cells, and the understanding of how progenitor cells differentiate and give rise to specialized lung cells, are vital for lung organogenesis [13, 15]. Upon completion of this initial developmental phase, the structures of the lower internal respiratory tract become much more apparent and tree-like. The lower respiratory tract contains the main structures of the trachea, bronchi, bronchioles, and alveoli.

The trachea has the largest cross-sectional area and both bronchi, left and right, are classically viewed as large airway structures (Fig. 1.1, top). Within the trachea, a significant amount of cartilaginous tissue makes the airway extremely rigid when compared to the rest of the airway [16, 17]. Additionally, along the trachea, there is a high proportion of ciliated epithelial cells, goblet cells, and basal cells (Fig. 1.1, top) [18, 19].

Basal cells are capable of being differentiated into all types of luminal cells [20]. Basal cells secrete mucus, while more distal alveolar epithelial type II cells secrete surfactants which both line the airways and can be overproduced when dysfunction occurs. In the case of SARS-CoV-2 infection, the upper airway is considered a principal entry and target of SARS-CoV-2 and commonly results in inflammation, neutrophilia, and cytokine release, and epithelial barrier damage ensues (Fig. 1.1, top right) [21–23]. Then, moving toward the end of the trachea and along the bronchi, the amount of cartilage decreases. While there are still ciliated epithelial cells, the predominant other cell types found here are the club cells. The main function of club cells is to protect the epithelium by secreting protein and a component of surfactant that can detoxify harmful substances that were inhaled before they pass onto more sensitive areas [24].

Next, moving distally along the respiratory tract ultimately leads to the parenchyma and mesenchyme (Fig. 1.1, middle). The parenchyma refers to the region of the lungs involved in gas exchange. It includes the alveoli that facilitate gas exchange

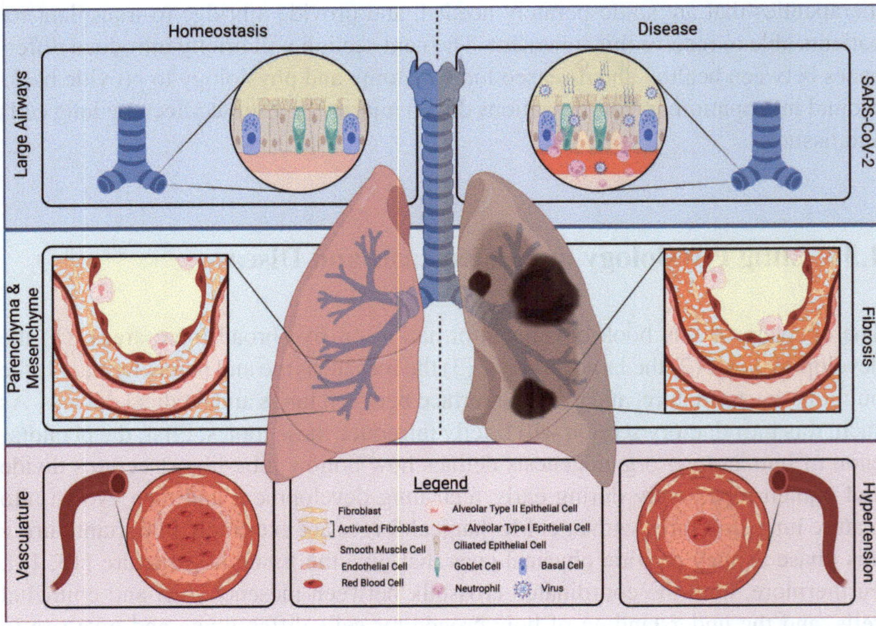


**Fig. 1.1** Schematic highlighting basic lung biology within the large airways (top), parenchyma and mesenchyme (middle), and vasculature (bottom), both at homeostasis (left side) and during disease (right side), including acute SARS-CoV-2 infection, fibrosis, and hypertension. (Created with BioRender.com)

across membranes one cell layer thick, and the extracellular matrix (ECM) [25]. Within the alveoli, the main function of the lungs takes place: gas exchange. Here, diffusion of carbon dioxide and oxygen occurs across a thin membrane to oxygenate the body. Meanwhile, the ECM provides both biochemical and biophysical cues to the cells and is composed of over 150 different types of proteins, enzymes, growth factors, and proteoglycans [26]. Additionally, the ECM provides structure to the lungs and experiences continual remodeling, often being referred to as the cells' microenvironment. Normally, when an injury or insult occurs, fibroblast cells activate and differentiate into myofibroblast cells that start the normal wound healing cascade to repair that site. When this healing process does not resolve, the process of fibrosis begins. Myofibroblast cells deposit excess ECM proteins, particularly collagen, that thicken the alveolar walls and leads to ineffective gas exchange and decreased ease of lung expansion (Fig. 1.1, middle right) [27]. Then, surrounding fibroblasts sense the increased tissue stiffness and a positive and viscous feedback loop begins, activating more fibroblasts and furthering the aberrant wound healing process of fibrosis [27].

Furthermore, the mesenchyme refers to a wide array of cells including lymphatic cells, pericytes, fibroblast and myofibroblasts, endothelial cells, mesenchymal stromal cells, and smooth muscle cells [28]. While all these cells contribute to the functions of the mesenchyme, much of the mesenchymal regulatory mechanisms involved in directing airway and alveoli development, as well as the composition of the mesenchyme and the relationship and interactions with the pulmonary epithelium, are still not well understood [29]. Thus, understanding better the relationship and interaction between the mesenchyme and the pulmonary epithelium is notably important during each step of lung development, as it serves as a central factor in the creation of functional lung tissue [29].

Lastly, the lung vasculature (Fig. 1.1, bottom) is made up of arteries, veins, and capillaries that run along lung structures and are most critical near the alveolar region. Deoxygenated blood is transported from the right side of the heart to the lungs by the pulmonary arteries. Three anatomical layers make up the pulmonary arteries: the intima (inner-most layer), the media (middle layer), and the adventitia (outer layer). Then, the capillary beds facilitate the diffusion of carbon dioxide and oxygen within the alveolar region while the pulmonary veins carry oxygenated blood back to the heart from the lungs. Pulmonary endothelial cells line the vasculature and are found at the air-liquid interface. Pulmonary hypertension (PH) is one a disease that affects the pulmonary circulatory system and often develops secondarily in IPF patients [30]. This disease is characterized by blood vessel thickening that narrows and blocks blood flow, making it harder for the heart to pump the same amount of volume of blood to the body (Fig. 1.1, bottom right) [30]. Hence, lung-tissue stiffness increase in patients suffering from pulmonary hypertension and pulmonary fibrosis. The opposite trend in tissue stiffness occurs in COPD. Evaluation of COPD patients lungs has revealed increased compliance and decreased tissue stiffness [31].

In summary, disease often begins by an initial or acute injury (e.g., inhaled irritant, lung puncture, or viral infection). When the injury does not resolve following a normal healing process, it often leads to irreversible damage. Both mechanical and compositional changes contribute to disease progression, yet for many diseases, it is still unknown which factor has a greater impact. Regardless, there are similarities among many conditions such as inflammation, changes in cellular phenotype, the extracellular matrix, and the cell-cell and cell-matrix interactions, which all have repercussions for gas exchange. Thus, it is important to consider disease-specific factors and tailor experimental designs to include these design parameters to achieve the most accurate ex vivo system.

## 1.4 Engineering Translational Models of Lung Homeostasis and Disease

Part I (Chaps. 2, 3 and 4) focuses on Engineering and Modeling the Developing Lung. In Chap. 2, Charlotte Dean and Sek-Shir Cheong start off by introducing simple non-engineered models of lung development that showcase multiple technical modalities (e.g., in vitro, ex vivo, and in silico). This chapter facilitates discussion about advantages and disadvantages for each approach, which is beneficial to readers considering their own experimental designs. Within the chapter, the authors systematically review the work that currently exists for recapitulating the five stages of lung development: embryonic, pseudoglandular, canalicular, saccular, and alveolar. The use of explanted cultures, organoids, and precision-cut lung slice models are a common thread between the different models.

In Chap. 3, Brea Chernokal, Cailin Gonyear, and Jason Gleghorn focus on employing primarily in vitro techniques to study lung development. They highlight the significant amount of work that has come from differentiating iPSCs into a variety of mature lung cells, including bronchospheres (proximal) and alveolospheres (distal). Time, three-dimensional organization, and mechanical forces all play a crucial role in model design for developing lungs and, therefore, are a major emphasis of this chapter. Additionally, this chapter demonstrates how simple models of lung development can be transitioned into more complex physiologic models by implementing microfluidic devices.

Laertis Ikonomou, Maria Yampolskaya, and Pankaj Mehta continue the discussion in Chap. 4. The authors highlight the benefit of multipotent embryonic lung progenitor cells, both human- and animal-derived, to create distinct lung models of organogenesis. Numerous cell markers and signaling pathways for a variety of different cell types and stages are presented, which provides a great starting place for researchers to explore and use as validation within their own models. Lastly, the authors mention the imperative need for computing techniques, such as RNA sequencing, and leave readers with concluding remarks about how lung progenitor-based organoids are likely to become well-established norms in both the medical and developmental biology fields.

Next, Chap. 5 by Lalit Gautam, Noa Harriott, Adrian Caceres, and Amy Ryan marks the beginning of Part II (Chaps. 5, 6 and 7): Engineering and Modeling Large Airways. Within this chapter, the authors discuss the crucial role of the airway epithelium as one of the first defense mechanisms that protects the lungs from irritants and infiltrates. Additionally, since the upper airways provide both physical and biochemical barriers, the chapter also introduces the wide variety of basic cellular and biological components present. Readers can refer to this chapter to gain a broad understanding about the structure-function relationships of goblet, club, basal, and other less prominent secretory and ciliated cells found within this region.

In Chap. 6, Rebecca Heise continues to discuss the fundamental biology and components that play pivotal roles in the large airway. The structure-function relationships between cells and their surrounding environment are introduced and provide context for why investigating the mechanical properties of the large airways has been proven very insightful. Then, multiple pathologies known to impact the large airways are briefly discussed, to highlight a need for computational and physiological models to better understand the disease mechanisms at play.

Finally, Chap. 7 written by Tehreem Khalid and Cian O'Leary wraps up the discussion of engineering large airways. This chapter begins by providing a detailed explanation of the clinical need for trachea replacements and discusses the mechanical forces involved in respiration. Then, the complex design requirements are individually considered, such as withstanding different pressure gradients, developing a non-uniform material that captures the anisotropic features of the native tissue, and ability to rapidly deform and expand. The mechanical properties of tracheal cartilage, muscles, and whole tissue are tabulated in a concise manner. Additionally, the methods employed to measure the mechanical properties are thoroughly presented and serve as a great introduction into how to implement these techniques.

Part III (Chaps. 8, 9, 10 and 11) – Engineering and Modeling the Mesenchyme and Parenchyma – starts with Jarod Zepp and colleagues presenting an overview of engineering and modeling the mesenchyme in Chap. 8. This content describes the structure and components that make up the mesenchyme, focusing on the extracellular matrix and mesenchymal cell types. The topics of fibroblast discovery, heterogeneity, and organization are progressively discussed. Then, the authors transition into describing complex fibroblast behaviors, in disease progression and how to model mesenchymal-epithelial interactions. Lastly, this chapter ends with a promising approach for how to target fibroblasts with nanoparticles as a strategy for intervening in disease progression.

Next, Chelsea Magin et al. build on this discussion in Chap. 9 by highlighting some of their exciting work developing dynamic 3D lung models. Within this chapter, they present their unique perspectives on the advantages, limitations, and potential future directions of the use of both synthetic and naturally derived hydrogel materials within precision-cut lung slice, organoid, and lung-on-a-chip models. Pulmonary fibrosis is introduced and interwoven into the discussion to showcase differences between healthy and diseased lung models. Furthermore, a detailed explanation of click chemistry is showcased for thiol-ene and thiol-yne reactions. Multiple techniques of using magnetically labeling microspheres, incorporating photodegradable microspheres, and functionalizing microspheres to ultimately engineer and mimic distal lung geometry are all interesting concepts to explore when reading this chapter [31].

Similarly, lung-on-a-chip models for the lung parenchyma are discussed in-depth within Chap. 10, written by Olivier Guenat and colleagues. Significant improvements have been made in the functionality and complexity of this technique since its inception in 2010. The authors highlight the importance of including aspects such as flow, an air-liquid interface, and multiple cell types, particularly to develop the most physiologically relevant alveolar-focused lung-on-a-chip model. Readers not as familiar with lung-on-a-chip systems should utilize this thorough overview.

Chapter 11 by Claudia Staab-Weijnitz, Ceylan Onursal, Deepika Nambiar, and Roberto Vanacore rounds out the discussion for this section. This chapter details the significance of assessing collagen formation, remodeling, and types within translational lung models. If readers are interested in learning immune-or mass-spectrometry-based collagen quantification and visualization techniques, this chapter serves as a great starting point that discusses state-of-the-art research. Advantages and disadvantages are compared for multiple techniques, and implications of collagen changes are reviewed within a disease context. This chapter also includes tabulated information on collagen types and which protease is associated with those collagen forms.

Chapter 12 by Wai Hoe Ng, Barbie Varghese, and Xi (Charlie) Ren begins Part IV: Engineering and Modeling the Pulmonary Vasculature. Some basic biology and important features of blood circulation are discussed to lay the foundation for how complex it truly is to recreate blood vessels ex vivo. Research studies using a variety of engineering approaches toward vascularization of models are thoroughly reviewed, including co-differentiated organoids, 3D printing, microfluidic devices, and recellularization techniques. This chapter will help readers better appreciate the role of endothelial cells, cellular cross talk, and complex vascularization, while highlighting significant new ways to engineer vasculature.

Likewise, within Chap. 13, Trieu Nguyen and Fakhrul Ahsan elaborate on lung-on-a-chip models, employing microfluidic devices that lend themselves to recapitulate the pulmonary vasculature. This method serves as an alternative to animal models and has tremendous power for drug screening. As such, new drugs can be potentially identified before clinical trials begin to treat pulmonary vasculature diseases, such as pulmonary arterial hypertension. The materials commonly chosen for microfluidic devices and the addition of patterning to create microvascular networks particularly stand out within this chapter.

Similarly, many of the same approaches continue to be discussed in Chap. 14 by Yifan Yuan. However, within this chapter, more specific clinical details are included that enable readers to incorporate them as design parameters, such as typical arterial pressure values, volumes, and symptoms. Sections on extracellular substrates, cell-cell cross talk, and mechanical shear and stretch highlight a variety of work across many different research groups. Yuan ends the chapter by suggesting that implementing multi-omics will drastically improve our understanding of transcriptomics and proteomics, and by transitioning away from using animal-derived materials and instead using human derived material, researchers will gain further insight into critical human mechanisms.

Finally, Chaps. 15 and 16 make up the remaining section of this book, Part V: Engineering and Modeling the Interface Between Devices and the Lung. These two chapters contain the most clinically dense and relevant information in terms of extracorporeal membrane oxygenation (ECMO) as a therapy. Chapter 15 is written by Anna Niroomand, Franziska Olm, and Sandra Lindstedt, while Chap. 16 is written by Deniz Bölükbas and Sinem Tas. Both chapters introduce the technique of ECMO and how the system interfaces with the body when in use, but Chap. 14 provides a more macroscopic view about the concerns of thrombosis and bleeding. Meanwhile, Chap. 15 delves into each component that makes up the machine, provides justification for each part, and addresses how the body responds to the materials it is exposed to. This therapy is very invasive and significant efforts are scaling the machines down to improve the portability and success in bridging patients until they receive transplants.

## 1.5 Conclusion

In the past, researchers often focused on one aspect of the lungs (e.g., a specific cell type, a cell signaling pathway, or a specific biomarker) and worked to develop a successful model that provided insight with regard to that feature. However, with the improvements in current technology and techniques available, standards are much higher today. Multidisciplinary approaches are becoming more widely recognized as crucial to the production of impactful work, so it is common to find biologists collaborating with biomaterial scientists and engineers and/or clinicians. Additionally, authors are asked to verify their results using several different modalities to comprehensively support their claims with rigorous and reproducible evidence. As such, individuals must rise and adapt to the quickening pace of these advancements, incorporating new technologies as well as multidisciplinary approaches into their own work, or they risk falling behind in the quest for translational models that can ultimately identify new ways to combat lung diseases.

Reducing our reliance on materials derived from animals, such as Matrigel and transitioning to engineered biomaterials will create well-defined cell-culture platforms that can investigate factors systematically. Also, there has historically been a lack of focus on sex-specific [33–35] and patient-specific [36, 37] considerations. It would be very practical to develop iPSC cell lines from both male and female patients to study the role sex has within the context of disease. Individuals should work to incorporate these two considerations into 3D lung models coupled with the use of 3D approaches, including lung-on-a-chip, to study flow and the air-liquid interface, bioprinted scaffolds capable of recreating lung geometry, and stretchable membranes that can deform and simulate ventilation patterns. Furthermore, support should continue for exploring alternative methods, such as refurbishing lungs previously deemed as unsuitable for organ transplants, to increase the number of patients that benefit from this treatment. Gordana Vunjak-Novakovic and colleagues performed interventional cross-circulation using prolonged extracorporeal support to improve lung function of damaged but living pig lungs to a level that met transplantation criteria [38]. Efforts like these have tremendous potential to improve the outcomes of patients with end-stage lung diseases who would have otherwise not have received lungs.

Leveraging partnerships between academia and industry will also be key to successfully translating the results obtained in engineered models of lung homeostasis and disease into clinical use. Recently, Volumetric, a 3D printing company started by Jordan Miller and colleagues from Rice University, was acquired by 3D Systems and began a partnership with United Therapeutics toward building new lungs [39–41]. This example highlights a true paradigm shift in the translation of engineered lung models and a new approach to solving the shortage of transplantable organs. Figure 1.2 showcases how truly novel advances in personalized medicine, 3D technologies, and biomaterial and co-culture systems will revolutionize pulmonary research and treatment of respiratory diseases in coming years. Great work

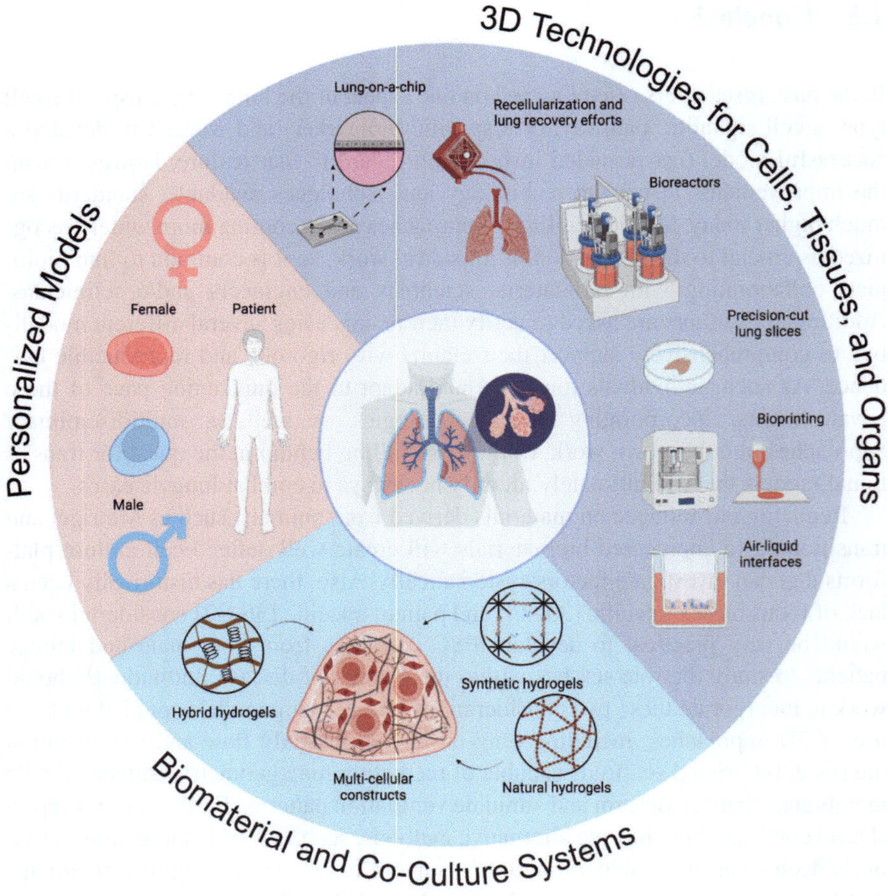


**Fig. 1.2** Schematic showcasing promising techniques and areas of focus for improved lung models. Personalized models, biomaterial and co-culture systems, and 3D technologies are specific areas to explore for future advances. (Created with BioRender.com)

progresses in engineering translational models of lung homeostasis and disease. Readers are encouraged to read every chapter of this book to further their knowledge of the basic science and exciting new technologies emerging in this area so that they may start a new collaboration to improve patient health and treatment options for the millions of people worldwide impacted by respiratory diseases.

## References

1. Forum of International Respiratory Societies. The Global Impact of Respiratory Disease – Second Edition. Sheffield, European Respiratory Society, 2017.
2. National Center for Health Statistics. Underlying Cause of Death 1999–2019 on CDC WONDER Online Database, released in 2020.

3. Deterding, R. R., DeBoer, E. M., Cidon, M. J., Robinson, T. E., Warburton, D., Deutsch, G. H., & Young, L. R. (2019). Approaching Clinical Trials in Childhood Interstitial Lung Disease and Pediatric Pulmonary Fibrosis. *American Journal of Respiratory and Critical Care Medicine*, *200*(10), 1219–1227. https://doi.org/10.1164/rccm.201903-0544CI
4. George, P. M., Wells, A. U., & Jenkins, R. G. (2020). Pulmonary fibrosis and COVID-19: The potential role for antifibrotic therapy. *The Lancet. Respiratory Medicine*, *8*(8), 807–815. https://doi.org/10.1016/S2213-2600(20)30225-3
5. Michalski, J. E., Kurche, J. S., & Schwartz, D. A. (2022). From ARDS to pulmonary fibrosis: The next phase of the COVID-19 pandemic? *Translational Research*, *241*, 13–24. https://doi.org/10.1016/j.trsl.2021.09.001
6. Gibson, P. G., Qin, L., & Puah, S. H. (2020). COVID-19 acute respiratory distress syndrome (ARDS): Clinical features and differences from typical pre-COVID-19 ARDS. *The Medical Journal of Australia*, https://doi.org/10.5694/mja2.50674. https://doi.org/10.5694/mja2.50674
7. Ley, B., Collard, H. R., & King, T. E. (2011). Clinical Course and Prediction of Survival in Idiopathic Pulmonary Fibrosis. *American Journal of Respiratory and Critical Care Medicine*, *183*(4), 431–440. https://doi.org/10.1164/rccm.201006-0894CI
8. King, T. E., Bradford, W. Z., Castro-Bernardini, S., Fagan, E. A., Glaspole, I., Glassberg, M. K., Gorina, E., Hopkins, P. M., Kardatzke, D., Lancaster, L., Lederer, D. J., Nathan, S. D., Pereira, C. A., Sahn, S. A., Sussman, R., Swigris, J. J., Noble, P. W., & ASCEND Study Group. (2014). A phase 3 trial of pirfenidone in patients with idiopathic pulmonary fibrosis. *The New England Journal of Medicine*, *370*(22), 2083–2092. https://doi.org/10.1056/NEJMoa1402582
9. Richeldi, L., du Bois, R. M., Raghu, G., Azuma, A., Brown, K. K., Costabel, U., Cottin, V., Flaherty, K. R., Hansell, D. M., Inoue, Y., Kim, D. S., Kolb, M., Nicholson, A. G., Noble, P. W., Selman, M., Taniguchi, H., Brun, M., Le Maulf, F., Girard, M., … Collard, H. R. (2014). Efficacy and Safety of Nintedanib in Idiopathic Pulmonary Fibrosis. *New England Journal of Medicine*, *370*(22), 2071–2082. https://doi.org/10.1056/NEJMoa1402584
10. Saidi, R. F., & Hejazii Kenari, S. K. (2014). Challenges of organ shortage for transplantation: Solutions and opportunities. *International Journal of Organ Transplantation Medicine*, *5*(3), 87–96.
11. Health Resources & Services Administration. (2022). *Organ Donation Statistics | organdonor.gov*. https://www.organdonor.gov/learn/organ-donation-statistics
12. Miller, A. J., & Spence, J. R. (2017). In Vitro Models to Study Human Lung Development, Disease and Homeostasis. *Physiology*, *32*(3), 246–260. https://doi.org/10.1152/physiol.00041.2016
13. Hogan, B. L. M. (2018). Integrating Mechanical Force into Lung Development. *Developmental Cell*, *44*(3), 273–275. https://doi.org/10.1016/j.devcel.2018.01.015
14. Li, J., Wang, Z., Chu, Q., Jiang, K., Li, J., & Tang, N. (2018). The Strength of Mechanical Forces Determines the Differentiation of Alveolar Epithelial Cells. *Developmental Cell*, *44*(3), 297-312.e5. https://doi.org/10.1016/j.devcel.2018.01.008
15. Warburton, D., El-Hashash, A., Carraro, G., Tiozzo, C., Sala, F., Rogers, O., De Langhe, S., Kemp, P. J., Riccardi, D., Torday, J., Bellusci, S., Shi, W., Lubkin, S. R., & Jesudason, E. (2010). Lung organogenesis. *Current Topics in Developmental Biology*, *90*, 73–158. https://doi.org/10.1016/S0070-2153(10)90003-3.
16. Sicard, D., Haak, A. J., Choi, K. M., Craig, A. R., Fredenburgh, L. E., & Tschumperlin, D. J. (2018). Aging and anatomical variations in lung tissue stiffness. *American Journal of Physiology - Lung Cellular and Molecular Physiology*, *314*(6), L946–L955. https://doi.org/10.1152/ajplung.00415.2017
17. Safshekan, F., Tafazzoli-Shadpour, M., Abdouss, M., Behgam Shadmehr, M., & Ghorbani, F. (2017). Investigation of the Mechanical Properties of the Human Tracheal Cartilage. *Tanaffos*, *16*(2), 107–114.
18. Kia'i, N., & Bajaj, T. (2022). Histology, Respiratory Epithelium. In *StatPearls*. StatPearls Publishing. http://www.ncbi.nlm.nih.gov/books/NBK541061/
19. Gartner, L. (2021). Respiratory System. *Textbook of Histology*, 15, 355-378.e2

20. Busch, S. M., Lorenzana, Z., & Ryan, A. L. (2021). Implications for Extracellular Matrix Interactions with Human Lung Basal Stem Cells in Lung Development, Disease, and Airway Modeling. *Frontiers in Pharmacology*, *12*, 645858. https://doi.org/10.3389/fphar.2021.645858
21. Yuen, E., Gudis, D. A., Rowan, N. R., Nguyen, S. A., & Schlosser, R. J. (2021). Viral Infections of the Upper Airway in the Setting of COVID-19: A Primer for Rhinologists. *American Journal of Rhinology & Allergy*, *35*(1), 122–131. https://doi.org/10.1177/1945892420947929
22. Calvert, B. A., Quiroz, E. J., Lorenzana, Z., Doan, N., Kim, S., Senger, C. N., Wallace, W. D., Salomon, M. P., Henley, J., & Ryan, A. L. (2022). Neutrophilic inflammation promotes SARS-CoV-2 infectivity and augments the inflammatory responses in airway epithelial cells. *BioRxiv: The Preprint Server for Biology*, 2021.08.09.455472. https://doi.org/10.1101/2021.08.09.455472
23. Wang, R., Hume, A. J., Beermann, M. L., Simone-Roach, C., Lindstrom-Vautrin, J., Le Suer, J., Huang, J., Olejnik, J., Villacorta-Martin, C., Bullitt, E., Hinds, A., Ghaedi, M., Rollins, S., Werder, R. B., Abo, K. M., Wilson, A. A., Mühlberger, E., Kotton, D. N., & Hawkins, F. J. (2022). Human airway lineages derived from pluripotent stem cells reveal the epithelial responses to SARS-CoV-2 infection. *American Journal of Physiology - Lung Cellular and Molecular Physiology*, *322*(3), L462–L478. https://doi.org/10.1152/ajplung.00397.2021
24. Widdicombe, J. G., & Pack, R. J. (1982). The Clara Cell. *European Journal of Respiratory Diseases*, *63*(3), 202–220.
25. Suki, B., Stamenovic, D., & Hubmayr, R. (2011). Lung Parenchymal Mechanics. *Comprehensive Physiology*, *1*(3), 1317–1351. https://doi.org/10.1002/cphy.c100033
26. Burgstaller, G., Oehrle, B., Gerckens, M., White, E. S., Schiller, H. B., & Eickelberg, O. (2017). The instructive extracellular matrix of the lung: Basic composition and alterations in chronic lung disease. *European Respiratory Journal*, *50*(1), 1601805. https://doi.org/10.1183/13993003.01805-2016
27. Richeldi, L., Collard, H. R., & Jones, M. G. (2017). Idiopathic pulmonary fibrosis. *The Lancet*, *389*(10082), 1941–1952. https://doi.org/10.1016/S0140-6736(17)30866-8
28. Nasri, A., Foisset, F., Ahmed, E., Lahmar, Z., Vachier, I., Jorgensen, C., Assou, S., Bourdin, A., & De Vos, J. (2021). Roles of Mesenchymal Cells in the Lung: From Lung Development to Chronic Obstructive Pulmonary Disease. *Cells*, *10*(12), 3467. https://doi.org/10.3390/cells10123467
29. McCulley, D., Wienhold, M., & Sun, X. (2015). The Pulmonary Mesenchyme Directs Lung Development. *Current Opinion in Genetics & Development*, *32*, 98–105. https://doi.org/10.1016/j.gde.2015.01.011
30. Ruffenach, G., Hong, J., Vaillancourt, M., Medzikovic, L., & Eghbali, M. (2020). Pulmonary hypertension secondary to pulmonary fibrosis: Clinical data, histopathology, and molecular insights. *Respiratory Research*, *21*(1), 303. https://doi.org/10.1186/s12931-020-01570-2
31. Burgess, J. K., Mauad, T., Tjin, G., Karlsson, J. C. (2016). Westergren-Thorsson, G. The extracellular matrix – the under-recognized element in lung disease? *The Journal of Pathology*, *240*(4), 397–409. https://doi.org/10.1002/path.4808
32. Caracena, T., Blomberg, R., Hewawasam, R. S., Fry, Z. E., Riches, D. W. H., & Magin, C. M. (2022). Alveolar epithelial cells and microenvironmental stiffness synergistically drive fibroblast activation in three-dimensional hydrogel lung models. *Biomaterials Science*. https://doi.org/10.1039/D2BM00827K
33. Studies, I. of M. (US) C. on E. and L. I. R. to the I. of W. in C., Mastroianni, A. C., Faden, R., & Federman, D. (1994). NIH Revitalization Act of 1993 Public Law 103-43. In *Women and Health Research: Ethical and Legal Issues of Including Women in Clinical Studies: Volume I*. National Academies Press (US). http://www.ncbi.nlm.nih.gov/books/NBK236531/
34. Shah, K., McCormack, C. E., & Bradbury, N. A. (2014). Do you know the sex of your cells? *American Journal of Physiology – Cell Physiology*, *306*(1), C3–C18. https://doi.org/10.1152/ajpcell.00281.2013

35. James, B. D., & Allen, J. B. (2021). Sex-Specific Response to Combinations of Shear Stress and Substrate Stiffness by Endothelial Cells In Vitro. *Advanced Healthcare Materials*, *10*(18), 2100735. https://doi.org/10.1002/adhm.202100735
36. Alysandratos, K.-D., Russo, S. J., Petcherski, A., Taddeo, E. P., Acín-Pérez, R., Villacorta-Martin, C., Jean, J. C., Mulugeta, S., Rodriguez, L. R., Blum, B. C., Hekman, R. M., Hix, O. T., Minakin, K., Vedaie, M., Kook, S., Tilston-Lunel, A. M., Varelas, X., Wambach, J. A., Cole, F. S., … Kotton, D. N. (2021). Patient-specific iPSCs carrying an SFTPC mutation reveal the intrinsic alveolar epithelial dysfunction at the inception of interstitial lung disease. *Cell Reports*, *36*(9), 109636. https://doi.org/10.1016/j.celrep.2021.109636
37. Aguado, B. A., Grim, J. C., Rosales, A. M., Watson-Capps, J. J., & Anseth, K. S. (2018). Engineering precision biomaterials for personalized medicine. *Science Translational Medicine*, *10*(424), eaam8645. https://doi.org/10.1126/scitranslmed.aam8645
38. Guenthart, B. A., O'Neill, J. D., Kim, J., Queen, D., Chicotka, S., Fung, K., Simpson, M., Donocoff, R., Salna, M., Marboe, C. C., Cunningham, K., Halligan, S. P., Wobma, H. M., Hozain, A. E., Romanov, A., Vunjak-Novakovic, G., & Bacchetta, M. (2019). Regeneration of severely damaged lungs using an interventional cross-circulation platform. *Nature Communications*, *10*, 1985. https://doi.org/10.1038/s41467-019-09908-1
39. *3D Systems Announces Acquisition of Volumetric Biotechnologies*. (2021, October 27). 3D Systems. https://www.3dsystems.com/press-releases/3d-systems-announces-acquisition-volumetric-biotechnologies
40. Listek, V. (2021, November 1). *3D Systems Pursuing Breakthrough Advances in Bioprinting, Acquires Volumetric*. 3DPrint.Com | The Voice of 3D Printing/Additive Manufacturing. https://3dprint.com/286334/3d-systems-pursuing-breakthrough-advances-in-bioprinting-buys-jordan-millers-volumetric/
41. *Volumetric to Be Acquired by 3D Systems to Advance Tissue and Organ Manufacturing*. (2021, October 27). Business Wire. https://www.businesswire.com/news/home/20211027006115/en/Volumetric-to-Be-Acquired-by-3D-Systems-to-Advance-Tissue-and-Organ-Manufacturing

# Part I
# Engineering and Modeling the Developing Lung

# Chapter 2
# Simple Models of Lung Development

**Charlotte H. Dean and Sek-Shir Cheong**

## 2.1 Introduction

### *2.1.1 Basics of Lung Development*

Our understanding of lung organogenesis has progressed in parallel with the generation of models designed to investigate the different stages of development [44]. The lungs are generated from the foregut endoderm; two initial buds give rise to the primordial buds of each future lung lobe. This is followed by secondary branching, a series of reiterative rounds of branching that generate the network of airways known as the respiratory tree. Subsequently, the gas-exchanging region of the lungs is formed from the terminal airway branches by thinning of interstitial tissue and widening of the airspaces to form clusters of epithelial sacs followed by the final stage of development, alveologenesis [44].

In all there are five stages of lung development, each of which relates to the morphological changes that are occurring as organogenesis proceeds. These five stages, embryonic, pseudoglandular, canalicular, saccular and alveolar, have been well documented [17, 28, 43]. The focus of these stages of organogenesis is on the epithelial component of the lungs, but the vasculature develops alongside the epithelium and has an equally important role to play in both organ development and function, since gas exchange takes place between the capillary endothelium and epithelium in the alveoli [21] (Fig. 2.1).

In addition to the endothelium and epithelium, the lungs are comprised of a myriad of other cell types, including sub-populations of fibroblasts, neuroendocrine cells and both resident and recruited immune cells, all of which contribute to the

C. H. Dean (✉) · S.-S. Cheong
National Heart and Lung Institute, Imperial College London, London, UK
e-mail: c.dean@imperial.ac.uk; sek-shir.cheong@imperial.ac.uk



C. M. Magin (ed.), *Engineering Translational Models of Lung Homeostasis and Disease*, Advances in Experimental Medicine and Biology 1413,
https://doi.org/10.1007/978-3-031-26625-6_2

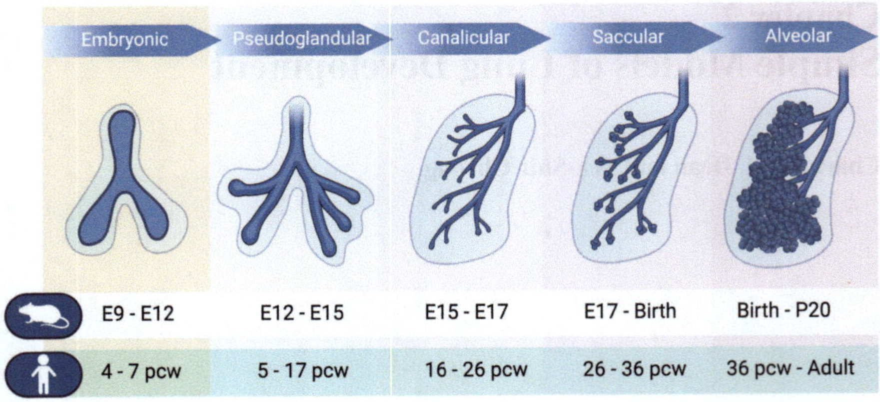


**Fig. 2.1** Stages of lung development. Schematic shows the five stages of lung development along with the period of human or mouse development when they occur. N.B. Alveologenesis takes place after birth in mouse

organ function. These other cell types are less well-represented in most of the commonly used models [6].

Many of the simple models of lung development were generated from mouse tissue or cells, but a growing number of human lung models have also been established. The existence of human models is important because despite the similarities between mouse and human development, there are also differences. This includes differences in the cellular composition of the airways and in the timing of alveologenesis, which begins prenatally in humans but takes place postnatally in the mouse [26, 35]. Molecular differences have also been found between mouse and human lung generation [31]. Knowledge of the morphological and molecular differences between mouse and human lung is of key importance as we move towards smarter models and effective translation of data obtained from animal studies.

## 2.2 Models to Study Lung Development

Models are a necessity to study the lungs since their deep position within the body and requirement for life makes this organ very difficult to study in situ. The generation of models to study lung organogenesis has broadly followed the stages of lung development with early models mainly focused on the pseudoglandular phase followed by more recent models of saccular and alveolar development.

This chapter looks at 'simple' (non-engineered) cell- and tissue-derived models used to represent lung development. The breadth of these lung development models has greatly increased over the past 15 years so that we are now in the position of having several models available for each stage of development. Moreover, the combination of bioengineering approaches with cell- and tissue-based models allows researchers to choose models that can be 'tuned' to represent the native biomechanics of the lungs. These more complex, engineered models of lung development are covered in Chaps. 3 and 4. Ultimately, the optimal choice of model(s) depends on the specific questions being addressed.

The primary functions of the airways and the alveoli are very different. Airways conduct the air down to the gas-exchanging region of the lungs, and they have a key role in keeping [33] noxious/injurious substances away from the delicate alveoli in the distal lungs, where gas exchange takes place. These alveoli are comprised of a thin epithelial layer adjacent to capillary vasculature so that gaseous diffusion can easily occur. Because they are particularly delicate, the alveoli often become damaged due to acute or chronic lung disease, and this in turn compromises breathing. Both the airways and the distal lung contain resident macrophages that play a key role in preventing noxious substances, such as pollutants or smoke, from damaging the structural components of the lungs. In addition, immune cells, such as monocytes and neutrophils, are recruited from the circulation when required, e.g. to fight infection.

The discovery that both rodent and human lungs contain multiple tissue resident stem cell populations and have substantial innate capacity for repair [3, 10] offers the possibility that regenerative biology could hold the key for new treatments for many lung diseases [34]. See Chap. 3 for further details of lung stem and progenitor cells.

Many models of lung development have focused on the epithelium, but the capillary network is equally important for lung development and function. Investigation of the fibroblast, neuroendocrine and immune cell populations of the lungs has also revealed critical roles for these cell populations in homeostasis, disease and repair, providing an increasingly complex picture of lung biology. Rapid advances in our understanding of different lung cell populations and their functions have come from using advanced transcriptomic approaches, including RNA-seq as well as a variety of different models.

## 2.3 Models of Early Lung Development (Airways)

Many early models of lung development focused on the airways partly because airway generation occurs throughout the majority of in utero development. In addition, techniques to isolate and image the lungs are easier in the early stages when only the airways are present in comparison to later saccular and alveolar stages when the organ is much more solid and no longer translucent. Airway generation occurs by branching morphogenesis, a series of re-iterative rounds of branching of the airway epithelium that generates the respiratory tree between E12.5 and E17.5 in mouse and 5 and 26 weeks in humans [16]. This phase of development has been studied extensively using a combination of explant cultures, mathematical models, spatial mapping and imaging modalities.

### 2.3.1 Explant Cultures

A simple and effective way to model early lung development is to extract the intact organ for explant culture. The lungs isolated between E11.5 and E13.5 are placed on a filter at the air-liquid interface (ALI) and cultured for up to 96 hours [12, 46]. Keeping the explants at the ALI facilitates continued development. A limitation of this model is the size of the tissue as nutrients need to diffuse through the entire explant; consequently, this technique is not suitable for later stages of lung development. The advantage of explants is that the native physiological interactions between different cell types in the organ are retained and they are particularly useful for studying morphological events during organ development. Because of these attributes, explant cultures have been used extensively to identify the molecular signals that contribute to branching morphogenesis [42]. Manipulation of oxygen concentration in explant cultures has shown that low oxygen concentration, which mimics the in utero environment (3%), facilitates increased terminal branching and cell proliferation [15].

In addition to whole lung explants, cultures of separated epithelium and mesenchyme have provided important information about lung development. Experiments where these compartments have been separated and then cultured either alone or with both added together have proved the importance of reciprocal signalling interactions to drive lung development [18]. Interestingly, explanted epithelium can undergo branching in the absence of mesenchyme, if cultured in the right conditions, such as embedded within a basement membrane matrix [32].

Although the majority of explant models use rodent tissue, other strategies have been used to provide models of human airway branching. Co-culture of the human basal-like airway epithelial cell line (VA-10) with human umbilical vein endothelial cells (HUVEC) in a 3D environment results in the formation of complex branching structures that resemble developing bronchoalveolar units [14]. Interestingly, this model also highlights the key role of the vascular endothelium in lung development [21].

### 2.3.2 2D and 3D Imaging of Branching Morphogenesis

Branching morphogenesis has typically been studied using 2D microscopy to capture the process of branching over the desired time period, and explant cultures have been particularly useful in this regard [12]. However, 2D methods cannot fully recapitulate a 3D process and are susceptible to the introduction of artefacts. This has led to the development of increasingly sophisticated 3D models of lung development.

Despite these advances, for some research questions, simpler models can prove more informative. For example, Yates at al. combined organotypic culture of E11.5 mouse lung with antisense morpholino knockdown to show that the polarity protein, Scribble, has a key role in lumen formation [45]. The dissociated lung cells formed epithelial cysts surrounded by mesenchymal cells, whereas cysts cultured with Scribble morpholino failed to form lumen. The simple cyst-like structures that form in these cultures can be immunostained and imaged to obtain both mechanistic and morphological information about the polarity of airway epithelial cells and lumen formation that are not easily discerned from more complex models.

To decipher the normal pattern of branching morphogenesis in whole mouse lungs, Mezger et al. undertook a comprehensive analysis of the architecture of lung branching morphogenesis using fixed, immunostained lungs to develop a formal model of the branching programme. This study mapped the pattern of branching from the lungs collected between E11 and E15, from which the authors identified three different modes of branching: domain branching, planar bifurcation and orthogonal bifurcation [25].

Imaging techniques, such as optical projection tomography (OPT), enable visualisation of whole intact organs in 3D [8, 37]. OPT imaging of the lungs and kidneys was used to generate data sets for accurate spatial mapping of branching morphogenesis using the Tree Surveyor software programme, which accurately quantifies branching morphogenesis and reveals any disruptions caused by genetic mutations or environmental changes [37]. Other mathematical models have proved highly valuable for predicting mechanistic information about lung branching morphogenesis [11, 24]. The re-iterative nature of branching morphogenesis is an attractive process to model in silico. 3D data sets can be generated from the embryonic lungs obtained at precise time points, and these can then be compared to predictions obtained from an in silico model of branching to determine whether the model does or doesn't reproduce the in vivo pattern of development. Fractal models of lung branching are one of the simplest types of in silico model applied to lung branching. Fractals are complex structures formed by repetitive application of a set of simple rules, which, in theory, describes branching morphogenesis very well. However, in vivo, the lungs are shaped by multiple factors, including mechanical forces, and the diffusion of spatially restricted molecular signals, such as fibroblast growth factor 10 (FGF10). Therefore, more complex models have been built to take account of different aspects of lung development [20, 24].

### 2.3.3 Time-Lapse Imaging

Repeated imaging of live lung explants over extended time periods (time-lapse) provides another level of information beyond that gained from imaging at a single time point. 2D time-lapse imaging of isolated epithelial explants was utilised to study how clefting and budding contribute to branching morphogenesis [27]. 3D time-lapse confocal imaging of E11.5 whole lung explants was employed to study cellular and molecular mechanisms of planar bifurcation and domain branching identified by Mezger et al. [36]. Explants were imaged repeatedly in the same positions approximately every 10 minutes for up to 48 hours and compared to freshly isolated lungs from the equivalent stage. This study showed that orientated cell divisions and myosin-dependent forces are important for the epithelium to form 3D branches. In a separate manuscript, the same 3D imaging technique was used to show that genetic manipulation of the polarity protein, Scribble, leads to a reduced number of airways that are misshaped [45].

### *2.3.4 Organoids*

The paucity of human models available to study lung development led to the establishment of new organoid models, spheroids, usually formed from a stem or progenitor cell population present in the lungs [4]. Organoids are a powerful tool for research, and they are amenable to imaging and can be produced in large numbers making them useful for screening purposes. Stem cell-derived organoids have greatly contributed to our knowledge about airway regeneration. This discovery research is crucial for development of successful regenerative medicine treatments. One important consideration when using organoids is that the read-outs should be tailored to the specific investigation rather than relying solely on generic read-outs, such as size and number.

Nikolic and colleagues established human organoids from distal tip cells of human embryonic airways. These organoids are amenable to long-term culture and provide a valuable tool to study aspects of human lung development [29]. Human airway basal epithelial cells are multipotent tissue stem cells found in adult airways that can give rise to multiple differentiated cell types present in the mature pseudostratified airway epithelium and are therefore useful material to provide airway-based organoids. Human basal epithelial cells obtained from endobronchial biopsies can be expanded by culture on a 3T3 fibroblast feeder layer in the presence of the ROCK inhibitor to produce a large number of cells. When these cells are detached from their feeder layer and cultured on Matrigel, they form tracheospheres that contain differentiated airway cell types [19]. The protocol used in Hynds et al. to expand the primary basal cells has also been used for human airway tissue engineering approaches aimed at tracheal replacement [9]. Murine-derived tracheospheres have also shed light on the dynamics and molecular drivers of airway regeneration [38–40]. The methods and uses of lung cell-derived organoids are reviewed in Barkauskas et al. [5]. Other types of organoids established from more than one cell type or derived from induced pluripotent stem cells (iPSCs) are covered elsewhere (Chaps. 3 and 4) (Fig. 2.2).

## 2.4 Models of Late Lung Development

As mentioned earlier in this chapter, models of later stages of lung development have been established much more recently than airway models.

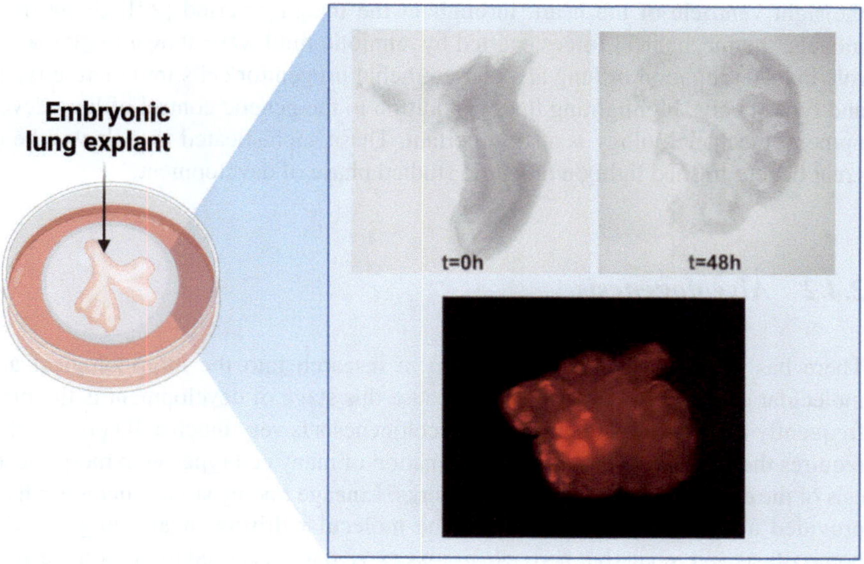


**Fig. 2.2** Lung explant culture. Lungs are explanted and placed on a filter at the air-liquid interface. Phase contrast images show an E11.5 mouse left lung lobe at 0 hour (top left) and 48 hours in culture (top right). The bottom image shows E11.5 murine explanted lungs after 48 hours in culture immunostained for pan-cytokeratin to highlight the epithelial branches

### *2.4.1 Saccular Phase Models*

During the saccular phase of development, the primitive air sacs form, and the epithelial cells differentiate to become squamous type I and cuboidal type II cells. The interstitium narrows due to differentiation of mesenchymal cells, and the capillaries rapidly expand to form a double parallel capillary network that surrounds the saccules. In humans this phase occurs between 26 weeks and the late foetal period, approximately 36 weeks, when alveologenesis begins, but in mouse the saccular phase begins in utero, around E17.5, and extends into the first few days after birth [43]. To investigate the mechanisms underlying the maturation of epithelial cells during the saccular phase, Li et al. established two models, both of which utilised live imaging to capture the lungs in real time. Firstly, an in vivo model was adapted from intravital microscopy of adult mice [23]. E16.5 mouse embryos were removed from the mother's abdomen but remained connected to the maternal circulatory system. The embryonic lungs were exposed and attached to a glass coverslip held in place by a gentle vacuum to enable confocal imaging for up to 3 hours in living embryos. Secondly, to monitor lung development for up to 10 hours, an ex vivo system was established where mouse embryos were removed from the uterus into a dish and the lungs were exposed to facilitate imaging but retained inside the body cavity. Media and oxygen were pumped into the pulmonary circulatory system via the right ventricle of the heart, throughout the imaging period [22]. Using these models, the mechanical forces exerted by amniotic fluid were shown to play a key role in differentiation of lung alveolar epithelial progenitor cells into mature type I and type II cells, highlighting that, in addition to the genetic control of lung development, mechanobiology is also important. These sophisticated models will be of great benefit to shed light on this little studied phase of development.

### 2.4.2 *Alveologenesis*

There has been a significant expansion in research into the morphological and molecular control of alveologenesis because this stage of development is the most frequently affected in lung diseases. Alveologenesis is very much a 3D process that requires the spatial and temporal co-ordination of many cell types with morphogenesis of the epithelium and capillary networks. Lineage tracing studies in mouse have provided a great deal of insight into the molecular drivers of alveologenesis in homeostasis and repair, but here we discuss ex vivo models used to investigate alveolar generation.

Alveolospheres are organoid models derived from alveolar type II (ATII) cells, which are the stem/progenitor cells of the distal lung [1, 13]. Alveolospheres are usually derived from ATII cells isolated from adult lungs or from iPSCs [4]. Whilst these organoid models are extremely useful for research and screening purposes, they do have limitations [30]. Organoids generated from ATII cells do not always produce differentiated ATI cells, and because they are usually comprised of one cell type, they don't reflect the complex environment of the distal lung [4].

### 2.4.3 *Other 3D Models of Alveologenesis*

Techniques developed to capture alveologenesis in 3D have contributed a great deal to our current understanding of alveologenesis and revealed artefacts from studying 2D sections [7]. Vibratome sections obtained from P0 to P15 lungs were used to generate 3D reconstructions of the lungs from confocal microscope images. Analysis of the reconstructed lungs showed that septal crests previously identified from 2D images were actually ridge-like structures that rise from the base of the existing alveolar pockets to subdivide them further and provide an increased surface area [7].

Another study used real-time imaging of precision-cut lung slices (PCLS) to reveal novel information about alveologenesis [2]. PCLS contain intact alveoli as well as all the lung-resident cell populations. Critically, they also retain the native architecture of the lungs. Live imaging of PCLS obtained from mouse lungs during alveologenesis revealed details about the cell behaviours, which contribute to alveologenesis, including cell clustering, hollowing and cell extension. These dynamic cell behaviours can only be captured using 3D live imaging techniques. These

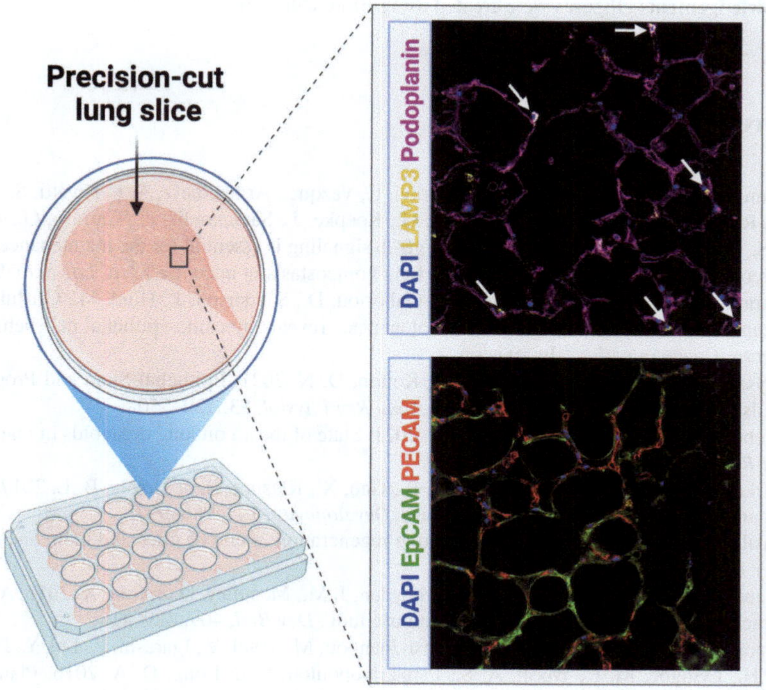


**Fig. 2.3** Precision-cut lung slice model. Precision-cut lung slices are 250–400-μm-thick slices of intact lung tissue. Images show immunostaining of lung parenchyma in PCLS for cell-type-specific markers. Podoplanin (pink) and Lamp3 (yellow) label type I and type II alveolar epithelial cells, respectively (top). EpCAM (green) and PECAM (red) label all epithelial cells and the capillary network, respectively (bottom)

findings from imaging of vibratome sections/slices are also consistent with the suggestion that the alveoli may form by expanding out from the lumen of an alveolar duct facilitated by a purse string-type structure at the mouth of the new alveolus like a gum bubble that pops out from a thin layer of gum [41] (Fig. 2.3).

## 2.5 Conclusion

The mature lungs are comprised of many different cell types as well as different tissue compartments. Our understanding of lung development has been greatly facilitated by the many different models that scientists have developed to study them. Model development that combines engineering approaches with either tissue-based models or those derived from multiple cell types will likely produce even better, more faithful models of the lungs, which will be of great benefit for respiratory research and to identify potential new treatments for a broad range of lung diseases.

**Acknowledgements** Figures were created using Biorender.com.

## References

1. Ahmadvand, N., Lingampally, A., Khosravi, F., Vazquez-Armendariz, A. I., Rivetti, S., Jones, M. R., Wilhelm, J., Herold, S., Barreto, G., Koepke, J., Samakovlis, C., Carraro, G., Zhang, J. S., Al Alam, D. & Bellusci, S. 2022. Fgfr2b signaling is essential for the maintenance of the alveolar epithelial type 2 lineage during lung homeostasis in mice. *Cell Mol Life Sci*, 79, 302.
2. Akram, K. M., Yates, L.L., Rothery, S., Gaboriou, D., Sanderson, J., Hind, M., Gfiffiths, M., Dean C.H. 2018. Live imaging of alveologenesis reveals dynamic epithelial cell behaviour. *Nature Communications*, In revision.
3. Alysandratos, K. D., Herriges, M. J. & Kotton, D. N. 2021. Epithelial Stem and Progenitor Cells in Lung Repair and Regeneration. *Annu Rev Physiol*, 83, 529–550.
4. Archer, F., Bobet-Erny, A. & Gomes, M. 2021. State of the art on lung organoids in mammals. *Vet Res*, 52, 77.
5. Barkauskas, C. E., Chung, M. I., Fioret, B., Gao, X., Katsura, H. & Hogan, B. L. 2017. Lung organoids: current uses and future promise. *Development*, 144, 986–997.
6. Basil, M. C. & Morrisey, E. E. 2020. Lung regeneration: a tale of mice and men. *Semin Cell Dev Biol*, 100, 88–100.
7. Branchfield, K., Li, R., Lungova, V., Verheyden, J. M., Mcculley, D. & Sun, X. 2016. A three-dimensional study of alveologenesis in mouse lung. *Dev Biol*, 409, 429–41.
8. Brzoska, H. L., d'Esposito, A. M., Kolatsi-Joannou, M., Patel, V., Igarashi, P., Lei, Y., Finnell, R. H., Lythgoe, M. F., Woolf, A. S., Papakrivopoulou, E. & Long, D. A. 2016. Planar cell polarity genes Celsr1 and Vangl2 are necessary for kidney growth, differentiation, and rostrocaudal patterning. *Kidney Int*, 90, 1274–1284.
9. Butler, C. R., Hynds, R. E., Gowers, K. H., Lee Ddo, H., Brown, J. M., Crowley, C., Teixeira, V. H., Smith, C. M., Urbani, L., Hamilton, N. J., Thakrar, R. M., Booth, H. L., Birchall, M. A., De Coppi, P., Giangreco, A., O'Callaghan, C. & Janes, S. M. 2016. Rapid Expansion of Human Epithelial Stem Cells Suitable for Airway Tissue Engineering. *Am J Respir Crit Care Med*, 194, 156–68.
10. Butler, J. P., Loring, S. H., Patz, S., Tsuda, A., Yablonskiy, D. A. & Mentzer, S. J. 2012. Evidence for adult lung growth in humans. *The New England journal of medicine*, 367, 244–7.
11. Celliere, G., Menshykau, D. & Iber, D. 2012. Simulations demonstrate a simple network to be sufficient to control branch point selection, smooth muscle and vasculature formation during lung branching morphogenesis. *Biol Open*, 1, 775–88.
12. Del Moral, P. M. & Warburton, D. 2010. Explant culture of mouse embryonic whole lung, isolated epithelium, or mesenchyme under chemically defined conditions as a system to evaluate the molecular mechanism of branching morphogenesis and cellular differentiation. *Methods Mol Biol*, 633, 71–9.
13. Ebisudani, T., Sugimoto, S., Haga, K., Mitsuishi, A., Takai-Todaka, R., Fujii, M., Toshimitsu, K., Hamamoto, J., Sugihara, K., Hishida, T., Asamura, H., Fukunaga, K., Yasuda, H., Katayama, K. & Sato, T. 2021. Direct derivation of human alveolospheres for SARS-CoV-2 infection modeling and drug screening. *Cell Rep*, 35, 109218.
14. Franzdottir, S. R., Axelsson, I. T., Arason, A. J., Baldursson, O., Gudjonsson, T. & Magnusson, M. K. 2010. Airway branching morphogenesis in three dimensional culture. *Respir Res*, 11, 162.
15. Gebb, S. A. & Jones, P. L. 2003. Hypoxia and lung branching morphogenesis. *Adv Exp Med Biol*, 543, 117–25.
16. Goodwin, K. & Nelson, C. M. 2020. Branching morphogenesis. *Development*, 147.

17. Herriges, M. & Morrisey, E. E. 2014. Lung development: orchestrating the generation and regeneration of a complex organ. *Development*, 141, 502–13.
18. Hines, E. A. & Sun, X. 2014. Tissue crosstalk in lung development. *J Cell Biochem*, 115, 1469–77.
19. Hynds, R. E., Butler, C. R., Janes, S. M. & Giangreco, A. 2019. Expansion of Human Airway Basal Stem Cells and Their Differentiation as 3D Tracheospheres. *Methods Mol Biol*, 1576, 43–53.
20. Iber, D. & Menshykau, D. 2013. The control of branching morphogenesis. *Open Biol*, 3, 130088.
21. Kina, Y. P., Khadim, A., Seeger, W. & El Agha, E. 2020. The Lung Vasculature: A Driver or Passenger in Lung Branching Morphogenesis? *Front Cell Dev Biol*, 8, 623868.
22. Li, J., Wang, Z., Chu, Q., Jiang, K., Li, J. & Tang, N. 2018. The Strength of Mechanical Forces Determines the Differentiation of Alveolar Epithelial Cells. *Dev Cell*, 44, 297–312 e5.
23. Looney, M. R., Thornton, E. E., Sen, D., Lamm, W. J., Glenny, R. W. & Krummel, M. F. 2011. Stabilized imaging of immune surveillance in the mouse lung. *Nat Methods*, 8, 91–6.
24. Menshykau, D., Blanc, P., Unal, E., Sapin, V. & Iber, D. 2014. An interplay of geometry and signaling enables robust lung branching morphogenesis. *Development*, 141, 4526–36.
25. Metzger, R. J., Klein, O. D., Martin, G. R. & Krasnow, M. A. 2008. The branching programme of mouse lung development. *Nature*, 453, 745–50.
26. Miller, A. J. & Spence, J. R. 2017. In Vitro Models to Study Human Lung Development, Disease and Homeostasis. *Physiology (Bethesda)*, 32, 246–260.
27. Miura, T. & Shiota, K. 2000. Time-lapse observation of branching morphogenesis of the lung bud epithelium in mesenchyme-free culture and its relationship with the localization of actin filaments. *Int J Dev Biol*, 44, 899–902.
28. Mullassery, D. & Smith, N. P. 2015. Lung development. *Semin Pediatr Surg*, 24, 152–5.
29. Nikolic, M. Z., Caritg, O., Jeng, Q., Johnson, J. A., Sun, D., Howell, K. J., Brady, J. L., Laresgoiti, U., Allen, G., Butler, R., Zilbauer, M., Giangreco, A. & Rawlins, E. L. 2017. Human embryonic lung epithelial tips are multipotent progenitors that can be expanded in vitro as long-term self-renewing organoids. *Elife*, 6.
30. Nikolic, M. Z. & Rawlins, E. L. 2017. Lung Organoids and Their Use To Study Cell-Cell Interaction. *Curr Pathobiol Rep*, 5, 223–231.
31. Nikolic, M. Z., Sun, D. & Rawlins, E. L. 2018. Human lung development: recent progress and new challenges. *Development*, 145.
32. Nogawa, H. & Ito, T. 1995. Branching morphogenesis of embryonic mouse lung epithelium in mesenchyme-free culture. *Development*, 121, 1015–22.
33. Parekh, K. R., Nawroth, J., Pai, A., Busch, S. M., Senger, C. N. & Ryan, A. L. 2020. Stem cells and lung regeneration. *Am J Physiol Cell Physiol*, 319, C675-C693.
34. Rock, J. & Konigshoff, M. 2012. Endogenous lung regeneration: potential and limitations. *American journal of respiratory and critical care medicine*, 186, 1213–9.
35. Rock, J. R., Onaitis, M. W., Rawlins, E. L., Lu, Y., Clark, C. P., Xue, Y., Randell, S. H. & Hogan, B. L. 2009. Basal cells as stem cells of the mouse trachea and human airway epithelium. *Proc Natl Acad Sci U S A*, 106, 12771–5.
36. Schnatwinkel, C. & Niswander, L. 2013. Multiparametric image analysis of lung-branching morphogenesis. *Dev Dyn*, 242, 622–37.
37. Short, K., Hodson, M. & Smyth, I. 2013. Spatial mapping and quantification of developmental branching morphogenesis. *Development*, 140, 471–8.
38. Tadokoro, T., Gao, X., Hong, C. C., Hotten, D. & Hogan, B. L. 2016. Bmp signaling and cellular dynamics during regeneration of airway epithelium from basal progenitors. *Development*, 143, 764–73.
39. Tadokoro, T., Wang, Y., Barak, L. S., Bai, Y., Randell, S. H. & Hogan, B. L. 2014. IL-6/STAT3 promotes regeneration of airway ciliated cells from basal stem cells. *Proc Natl Acad Sci U S A*, 111, E3641–9.

40. Wansleeben, C., Bowie, E., Hotten, D. F., Yu, Y. R. & Hogan, B. L. 2014. Age-related changes in the cellular composition and epithelial organization of the mouse trachea. *PLoS One*, 9, e93496.
41. Warburton, D. 2021. Conserved Mechanisms in the Formation of the Airways and Alveoli of the Lung. *Front Cell Dev Biol*, 9, 662059.
42. Warburton, D., Bellusci, S., De Langhe, S., Del Moral, P. M., Fleury, V., Mailleux, A., Tefft, D., Unbekandt, M., Wang, K. & Shi, W. 2005. Molecular mechanisms of early lung specification and branching morphogenesis. *Pediatric research*, 57, 26R-37R.
43. Warburton, D., El-Hashash, A., Carraro, G., Tiozzo, C., Sala, F., Rogers, O., De Langhe, S., Kemp, P. J., Riccardi, D., Torday, J., Bellusci, S., Shi, W., Lubkin, S. R. & Jesudason, E. 2010a. Lung organogenesis. *Current topics in developmental biology*, 90, 73–158.
44. Warburton, D., El-Hashash, A., Carraro, G., Tiozzo, C., Sala, F., Rogers, O., De Langhe, S., Kemp, P. J., Riccardi, D., Torday, J., Bellusci, S., Shi, W., Lubkin, S. R. & Jesudason, E. 2010b. Lung organogenesis. *Curr Top Dev Biol*, 90, 73–158.
45. Yates, L. L., Schnatwinkel, C., Hazelwood, L., Chessum, L., Paudyal, A., Hilton, H., Romero, M. R., Wilde, J., Bogani, D., Sanderson, J., Formstone, C., Murdoch, J. N., Niswander, L. A., Greenfield, A. & Dean, C. H. 2013. Scribble is required for normal epithelial cell-cell contacts and lumen morphogenesis in the mammalian lung. *Dev Biol*, 373, 267–80.
46. Yeganeh, B., Bilodeau, C. & Post, M. 2018. Explant Culture for Studying Lung Development. *Methods Mol Biol*, 1752, 81–90.

# Chapter 3
# Lung Development in a Dish: Models to Interrogate the Cellular Niche and the Role of Mechanical Forces in Development

**Brea Chernokal, Cailin R. Gonyea, and Jason P. Gleghorn**

## 3.1 Introduction

Over the past decade, emphasis has been placed on recapitulating in vitro the architecture and multicellular interactions found in organs in vivo [1, 2]. Whereas traditional reductionist approaches to in vitro models enable teasing apart the precise signaling pathways, cellular interactions, and response to biochemical and biophysical cues, model systems that incorporate higher complexity are needed to ask questions about physiology and morphogenesis at the tissue scale. Significant advancements have been made in establishing in vitro models of lung development to understand cell-fate specification, gene regulatory networks, sexual dimorphism, three-dimensional organization, and how mechanical forces interact to drive lung organogenesis [3–5]. In this chapter, we highlight recent advances in the rapid development of various lung organoids, organ-on-a-chip models, and whole lung ex vivo explant models currently used to dissect the roles of these cellular signals and mechanical cues in lung development and potential avenues for future investigation (Fig. 3.1).

---

Brea Chernokal and Cailin R. Gonyea contributed equally with all other contributors.

---

The original version of this chapter was revised. The correction to this chapter is available at https://doi.org/10.1007/978-3-031-26625-6_17

---

B. Chernokal · C. R. Gonyea · J. P. Gleghorn (✉)
Department of Biomedical Engineering, University of Delaware, Newark, DE, USA
e-mail: gleghorn@udel.edu





C. M. Magin (ed.), *Engineering Translational Models of Lung Homeostasis and Disease*, Advances in Experimental Medicine and Biology 1413,
https://doi.org/10.1007/978-3-031-26625-6_3

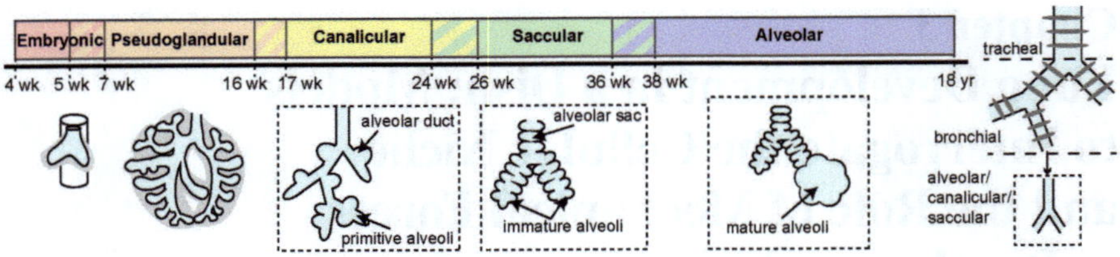


**Fig. 3.1** Stages of human lung development starting at 4 weeks postconception and continuing through 18 years post birth. This process builds the entire complexity of the mammalian lung

## 3.2 Self-Assembled Organoid and Spheroid Models

Self-assembled differentiated organoids have become increasingly important and powerful models for studying the development and disease of many organs and tissues. In particular, numerous organoid models have been generated for the lung. Current models encompass varying levels of cellular complexity and regional specificity, from early developmental stages with multipotent progenitor cells to differentiated organoids that reflect the regional cellular heterogeneity of the large airways and the alveoli. Whereas traditionally organoids are derived from primary cells, lung organoids have also been generated from stem cell populations, such as induced pluripotent stem cells (iPSCs) or embryonic stem cells (ESCs) [2, 6–8] (Fig. 3.2a). This section will provide a high-level summary of the current state of the art in organoid modeling of the lung airway and a description of the recent advancements and opportunities these models may offer to study lung development.

### *3.2.1 Creating Lung Organoid Models That Represent Regional Composition and Heterogeneity*

Since the initial 3D lung epithelial cultures were created [9–11], significant advancements have been made to determine cell source, culture medium formulation, and culture methods to generate region-specific organoid-like models, including tracheo- and tracheobronchospheres, bronchiolar organoids, and alveolospheres. As such, these models have enabled the dissection of cell-fate decision pathways, molecular mediators of morphogenesis, and intercellular and cell-extracellular matrix (ECM) interactions in the various lung niches. These models continue to advance our understanding of the importance and role of cellular ecology in development, homeostasis, and disease.

To model the tracheal niche, tracheospheres were initially generated from isolated mouse and, subsequently, human basal cells [12]. Adapting previous protocols for three-dimensional (3D) spheroid culture of other organ stem cells [13], isolated primary NGFR+ basal cells were FACS sorted and encapsulated as single cells in Matrigel and cultured over 2 weeks (Fig. 3.2a) [12]. These initial tracheospheres

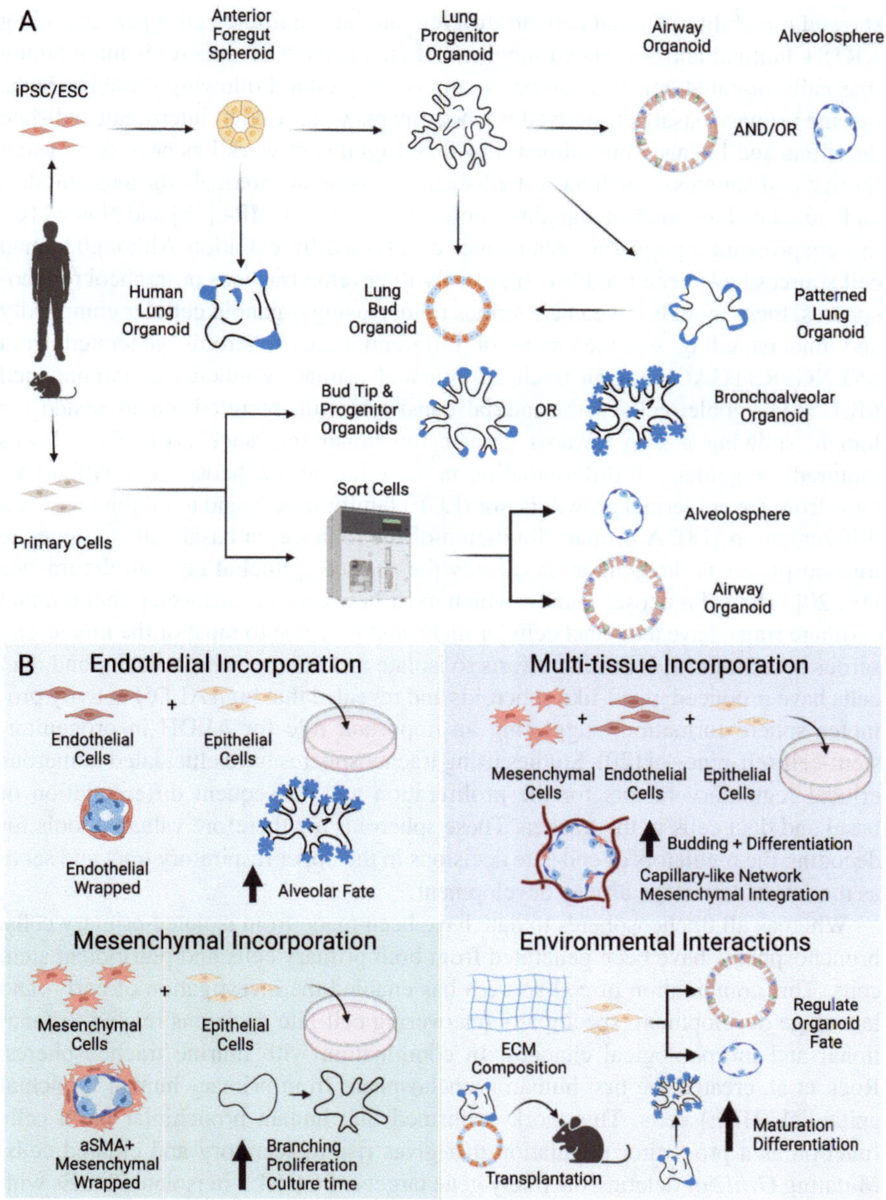


**Fig. 3.2** Organoid models for studying lung development. (**a**) Stem and primary cells are used to develop lung spheroids and organoids. Methodologies vary by cell source and depend on the lung region or cell type of interest. (**b**) Current techniques for increasing spheroid/organoid complexity

showed the ability of basal cells to differentiate into multiple cell types, including KRT8+ luminal and acetylated tubulin+ ciliated cells, with spheroids maintaining internally oriented apical polarization and beating cilia. Following these methods, murine primary basal cell-derived tracheospheres were used to interrogate cell-fate decisions and lineage commitment [14–17]. Significant work has been done using murine cell sources, which has established the essential protocols for these models and elucidated the roles of signaling molecules, such as BMP4 [15] and Notch [16], in stem progenitor population maintenance and cell differentiation. Although human cell sources have been used less frequently to develop tracheo- or tracheobronchospheres, these models have made strides in increasing organoid cellular complexity and understanding the mediators of differentiation. Spheroids generated from p63;NGFR;ITGA6+ human tracheal/bronchial primary epithelial cells contained MUC5AC+ goblet cells with material consistent with secreted mucus inside the lumen, showing a step towards greater functional relevance [18]. Notch2 was required for goblet cell differentiation in these human tracheospheres, while proteins from the epidermal growth factor (EGF) family were found to inhibit basal cell differentiation [18]. A primary limitation of the reliance on basal cells to generate tracheospheres is the generation of only the surface epithelial layer in the trachea [19, 20]. The submucosal glands, which exist between the surface epithelium and cartilage rings, have a distinct cellular niche and give rise to most of the mucus and serous secretions [19]. Further efforts to isolate and culture submucosal gland duct cells have produced gland-like spheroids and revealed that high ALDH activity promotes sphere formation, suggesting an important role for ALDH in progenitor/stem-cell self-renewal [20]. Studies using tracheospheres have elucidated numerous critical regulatory factors for the proliferation and subsequent differentiation of basal and duct cells in the trachea. These spheroids are therefore valuable tools for decoding the regulators of cell-fate decisions in the upper respiratory tract and serve as models of late-stage airway development.

Whereas all tracheospheres to date have been made from isolated primary cells, bronchospheres have been generated from both primary cells and pluripotent stem cells. This combination of cell sources has enabled the investigation of early- and late-stage development, specifically uncovering cell-fate decisions related to functional and morphological changes. In combination with murine tracheospheres, Rock et al. created the first human bronchosphere from primary human bronchial epithelial (HBE) cells. This work confirmed that human bronchiolar basal cells function as a progenitor population that gives rise to secretory and ciliated cells. Mutating *Grhl2* or deleting the likely gene targets of GRHL2 in isolated HBEs with CRISPR/Cas9 genome editing technology showed critical roles for GRHL2 and ZNF750 were identified for basal cell proliferation, barrier function, and ciliogenesis [14]. This highlights the potential of organoid models to be used for rapid screening of genes related to the regulation of morphogenesis and differentiation. As these primary cell bronchosphere cultures were generated from adult cells, they do not capture the full multipotency of progenitors in development nor provide information on when these cell types arise in the developing lung. However, embryonic day (E)12.5–14.5 murine lung epithelial progenitors expressing *Nkx2.1* have been used to generate spheroids that were positive for markers of basal and secretory cells and, to a lesser extent, ciliated and submucosal gland cell markers [21]. Spheroids showed expression of SCGB1A1, microvilli presence, and secretory granule production, indicative of club cell precursors. Additionally, there was diversity among these cells, with region-specific markers *Reg3g, Gabrp,* and *Upk3a* heterogeneously distributed within the spheroids. Interestingly, these club cells further indicated the presence of a distinct, multipotent cell type from mature secretory cells as they gave rise to basal and submucosal gland cells (KRT5;SMA+) in addition to club and ciliated cells [21].

The first airway spheroids generated from stem cells were mouse ESC-derived NKX2.1+ progenitors suspended in Matrigel and transplanted in vivo [22]. These spheroids differentiated to contain proximal airway epithelial cells, basal cells, Clara cells, ciliated cells, and secretory cells, thereby producing a greater range of mature cell types than in previous work. Building on this, human iPSCs were differentiated into NKX2.1+ progenitors with subpopulations of SOX2+ and SOX9+ cells, suggesting the presence of committed airway (NKX2.1;SOX2+) and multipotent progenitors (NKX2.1;SOX9+) [22]. This work provided a basis for generating stem cell-derived bronchospheres; therefore, subsequent work focused on uncovering the regulatory factors involved in cell differentiation, recapitulating cellular functions, and investigating the effects of environmental mechanics. iPSC-derived NKX2.1;SOX2+ proximal airway progenitor spheroids were able to grow and fuse into large amorphous structures but could not undergo further lineage commitment under normal culture conditions [23]. However, when subjected to primary bronchial epithelial media, spheroids differentiated into CHGA+ and SYP+ pulmonary neuroendocrine cells, SCGB1A1+ club cells, KRT5+ basal cells, and acetylated tubulin;FOXJ1+ ciliated cells, with acetylated tubulin+ cells closely aligned with mucus-secreting MUC5AC+ cells [23]. The localization of these various cell types resembled that of the human fetal lung, and beating cilia were visualized. Further interrogation of signaling pathways in proximalized airway organoids via Notch inhibition [23], Wnt up- or downregulation [24–26], or fibroblast growth factor (FGF) supplementation [24] has indicated the importance of these factors in epithelial differentiation and lineage commitment. Airway organoids were occasionally generated from hepatic- and gastric-like cells [26], underscoring the need for continued investigation of pathways responsible for cell-fate decisions.

To model the distal lung, alveolospheres were generated from primary human alveolar type 2 (AT2) cells cultured in Matrigel (Fig. 3.2a) [27, 28]. These methods created spheroids through migration and attachment rather than proliferation and apoptosis, as in previous organoid models. Additionally, they could differentiate into alveolar type 1 (AT1) cells, with human serum treatment inducing AGER+ cells and forskolin-induced swelling resulting in a shift toward a flattened morphology [27, 28]. This phenotypic shift was thought to be due to increased secretion and pressure within the spheroid, aligning with previous reports of increased mechanical tension in epithelial acini driving morphogenesis and homeostasis [29]. Using similar AT2 spheroids, it was found that depleting EGF led to differentiation of AT2s into the newly discovered alveolar type 0 (AT0) cell, an intermediate progenitor cell [30]. ATOs could then be differentiated into AT1s with the addition of serum or into terminal and respiratory bronchial secretory cells by continued depletion of EGF [30]. To further confirm the differentiation capability of AT2 to AT1 cells, alveolar epithelial progenitor cells were directly differentiated from human pluripotent stem cells (hPSCs) into spheroids positive for AT1 and AT2 cell markers and distal lung lineage markers [31]. Adapting a 2D lung cell differentiation protocol, human iPSC alveolar organoids were generated containing functional lamellar bodies and clustered transcriptomically with primary fetal lung cells without the need for a feeder cell population [32–34]. These organoids were used to investigate the differentiation program of AT2 and AT1 cells, showing the importance of Wnt downregulation and the existence of an intermediate IGFBP2− cell, confirming in vivo lineage tracing results [34, 35]. Looking upstream of AT2s, a newly identified SCGB3A2+ respiratory airway secretory (RAS) cell was found to localize between alveoli [36]. Human ESC-derived RAS cell organoids supported Wnt and Notch regulation of AT2 differentiation [33, 36]. Further culture of RAS-derived AT2-committed organoids in airway media did not rescue *Scgb3a2* expression, demonstrating this is a unidirectional differentiation process [36]. While most alveolosphere culture methods use Matrigel, initial steps have been taken to use other hydrogel formulations with known compositions and tunable biophysical or biochemical properties [37]. For example, the culture of iPSC-derived AT2 alveolospheres in various 3D hyaluronic acid hydrogels showed that chemical and mechanical properties influence spheroid formation [38]. The incorporation of hyaluronic acid microwells further enabled spatial control of alveolospheres and maintained a larger proportion of cells with AT2 markers than Matrigel culture [38].

### 3.2.2 Advancing the Complexity of Organoids to Investigate Tissue Crosstalk

Cell-cell interactions across tissue types are known to be essential to lung development [39–42]. More complex organoids, incorporating different tissue compartments, including immune, endothelial, and other stromal cell populations, have been created for investigating and elucidating reciprocal tissue signaling mechanisms necessary for lung development. Adding mesenchymal cells, either as a feeder layer or as part of the organoid, was one of the first steps toward organoid models for intertissue interactions (Fig. 3.2b). Recent studies have begun investigating the effects of stromal cell populations and other hydrogel compositions on tracheal and basal cell spheroids. Tracheospheres cultured with a mitotically inactivated 3T3 fibroblast feeder layer and Rho-associated protein kinase (ROCK) inhibitor Y-27632 increased proliferation and maintained mucosecretory and ciliated cell fate [43]. Conditionally reprogrammed isolated human primary basal cells cultured with 3T3 fibroblast feeder cells and ROCK inhibition developed tracheobronchospheres with externally oriented apical polarization, providing the ability to test therapeutic delivery [44]. Additionally, investigation into matrix composition and signaling pathways found that a mixture of collagen I, collagen III, and Matrigel produced significantly more organoids than Matrigel alone, while supplementation of R-spondin 2 (Wnt agonist) and Noggin (BMP antagonist) further stimulated growth and differentiation [44]. Coculture of human lung-derived primary EPCAM+ epithelial cells and EPCAM-;SCA-1+ mesenchymal cells in Matrigel generated epithelial organoids that were wrapped in aSMA+ mesenchymal cells, mimicking in vivo interactions in the large airways [45]. Investigation of mesenchymal factors responsible for epithelial organoid formation revealed FGF10 and HGF supplementation could facilitate organoid formation without coculture of mesenchymal cells, suggesting the EPCAM-;SCA-1+ mesenchymal cells secrete these factors to direct epithelial proliferation and differentiation [45].

Coculture of VA10 human bronchial cells with endothelial cells led to branching organoids with alveolar-like buds that included tertiary level branches (Fig. 3.2b) [46]. Branching was endothelial mediated and relied on FGF signaling [46]. In other studies, upregulation of FGFR1 and VEGFR2 led to murine pulmonary capillary endothelial cell-mediated generation of alveolar saclike organoids [47]. Treatment with various inhibitors and agonists revealed that VEGFR2 and FGFR1 signaling triggered pulmonary capillary endothelial cells to produce MMP14, which was necessary for lung progenitor activation and alveolar-sac organoid formation [47]. Further investigation with epithelial-endothelial coculture organoids revealed a BMP4-controlled NFATC1-TSP1 axis in endothelial cells, which directs bronchoalveolar stem cell differentiation toward an alveolar fate [48]. Whereas the incorporation of multiple cell types within a single 3D culture model has been achieved with some cell types, a persistent challenge to the widespread incorporation of multiple cell types, particularly epithelial and endothelial cells, is the lack of an optimized culture medium to support both populations, even though relevant morphological structures can be generated in vitro with each cell type individually [49, 50].

Perhaps the most complex organoids have combined epithelial, mesenchymal, and endothelial interactions (Fig. 3.2b). Embryonic day 17.5 murine fetal pulmonary cells were cultured in collagen to generate lung organoids containing epithelial, endothelial, and mesenchymal cells [51]. Distinct roles of FGFs were found, with FGF2 supplementation leading to increased mesenchymal proliferation and endothelial network formation, whereas FGF7 and 10 increased epithelial and mesenchymal proliferation, with FGF10 specifically increasing bud formation [51]. Comparing various supplementation schemes suggested that epithelial proliferation and budding positively affect vascular development, likely through increased proangiogenic paracrine signaling [51]. Utilizing the same model, crosstalk between the FGF, SHH, and VEGF-A signaling pathways was observed [52]. Specifically, it was suggested that exogenous FGFs begin a cascade of many endogenous mediators, including SHH and VEGF-A, which affect epithelial and endothelial development in the lung [52]. Further, these signaling pathways regulated extracellular matrix ligand tenascin-C deposition, and tenascin-C patterns were important for endothelial network morphology and epithelial saccularization in the organoids [52]. In other studies, human bronchial epithelial cells were cultured with human lung microvascular endothelial and mesenchymal cells in soluble ECM-supplemented media atop a Matrigel layer to form branched and budding airway organoids [53]. The multiple cell types self-organized to create tubular structures within the organoid that recapitulated the basic structure observed in lung development, although mature airway and alveolar cells were not observed [53]. Generation of these tubular structures was dependent on myosin 2 contraction in fibroblasts and YAP signaling, respectively shown by attenuation of tubules with blebbistatin treatment or fibroblast-free culture and YAP localization in tubule tips with knockdown. Resulting in reduced tubular formation and disorganization of the epithelium, mesenchyme, and fibronectin deposition [53]. Multicell coculture organoids, therefore, provide a robust model for lung development and enable the investigation of various signaling pathways involved in cell differentiation and organ morphogenesis.

### *3.2.3 Induction of Lung Organoids to Create Multiple Tissue Compartments*

hPSCs have been used to develop multi-tissue compartment organoids with different differentiation schemes and medium supplements directing organoid composition (Fig. 3.2a). hPSCs treated with activin A and a combination of Noggin, SB431542, FGF4, smoothened agonist (SAG), and CHIR99201 generated medium-suspended organoids that were subsequently embedded in Matrigel and treated with FGF10 to form human lung organoids (HLOs), which mimicked early fetal murine lung development [54]. These HLOs included epithelial cell populations with proximal and distal airway-like structures surrounded by mesenchymal cells. Seeding of HLOs onto a decellularized lung matrix (Fig. 3.2b) resulted in the differentiation of multiciliated cells, reinforcing that extracellular matrix cues can regulate lung cell fate [54]. Other studies had demonstrated that when HLOs were seeded onto rigid microporous poly(lactide-co-glycolide) scaffolds and implanted for 4 weeks into the epididymal fat pad of mice, differentiated airway structures with beating multiciliated cells, a vasculature, the scattered presence of goblet and club cells, and surrounding smooth muscle, myofibroblasts, and cartilage were formed [55]. Using a slightly modified differentiation scheme of CHIR99021 followed by Noggin, FGF4, SB431542, and CHIR99021, before implantation of these HLOs into the kidney capsule of immunodeficient NSG mice, resulted in differentiation of bipotent alveolar progenitors, AT2, AT1, ciliated, basal, goblet, and club cells after 120 days [56]. These organoids had developed ACTA2+ vasculature and PGP9.5+ neuroendocrine cells, with transmission electron microscopy imaging revealing blood vessel and myelin sheath structures [56]. Although these HLOs achieved more mature cell fates, they never developed tubular structures. As such, HLOs have enabled the investigation of proximal and distal airway maturation and the mesenchyme's role in development. Using the same differentiation protocols, HLOs have been generated from iPSCs derived from human fetuses and neonates with normal lungs or congenital diaphragmatic hernia (CDH) [57]. This work highlights the potential of lung organoids to be used in the investigation of mechanisms underlying developmental diseases, such as CDH, and to study the saccularization and alveolarization stages of lung development [58].

Like HLOs, hPSC-derived lung bud organoids (LBOs) with mesoderm and endoderm develop structures upon transplantation and Matrigel culture [59]. Development of LBOs slightly differed from HLOs, with BMP4, FGF10, FGF7, retinoic acid, and CHIR99201 supplementation of the media. The resulting LBOs had no markers of maturity present other than p63, indicating their progenitor state [59]. Implantation of these organoids under the kidney capsule of immunodeficient NSG mice led to the development of proximal airway epithelium, branched structures with surrounding mesenchymal cells, cells positive for markers of later stage proximal (FOXJ1, CC10, and mucins) and distal (SFTPC and SFTPB) cell types, the presence of submucosal gland structures, neuroepithelial body-like structures, and networks of thin cell layers with AT1 and AT2 markers [59]. Interestingly, LBOs did not require an implantation scaffold for maturation and could develop into patterned lung-like structures, while HLOs only developed into airway-like structures. LBOs could also be cultured in Matrigel in vitro to generate branched structures with dilated tips, and RNAseq analysis of these organoids was found to match late second-trimester fetal lungs. However, they were distally biased and did not achieve mature AT1 cells [59]. Further development and improvement of organoid models that combine multiple tissue compartments is therefore an exciting opportunity space for understanding signaling mechanisms that underlie lung development.

## 3.3 Microfluidic and Organ-on-a-Chip Models to Study Lung Development

Microfluidic organ-on-a-chip systems are advantageous in their ability to incorporate well-defined mechanical stimuli, such as fluid flow, pressure, stretch, substratum stiffness, and shear stress, into in vitro human cell culture systems. Traditionally, microfluidic devices consist of micron-sized cellular channels perfused with cell culture medium or other fluids (Fig. 3.3). The perfusion provides nutrient replenishment from medium circulation and, if desired, the application of fluid forces for long-term studies. Microfluidic platforms have advanced dramatically from the investigation of single cells on a glass substratum within a channel to entire tissue networks within 3D hydrogel scaffolds [60–62]. With the advancements in device design and fabrication, the field has increasingly incorporated the ability to build and test tissue and organ physiology. Much of the work developing microfluidic platforms to investigate lung biology, signaling, and physiology has been used for understanding adult homeostasis and disease. However, there is tremendous opportunity to use these platforms for investigating lung development. In particular, the embryonic and fetal lungs are fluid filled [63], which can easily be captured by the

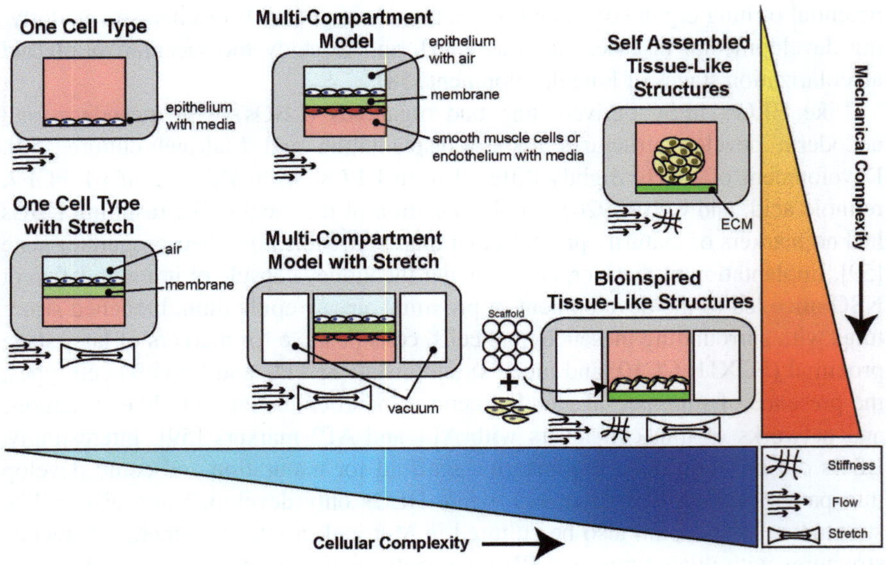


**Fig. 3.3** Diagram of organ-on-a-chip device designs illustrated based on cell complexity and the complexity of mechanical stimuli that can be achieved

small size and fluid control of these devices to answer questions of early lung development at a relevant length scale [64]. Similarly, scaling geometry and mechanical stimuli relevant to neonates provides significant opportunities for these models to advance our knowledge of late-stage lung development, which is directly translatable to the clinical management of preterm infants.

### *3.3.1 Moving Toward More Complex Physiology with Multiple Channels*

Beyond numerous studies using single fluidic channels, the introduction of multichannel devices for lung modeling improved the physiological relevance of the design by enabling controlled interrogation of one or more mechanical stimuli (Fig. 3.3). Takayama's group was among the first to use microfluidic platforms for modeling the lung and developed a compartmentalized microfluidic model of the airway epithelium [65, 66]. The structure and dimensions were designed to recapitulate in vivo airways and respiratory bronchioles. A two-chamber design created apical and basal compartments of the airway epithelium, separated by a thin "basement" membrane. Human primary small airway epithelial cells seeded in the top channel were subject to tunable airflows, with the bottom channel serving as a media reservoir. These mechanical forces led to cellular differentiation into secretory phenotypes [66]. Additional iterations of this device have been used to elucidate the protective role of surfactants from the damaging mechanical stresses caused by liquid-plug airway obstructions formed in premature infants or lungs with insufficient production of surfactants [67–70]. Since the initial development of these epithelial models, numerous designs have been used to study mechanical stimuli on airway epithelial function in various regions along the airway tree. In one such design, a two-channel microfluidic device was developed to understand how breathing-induced stretch affects AT2 cell phenotype and surfactant production [71]. A deformable membrane separated a medium-perfused channel from an air-filled pneumatic channel, and the epithelial cell monolayer on the membrane was deformed by pressurizing the pneumatic channel inducing stretch within the tissue in vitro. Loading via cyclic stretch to mimic breathing and ventilator-induced lung injury could be achieved, and by tuning the frequency and magnitude of cyclic stretch, healthy and pathological forces can be incorporated into the microfluidic model [71]. Importantly, although these models have been used to investigate adult airway mechanics, they have broad applicability. They specifically could be used to investigate airway development in utero and postnatally to model developmental effects and ventilator-induced lung injury in neonates.

Intercellular and intertissue signaling is essential to physiological function in homeostasis and disease. Likewise, complex signaling networks driven by soluble and mechanical cues exist in the developing lung that drive proliferation, differentiation, and ultimately morphogenesis. As such, multichannel microfluidic systems have sought to construct complex tissue environments to dissect the mechanisms of cellular crosstalk and understand the functional impacts on tissue physiology. Using similar approaches to the previously described multichannel devices, multi-tissue lung models have been developed (Fig. 3.3). Using a simple two-channel device, interactions can be studied between airway epithelium and smooth muscle cells separated by a thin hydrogel to mimic chronic lung disease [72]. Similar devices have also been used to investigate epithelial-endothelial crosstalk in the lung [73–75]. In one such device, the application of dynamic mechanical forces in a combined epithelial-endothelial device was introduced in a similar two-channel design with two lateral channels [73]. Air was applied in the epithelial channel and culture medium in the endothelial channel. With a dynamic vacuum force applied to the lateral channels, in-plane stretch is achieved on a porous elastic membrane that separates the epithelial and endothelial compartments. Further iterations of the device have enabled the application of breathing-like forces without the need for parallel chambers [75]. Seeding the epithelial channel with distal or proximal epithelial cells created an alveolar or airway model, respectively, which could be exposed to air to mimic adult and neonatal in vivo conditions. This model enables the investigation of epithelial-endothelial interactions in response to breathing-like forces. Additional variations on these device designs have enabled the study of signaling between tissue compartments [76–78] to decipher mechanisms of pathogenesis in adult lung models. However, these would be powerful models to investigate the relative contributions of tissue crosstalk and mechanical forces, including changes in ECM stiffness [79], during lung development. In addition to mechanistic insight, another important use of these in vitro models is for drug screening and understanding physiological transport across these barrier tissues [75, 78, 80–83], which would be directly translatable to developing new treatments for diseases associated with the neonatal lung and patient-specific therapeutic testing [84].

### *3.3.2 Integration of Dimensionality and Biomaterials into Organ-on-a-Chip Platforms*

It is well established that three-dimensionality, composition, and stiffness regulate gene expression, phenotype, proliferation, differentiation, and function of cells. As but one example, varying tissue structures were formed by human bronchiolar epithelial cells when cultured in 3D depending on whether they were cultured at the air-liquid interface on collagen, on Matrigel, or encapsulated within Matrigel [85]. Whereas air liquid interface (ALI) on collagen generated a monolayer, culture on top of Matrigel resulted in branched and budding tubule structures reminiscent of patterned lung organoids, and encapsulation in Matrigel resulted in spheroid formation. As such, in vitro lung models have sought to incorporate relevant geometric scales and material properties into microfluidic organ-on-a-chip devices. Strategies have ranged from biomimetic geometries to advanced materials, including hydrogels [86], microbeads [87], 3D printed plastics [88], and membranes [89]. Generally, the scaffolds are designed in the shape of the desired tissues to guide and direct cell growth, alignment, and differentiation. Circular or spherical geometries are often used to mimic alveolar structures, which are subsequently seeded with human primary alveolar epithelial cells to develop 3D alveolar-like constructs (Fig. 3.3). For example, a gold membrane with large hexagonal pores supported an elastin-collagen gel seeded with a mixture of AT1 and AT2 cells on the top and endothelial cells on the underside [89]. The membrane was stretched via negative pressure on the underside to mimic the distal airways during breathing. Similar strategies to model the distal airway include dome-shaped polycarbonate membranes [88] and porous gelatin methacryloyl hydrogel scaffolds [87]. Apart from these material approaches to generate relevant geometries and biophysical niches, methods have been developed to directly incorporate organoids within microfluidic systems [90] (Fig. 3.3). This latter approach may yield a powerful system wherein an organoid generates a complex cellular community that can be dissociated to serve as a cell source for these organ-on-a-chip systems. Subsequently, controlled biophysical cues, biochemical gradients, and defined geometric length scales not achievable in organoids can be used to investigate complex human cell populations within these microfluidic platforms. Combining these technologies is an exciting possibility as it will further develop the understanding of many mechanistic questions about the developing lung.

## 3.4 Whole Organ Models to Understand the Mechanics of Lung Development

Whereas much work has been done across a range of small and large animal models of lung development and diseases, including bronchopulmonary dysplasia (BPD) and CDH [91–95], the mouse has provided critical insights into the gene regulatory networks and cellular differentiation events that underpin mammalian lung morphogenesis. For decades the workhorse models used in lung developmental biology have been mesenchyme-free murine epithelial lung tips embedded in Matrigel [96] and ex vivo whole rodent lung culture on floating membranes at the air-liquid interface [97]. These approaches offer the opportunity to have the inherent complexity of the organ to recapitulate morphogenic processes in vitro. However, applying controlled mechanical forces to whole organs ex vivo remains a persistent challenge. In recent years, methods have been developed for lung culture models to determine how mechanical forces, including transmural fluid pressure, breathing movements, and airway smooth muscle contractions, are coupled to developmental programs to guide cell proliferation, differentiation, and airway morphogenesis [93, 98–100].

Indeed, work using whole mouse lung explants has revealed the critical roles of transmural fluid pressure in early embryonic airway branching in a microfluidic chest cavity model [101, 102]. Similarly, the trachea was intubated and pulmonary vasculature was perfused in E16.5–18.5 mouse embryos to investigate the role of fetal breathing movements and FGF10 signaling on distal airway development during saccularization [103]. These whole organ perfusion approaches are analogous to whole organ culture models in adult rodents, large animals, and human lung explants, which are often used for disease modeling and drug inhalation studies [104–108]. The adaptation of these techniques to mouse embryonic, fetal, and neonatal lung explants offers a powerful tool to interrogate how mechanical forces guide lung development in a genetically tractable lung model. Continued adaptation of these systems to murine models of late-stage development, encompassing the end of saccularization through alveolarization, is an important avenue for continued mechanistic investigation of normal development and disease.

## 3.5 Conclusion

The model systems reviewed herein represent significant and exciting advancements in our ability to add and interrogate complex cellular niches and physiological functions in the developing lung. These approaches have unique advantages, but the collective synergy of these models offers the ability to decouple tissue and organ-scale biochemical and biophysical signaling networks that drive and regulate lung organogenesis (Fig. 3.4). Organoids can generate cellular heterogeneity

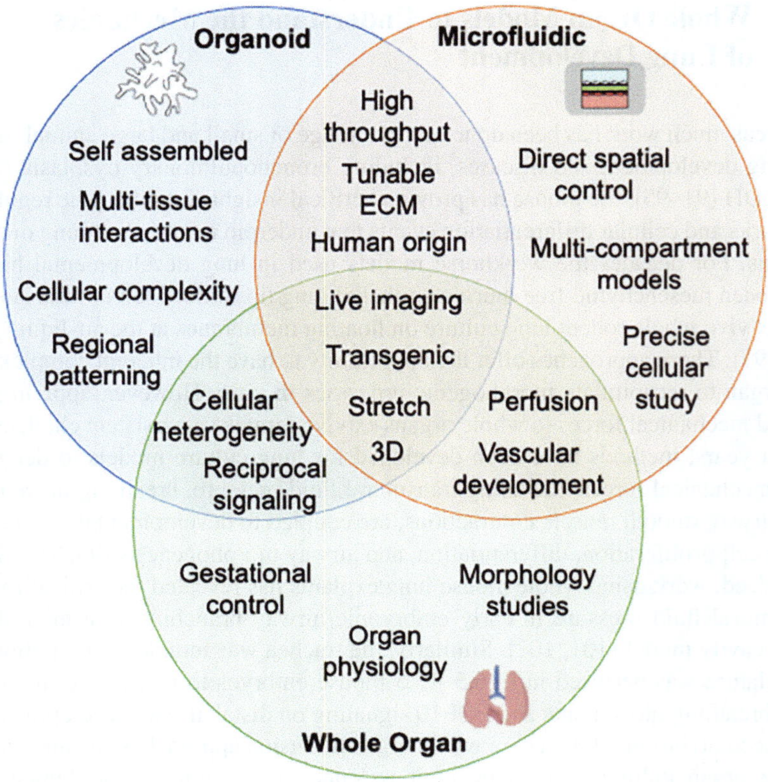


**Fig. 3.4** Capabilities and benefits of the three central model systems for building complexity into lung development studies.

and structural complexity that mimic native lung niches more closely. Moreover, given the ability to genetically modify the cells being used, organoids provide a powerful platform for screening signaling pathways and their impacts on morphology and function that are often intractable in humans. Organ-on-a-chip systems allow for the direct spatial control of cellular positioning and biochemical and biophysical cues. These systems provide geometric control with the ability to successfully integrate multiple tissue compartments to provide quantitative physiological and functional outputs from human cell sources. Whole organ culture captures the full complexity of the mammalian lung and gestational-stage control. Current efforts to advance these models and integrate them, leveraging the unique advantages of each, will enable the mechanistic understanding of lung development and provide insight into congenital disease mechanisms and translational therapies for newborns.

## References


1. R. R. Nadkarni, S. Abed, and J. S. Draper, "Organoids as a model system for studying human lung development and disease," *Biochemical and Biophysical Research Communications*, vol. 473, no. 3, pp. 675–682, May 2016, https://doi.org/10.1016/j.bbrc.2015.12.091.
2. J. van der Vaart and H. Clevers, "Airway organoids as models of human disease," *Journal of Internal Medicine*, vol. 289, no. 5, pp. 604–613, 2021, https://doi.org/10.1111/joim.13075.
3. Y. Zhang *et al.*, "MicroRNA-30a as a candidate underlying sex-specific differences in neonatal hyperoxic lung injury: implications for BPD," *American Journal of Physiology-Lung Cellular and Molecular Physiology*, vol. 316, no. 1, pp. L144–L156, 2019.
4. Y. Zhang, X. Dong, J. Shirazi, J. P. Gleghorn, and K. Lingappan, "Pulmonary endothelial cells exhibit sexual dimorphism in their response to hyperoxia," *American Journal of Physiology-Heart and Circulatory Physiology*, vol. 315, no. 5, pp. H1287–H1292, 2018.
5. Z. Jakus *et al.*, "Lymphatic function is required prenatally for lung inflation at birth," *J Exp Med*, vol. 211, no. 5, pp. 815–826, May 2014, https://doi.org/10.1084/jem.20132308.
6. A. J. Miller *et al.*, "Generation of lung organoids from human pluripotent stem cells in vitro," *Nat Protoc*, vol. 14, no. 2, pp. 518–540, Feb. 2019, https://doi.org/10.1038/s41596-018-0104-8.
7. A. J. Miller and J. R. Spence, "In Vitro Models to Study Human Lung Development, Disease and Homeostasis," *Physiology*, vol. 32, no. 3, pp. 246–260, May 2017, https://doi.org/10.1152/physiol.00041.2016.
8. H. Clevers, "Modeling Development and Disease with Organoids," *Cell*, vol. 165, no. 7, pp. 1586–1597, Jun. 2016, https://doi.org/10.1016/j.cell.2016.05.082.
9. H. Sugihara, S. Toda, S. Miyabara, C. Fujiyama, and N. Yonemitsu, "Reconstruction of alveolus-like structure from alveolar type II epithelial cells in three-dimensional collagen gel matrix culture.," *Am J Pathol*, vol. 142, no. 3, pp. 783–792, Mar. 1993.
10. W. H. J. Douglas, G. W. Moorman, and R. W. Teel, "The formation of histotypic structures from monodisperse fetal rat lung cells cultured on a three-dimensional substrate," *In Vitro Cell. Dev. Biol.-Plant*, vol. 12, no. 5, pp. 373–381, May 1976, https://doi.org/10.1007/BF02796315.
11. J. S. Paquette *et al.*, "PRODUCTION OF TISSUE-ENGINEERED THREE-DIMENSIONAL HUMAN BRONCHIAL MODELS," *In Vitro Cell Dev Biol Anim*, vol. 39, no. 5, p. 213, 2003, https://doi.org/10.1290/1543-706X(2003)039<0213:POTTHB>2.0.CO;2.
12. J. R. Rock *et al.*, "Basal cells as stem cells of the mouse trachea and human airway epithelium," *Proceedings of the National Academy of Sciences*, vol. 106, no. 31, pp. 12771–12775, Aug. 2009, https://doi.org/10.1073/pnas.0906850106.
13. D. A. Lawson, L. Xin, R. U. Lukacs, D. Cheng, and O. N. Witte, "Isolation and functional characterization of murine prostate stem cells," *Proc. Natl. Acad. Sci. U.S.A.*, vol. 104, no. 1, pp. 181–186, Jan. 2007, https://doi.org/10.1073/pnas.0609684104.
14. X. Gao, A. S. Bali, S. H. Randell, and B. L. M. Hogan, "GRHL2 coordinates regeneration of a polarized mucociliary epithelium from basal stem cells," *Journal of Cell Biology*, vol. 211, no. 3, pp. 669–682, Nov. 2015, https://doi.org/10.1083/jcb.201506014.
15. T. Tadokoro, X. Gao, C. C. Hong, D. Hotten, and B. L. M. Hogan, "BMP signaling and cellular dynamics during regeneration of airway epithelium from basal progenitors," *Development*, vol. 143, no. 5, pp. 764–773, Mar. 2016, https://doi.org/10.1242/dev.126656.
16. J. R. Rock, X. Gao, Y. Xue, S. H. Randell, Y.-Y. Kong, and B. L. M. Hogan, "Notch-Dependent Differentiation of Adult Airway Basal Stem Cells," *Cell Stem Cell*, vol. 8, no. 6, pp. 639–648, Jun. 2011, https://doi.org/10.1016/j.stem.2011.04.003.
17. T. Tadokoro, Y. Wang, L. S. Barak, Y. Bai, S. H. Randell, and B. L. Hogan, "IL-6/STAT3 promotes regeneration of airway ciliated cells from basal stem cells," *Proceedings of the National Academy of Sciences*, vol. 111, no. 35, pp. E3641–E3649, 2014.
18. H. Danahay *et al.*, "Notch2 Is Required for Inflammatory Cytokine-Driven Goblet Cell Metaplasia in the Lung," *Cell Reports*, vol. 10, no. 2, pp. 239–252, Jan. 2015, https://doi.org/10.1016/j.celrep.2014.12.017.

19. A. E. Hegab *et al.*, "Novel Stem/Progenitor Cell Population from Murine Tracheal Submucosal Gland Ducts with Multipotent Regenerative Potential," *Stem Cells*, vol. 29, no. 8, pp. 1283–1293, Aug. 2011, https://doi.org/10.1002/stem.680.
20. A. E. Hegab *et al.*, "Isolation and In Vitro Characterization of Basal and Submucosal Gland Duct Stem/Progenitor Cells from Human Proximal Airways," *Stem Cells Translational Medicine*, vol. 1, no. 10, pp. 719–724, Oct. 2012, https://doi.org/10.5966/sctm.2012-0056.
21. M. Bilodeau, S. Shojaie, C. Ackerley, M. Post, and J. Rossant, "Identification of a Proximal Progenitor Population from Murine Fetal Lungs with Clonogenic and Multilineage Differentiation Potential," *Stem Cell Reports*, vol. 3, no. 4, pp. 634–649, Oct. 2014, https://doi.org/10.1016/j.stemcr.2014.07.010.
22. H. Mou *et al.*, "Generation of Multipotent Lung and Airway Progenitors from Mouse ESCs and Patient-Specific Cystic Fibrosis iPSCs," *Cell Stem Cell*, vol. 10, no. 4, pp. 385–397, Apr. 2012, https://doi.org/10.1016/j.stem.2012.01.018.
23. S. Konishi *et al.*, "Directed Induction of Functional Multi-ciliated Cells in Proximal Airway Epithelial Spheroids from Human Pluripotent Stem Cells," *Stem Cell Reports*, vol. 6, no. 1, pp. 18–25, Jan. 2016, https://doi.org/10.1016/j.stemcr.2015.11.010.
24. "Efficient Derivation of Functional Human Airway Epithelium from Pluripotent Stem Cells via Temporal Regulation of Wnt Signaling," *Cell Stem Cell*, vol. 20, no. 6, pp. 844–857.e6, Jun. 2017, https://doi.org/10.1016/j.stem.2017.03.001.
25. M. Serra *et al.*, "Pluripotent stem cell differentiation reveals distinct developmental pathways regulating lung- versus thyroid-lineage specification," *Development*, vol. 144, no. 21, pp. 3879–3893, Nov. 2017, https://doi.org/10.1242/dev.150193.
26. K. B. McCauley *et al.*, "Single-Cell Transcriptomic Profiling of Pluripotent Stem Cell-Derived SCGB3A2+ Airway Epithelium," *Stem Cell Reports*, vol. 10, no. 5, pp. 1579–1595, May 2018, https://doi.org/10.1016/j.stemcr.2018.03.013.
27. W. Yu *et al.*, "Formation of Cysts by Alveolar Type II Cells in Three-dimensional Culture Reveals a Novel Mechanism for Epithelial Morphogenesis," *MBoC*, vol. 18, no. 5, pp. 1693–1700, May 2007, https://doi.org/10.1091/mbc.e06-11-1052.
28. H. Katsura *et al.*, "Human Lung Stem Cell-Based Alveolospheres Provide Insights into SARS-CoV-2-Mediated Interferon Responses and Pneumocyte Dysfunction," *Cell Stem Cell*, vol. 27, no. 6, pp. 890–904.e8, Dec. 2020, https://doi.org/10.1016/j.stem.2020.10.005.
29. V. Narayanan *et al.*, "Osmotic Gradients in Epithelial Acini Increase Mechanical Tension across E-cadherin, Drive Morphogenesis, and Maintain Homeostasis," *Current Biology*, vol. 30, no. 4, pp. 624–633.e4, Feb. 2020, https://doi.org/10.1016/j.cub.2019.12.025.
30. P. Kadur Lakshminarasimha Murthy *et al.*, "Human distal lung maps and lineage hierarchies reveal a bipotent progenitor," *Nature*, vol. 604, no. 7904, Art. no. 7904, Apr. 2022, https://doi.org/10.1038/s41586-022-04541-3.
31. "Generation of Alveolar Epithelial Spheroids via Isolated Progenitor Cells from Human Pluripotent Stem Cells," *Stem Cell Reports*, vol. 3, no. 3, pp. 394–403, Sep. 2014, https://doi.org/10.1016/j.stemcr.2014.07.005.
32. S. X. L. Huang *et al.*, "Efficient generation of lung and airway epithelial cells from human pluripotent stem cells," *Nat Biotechnol*, vol. 32, no. 1, pp. 84–91, Jan. 2014, https://doi.org/10.1038/nbt.2754.
33. "Differentiation of Human Pluripotent Stem Cells into Functional Lung Alveolar Epithelial Cells," *Cell Stem Cell*, vol. 21, no. 4, pp. 472–488.e10, Oct. 2017, https://doi.org/10.1016/j.stem.2017.08.014.
34. D. B. Frank *et al.*, "Emergence of a Wave of Wnt Signaling that Regulates Lung Alveologenesis by Controlling Epithelial Self-Renewal and Differentiation," *Cell Reports*, vol. 17, no. 9, pp. 2312–2325, Nov. 2016, https://doi.org/10.1016/j.celrep.2016.11.001.
35. "Pulmonary alveolar type I cell population consists of two distinct subtypes that differ in cell fate." https://www.pnas.org/doi/10.1073/pnas.1719474115 (accessed May 29, 2022).
36. M. C. Basil *et al.*, "Human distal airways contain a multipotent secretory cell that can regenerate alveoli," *Nature*, vol. 604, no. 7904, Art. no. 7904, Apr. 2022, https://doi.org/10.1038/s41586-022-04552-0.

37. C. S. Millar-Haskell, A. M. Dang, and J. P. Gleghorn, "Coupling synthetic biology and programmable materials to construct complex tissue ecosystems," *MRS Commun*, vol. 9, no. 2, pp. 421–432, Jun. 2019, https://doi.org/10.1557/mrc.2019.69.
38. C. Loebel *et al.*, "Microstructured hydrogels to guide self-assembly and function of lung alveolospheres," *Advanced Materials*, p. 2202992, May 2022, https://doi.org/10.1002/adma.202202992.
39. J. P. Gleghorn, J. Kwak, A. L. Pavlovich, and C. M. Nelson, "Inhibitory morphogens and monopodial branching of the embryonic chicken lung," *Developmental Dynamics*, vol. 241, no. 5, pp. 852–862, 2012, https://doi.org/10.1002/dvdy.23771.
40. K. Lingappan, B. Hayward-Piatkovskyi, and J. P. Gleghorn, "Neonatal Lung Disease: Mechanisms Driving Sex Differences," in *Sex-Based Differences in Lung Physiology*, P. Silveyra and X. T. Tigno, Eds. Cham: Springer International Publishing, 2021, pp. 115–144. https://doi.org/10.1007/978-3-030-63549-7_5.
41. "Sculpting Organs: Mechanical Regulation of Tissue Development | Annual Review of Biomedical Engineering." https://www.annualreviews.org/doi/10.1146/annurev-bioeng-071811-150043?url_ver=Z39.88-2003 & rfr_id=ori%3Arid%3Acrossref.org & rfr_dat=cr_pub++0pubmed (accessed Jul. 17, 2022).
42. J. T. Morgan, W. G. Stewart, R. A. McKee, and J. P. Gleghorn, "The Mechanosensitive Ion Channel TRPV4 is a Regulator of Lung Development and Pulmonary Vasculature Stabilization," *Cel. Mol. Bioeng.*, vol. 11, no. 5, pp. 309–320, Oct. 2018, https://doi.org/10.1007/s12195-018-0538-7.
43. C. R. Butler *et al.*, "Rapid Expansion of Human Epithelial Stem Cells Suitable for Airway Tissue Engineering," *Am J Respir Crit Care Med*, vol. 194, no. 2, pp. 156–168, Jul. 2016, https://doi.org/10.1164/rccm.201507-1414OC.
44. C. A. Boecking *et al.*, "A simple method to generate human airway epithelial organoids with externally orientated apical membranes," *American Journal of Physiology-Lung Cellular and Molecular Physiology*, vol. 322, no. 3, pp. L420–L437, Mar. 2022, https://doi.org/10.1152/ajplung.00536.2020.
45. "Evidence of an epithelial stem/progenitor cell hierarchy in the adult mouse lung." https://www.pnas.org/doi/10.1073/pnas.0909207107 (accessed May 30, 2022).
46. S. R. Franzdóttir, I. T. Axelsson, A. J. Arason, Ó. Baldursson, T. Gudjonsson, and M. K. Magnusson, "Airway branching morphogenesis in three dimensional culture," *Respiratory Research*, vol. 11, no. 1, p. 162, Nov. 2010, https://doi.org/10.1186/1465-9921-11-162.
47. B.-S. Ding *et al.*, "Endothelial-Derived Angiocrine Signals Induce and Sustain Regenerative Lung Alveolarization," *Cell*, vol. 147, no. 3, pp. 539–553, Oct. 2011, https://doi.org/10.1016/j.cell.2011.10.003.
48. J.-H. Lee *et al.*, "Lung Stem Cell Differentiation in Mice Directed by Endothelial Cells via a BMP4-NFATc1-Thrombospondin-1 Axis," *Cell*, vol. 156, no. 3, pp. 440–455, Jan. 2014, https://doi.org/10.1016/j.cell.2013.12.039.
49. J. Shirazi, J. T. Morgan, E. M. Comber, and J. P. Gleghorn, "Generation and morphological quantification of large scale, three-dimensional, self-assembled vascular networks," *MethodsX*, vol. 6, pp. 1907–1918, Jan. 2019, https://doi.org/10.1016/j.mex.2019.08.006.
50. J. T. Morgan, J. Shirazi, E. M. Comber, C. Eschenburg, and J. P. Gleghorn, "Fabrication of centimeter-scale and geometrically arbitrary vascular networks using in vitro self-assembly," *Biomaterials*, vol. 189, pp. 37–47, Jan. 2019, https://doi.org/10.1016/j.biomaterials.2018.10.021.
51. M. J. Mondrinos, S. Koutzaki, P. I. Lelkes, and C. M. Finck, "A tissue-engineered model of fetal distal lung tissue," *American Journal of Physiology-Lung Cellular and Molecular Physiology*, vol. 293, no. 3, pp. L639–L650, Sep. 2007, https://doi.org/10.1152/ajplung.00403.2006.
52. M. J, J. L, F. M, and L. I, "Engineering De Novo Assembly of Fetal Pulmonary Organoids," *Tissue Engineering Part A*, Jun. 2014, https://doi.org/10.1089/ten.tea.2014.0085.
53. Q. Tan, K. M. Choi, D. Sicard, and D. J. Tschumperlin, "Human airway organoid engineering as a step toward lung regeneration and disease modeling," *Biomaterials*, vol. 113, pp. 118–132, Jan. 2017, https://doi.org/10.1016/j.biomaterials.2016.10.046.

54. B. R. Dye *et al.*, "In vitro generation of human pluripotent stem cell derived lung organoids," *eLife*, vol. 4, p. e05098, Mar. 2015, https://doi.org/10.7554/eLife.05098.
55. B. R. Dye *et al.*, "A bioengineered niche promotes in vivo engraftment and maturation of pluripotent stem cell derived human lung organoids," *eLife*, vol. 5, p. e19732, Oct. 2016, https://doi.org/10.7554/eLife.19732.
56. Chen Yong *et al.*, "Long-Term Engraftment Promotes Differentiation of Alveolar Epithelial Cells from Human Embryonic Stem Cell Derived Lung Organoids," *Stem Cells and Development*, Sep. 2018, https://doi.org/10.1089/scd.2018.0042.
57. S. M. Kunisaki *et al.*, "Human induced pluripotent stem cell-derived lung organoids in an ex vivo model of the congenital diaphragmatic hernia fetal lung," *Stem Cells Transl Med*, vol. 10, no. 1, pp. 98–114, Sep. 2020, https://doi.org/10.1002/sctm.20-0199.
58. O. O. Olutoye II *et al.*, "The Cellular and Molecular Effects of Fetoscopic Endoluminal Tracheal Occlusion in Congenital Diaphragmatic Hernia," *Frontiers in Pediatrics*, vol. 10, p. 925106, 2022.
59. Y.-W. Chen *et al.*, "A three-dimensional model of human lung development and disease from pluripotent stem cells," *Nat Cell Biol*, vol. 19, no. 5, Art. no. 5, May 2017, https://doi.org/10.1038/ncb3510.
60. M. Cabodi, N. W. Choi, J. P. Gleghorn, C. S. Lee, L. J. Bonassar, and A. D. Stroock, "A microfluidic biomaterial," *Journal of the American Chemical Society*, vol. 127, no. 40, pp. 13788–13789, 2005.
61. S. Manivannan, J. P. Gleghorn, and C. M. Nelson, "Engineered tissues to quantify collective cell migration during morphogenesis," in *Kidney Development*, Springer, 2012, pp. 173–182.
62. S. Takayama, E. Ostuni, P. LeDuc, K. Naruse, D. E. Ingber, and G. M. Whitesides, "Subcellular positioning of small molecules," *Nature*, vol. 411, no. 6841, pp. 1016–1016, Jun. 2001, https://doi.org/10.1038/35082637.
63. R. M. Gilbert, J. T. Morgan, E. S. Marcin, and J. P. Gleghorn, "Fluid mechanics as a driver of tissue-scale mechanical signaling in organogenesis," *Current pathobiology reports*, vol. 4, no. 4, pp. 199–208, 2016.
64. J. P. Gleghorn, S. Manivannan, and C. M. Nelson, "Quantitative approaches to uncover physical mechanisms of tissue morphogenesis," *Current opinion in biotechnology*, vol. 24, no. 5, pp. 954–961, 2013.
65. D. Huh, Y. Kamotani, J. B. Grotberg, and S. Takayama, "Compartmentalized microfluidic lung epithelial cell culture device for pulmonary mechanotransduction studies," *SPECIAL PUBLICATION-ROYAL SOCIETY OF CHEMISTRY*, vol. 297, pp. 282–284, 2004.
66. D. Huh *et al.*, "Acoustically detectable cellular-level lung injury induced by fluid mechanical stresses in microfluidic airway systems," *Proceedings of the National Academy of Sciences*, vol. 104, no. 48, pp. 18886–18891, 2007.
67. H. Tavana, P. Zamankhan, P. J. Christensen, J. B. Grotberg, and S. Takayama, "Epithelium damage and protection during reopening of occluded airways in a physiologic microfluidic pulmonary airway model," *Biomedical microdevices*, vol. 13, no. 4, pp. 731–742, 2011.
68. Y. Hu *et al.*, "A microfluidic model to study fluid dynamics of mucus plug rupture in small lung airways," *Biomicrofluidics*, vol. 9, no. 4, p. 044119, 2015.
69. N. J. Douville *et al.*, "Combination of fluid and solid mechanical stresses contribute to cell death and detachment in a microfluidic alveolar model," *Lab on a Chip*, vol. 11, no. 4, pp. 609–619, 2011.
70. H. Tavana *et al.*, "Dynamics of liquid plugs of buffer and surfactant solutions in a micro-engineered pulmonary airway model," *Langmuir*, vol. 26, no. 5, pp. 3744–3752, 2010.
71. V. Kumar *et al.*, "An in vitro microfluidic alveolus model to study lung biomechanics," *Frontiers in bioengineering and biotechnology*, p. 166, 2022.
72. M. Humayun, C.-W. Chow, and E. W. Young, "Microfluidic lung airway-on-a-chip with arrayable suspended gels for studying epithelial and smooth muscle cell interactions," *Lab on a Chip*, vol. 18, no. 9, pp. 1298–1309, 2018.
73. D. Huh, B. D. Matthews, A. Mammoto, M. Montoya-Zavala, H. Y. Hsin, and D. E. Ingber, "Reconstituting organ-level lung functions on a chip," *Science*, vol. 328, no. 5986, pp. 1662–1668, 2010.

74. D. Huh and D. E. Ingber, "Lung organomimetic microdevice," presented at the Thirteenth International Conference on Miniaturized Systems for Chemistry and Life Sciences, 2009.
75. A. Jain *et al.*, "Primary human lung alveolus-on-a-chip model of intravascular thrombosis for assessment of therapeutics," *Clinical pharmacology & therapeutics*, vol. 103, no. 2, pp. 332–340, 2018.
76. H. Nam, Y. Choi, and J. Jang, "Vascularized lower respiratory-physiology-on-a-chip," *Applied Sciences*, vol. 10, no. 3, p. 900, 2020.
77. J. D. Stucki *et al.*, "Medium throughput breathing human primary cell alveolus-on-chip model," *Scientific reports*, vol. 8, no. 1, pp. 1–13, 2018.
78. B. F. Niemeyer, P. Zhao, R. M. Tuder, and K. H. Benam, "Advanced microengineered lung models for translational drug discovery," *SLAS DISCOVERY: Advancing Life Sciences R & D*, vol. 23, no. 8, pp. 777–789, 2018.
79. C. Lin, X. Zheng, S. Lin, Y. Zhang, J. Wu, and Y. Li, "Mechanotransduction Regulates the Interplays Between Alveolar Epithelial and Vascular Endothelial Cells in Lung," *Frontiers in Physiology*, p. 246, 2022.
80. L. Si, H. Bai, C. Y. Oh, L. Jin, R. Prantil-Baun, and D. E. Ingber, "Clinically relevant influenza virus evolution reconstituted in a human lung airway-on-a-chip," *Microbiology Spectrum*, vol. 9, no. 2, pp. e00257–21, 2021.
81. L. Si *et al.*, "A human-airway-on-a-chip for the rapid identification of candidate antiviral therapeutics and prophylactics," *Nature biomedical engineering*, vol. 5, no. 8, pp. 815–829, 2021.
82. K. H. Benam *et al.*, "Human small airway-on-a-chip: A novel microphysiological system to model lung inflammation, accelerate drug development and enable inhalational toxicoanalysis," 2016.
83. K. H. Benam *et al.*, "Small airway-on-a-chip enables analysis of human lung inflammation and drug responses in vitro," *Nature methods*, vol. 13, no. 2, pp. 151–157, 2016.
84. R. Plebani *et al.*, "Modeling pulmonary cystic fibrosis in a human lung airway-on-a-chip," *Journal of Cystic Fibrosis*, 2021.
85. O. Delgado *et al.*, "Multipotent capacity of immortalized human bronchial epithelial cells," *PloS one*, vol. 6, no. 7, p. e22023, 2011.
86. T. Caracena, R. Blomberg, R. S. Hewawasam, D. W. Riches, and C. M. Magin, "Transitional alveolar epithelial cells and microenvironmental stiffness synergistically drive fibroblast activation in three-dimensional hydrogel lung models," *bioRxiv*, 2022.
87. D. Huang *et al.*, "Reversed-engineered human alveolar lung-on-a-chip model," *Proceedings of the National Academy of Sciences*, vol. 118, no. 19, p. e2016146118, 2021.
88. D. Baptista *et al.*, "3D Lung-on-Chip Model Based on Biomimetically Microcurved Culture Membranes," *ACS Biomaterials Science & Engineering*, 2022.
89. P. Zamprogno *et al.*, "Second-generation lung-on-a-chip with an array of stretchable alveoli made with a biological membrane," *Communications biology*, vol. 4, no. 1, pp. 1–10, 2021.
90. T. H. Shin, M. Kim, C. O. Sung, S. J. Jang, and G. S. Jeong, "A one-stop microfluidic-based lung cancer organoid culture platform for testing drug sensitivity," *Lab on a Chip*, vol. 19, no. 17, pp. 2854–2865, 2019.
91. X. Zhang, X. Chu, B. Weng, X. Gong, and C. Cai, "An Innovative Model of Bronchopulmonary Dysplasia in Premature Infants," *Frontiers in Pediatrics*, vol. 8, 2020, Accessed: Jun. 21, 2022. [Online]. Available: https://www.frontiersin.org/article/10.3389/fped.2020.00271
92. N. Ambalavanan and R. E. Morty, "Searching for better animal models of BPD: a perspective," *Am J Physiol Lung Cell Mol Physiol*, vol. 311, no. 5, pp. L924–L927, Nov. 2016, https://doi.org/10.1152/ajplung.00355.2016.
93. R. M. Gilbert, L. E. Schappell, and J. P. Gleghorn, "Defective mesothelium and limited physical space are drivers of dysregulated lung development in a genetic model of congenital diaphragmatic hernia," *Development*, vol. 148, no. 10, p. dev199460, May 2021, https://doi.org/10.1242/dev.199460.
94. P. P. L. Chiu, "New Insights into Congenital Diaphragmatic Hernia – A Surgeon's Introduction to CDH Animal Models," *Frontiers in Pediatrics*, vol. 2, 2014, Accessed: Jun. 21, 2022. [Online]. Available: https://www.frontiersin.org/article/10.3389/fped.2014.00036
95. L. Sbragia *et al.*, "A novel surgical toxicological-free model of diaphragmatic hernia in fetal rats," *Pediatr Res*, pp. 1–7, Aug. 2021, https://doi.org/10.1038/s41390-021-01702-4.

96. V. D. Varner, J. P. Gleghorn, E. Miller, D. C. Radisky, and C. M. Nelson, "Mechanically patterning the embryonic airway epithelium," *Proceedings of the National Academy of Sciences*, vol. 112, no. 30, pp. 9230–9235, Jul. 2015, https://doi.org/10.1073/pnas.1504102112.
97. "Morphogenesis and morphometric scaling of lung airway development follows phylogeny in chicken, quail, and duck embryos | EvoDevo | Full Text." https://evodevojournal.biomedcentral.com/articles/10.1186/s13227-016-0049-3 (accessed Jul. 18, 2022).
98. E. El Agha, V. Kheirollahi, A. Moiseenko, W. Seeger, and S. Bellusci, "Ex vivo analysis of the contribution of FGF10+ cells to airway smooth muscle cell formation during early lung development," *Developmental Dynamics*, vol. 246, no. 7, pp. 531–538, 2017, https://doi.org/10.1002/dvdy.24504.
99. M. Unbekandt, P.-M. del Moral, F. G. Sala, S. Bellusci, D. Warburton, and V. Fleury, "Tracheal occlusion increases the rate of epithelial branching of embryonic mouse lung via the FGF10-FGFR2b-Sprouty2 pathway," *Mechanisms of Development*, vol. 125, no. 3, pp. 314–324, Mar. 2008, https://doi.org/10.1016/j.mod.2007.10.013.
100. L. E. Schappell, D. J. Minahan, and J. P. Gleghorn, "A Microfluidic System to Measure Neonatal Lung Compliance Over Late Stage Development as a Functional Measure of Lung Tissue Mechanics," *Journal of Biomechanical Engineering*, vol. 142, no. 10, Aug. 2020, https://doi.org/10.1115/1.4047133.
101. C. M. Nelson *et al.*, "Microfluidic chest cavities reveal that transmural pressure controls the rate of lung development," *Development*, p. dev.154823, Jan. 2017, https://doi.org/10.1242/dev.154823.
102. A. E. Stanton *et al.*, "Negative Transpulmonary Pressure Disrupts Airway Morphogenesis by Suppressing Fgf10," *Frontiers in Cell and Developmental Biology*, vol. 9, 2021.
103. J. Li, Z. Wang, Q. Chu, K. Jiang, J. Li, and N. Tang, "The Strength of Mechanical Forces Determines the Differentiation of Alveolar Epithelial Cells," *Developmental Cell*, vol. 44, no. 3, pp. 297–312.e5, Feb. 2018, https://doi.org/10.1016/j.devcel.2018.01.008.
104. J. B. Gordon and M. L. Tod, "Effects of N omega-nitro-L-arginine on total and segmental vascular resistances in developing lamb lungs," *Journal of Applied Physiology*, vol. 75, no. 1, pp. 76–85, Jul. 1993, https://doi.org/10.1152/jappl.1993.75.1.76.
105. H. Y. Yoo *et al.*, "Optimization of Isolated Perfused/Ventilated Mouse Lung to Study Hypoxic Pulmonary Vasoconstriction," *Pulm Circ*, vol. 3, no. 2, pp. 396–405, Apr. 2013, https://doi.org/10.4103/2045-8932.114776.
106. J. S. Torday, E. B. Olson, and N. L. First, "Production of cortisol from cortisone by the isolated, perfused fetal rabbit lung," *Steroids*, vol. 27, no. 6, pp. 869–880, Jun. 1976, https://doi.org/10.1016/0039-128X(76)90145-8.
107. M. Sakagami *et al.*, "Expression and Transport Functionality of FcRn within Rat Alveolar Epithelium: A Study in Primary Cell Culture and in the Isolated Perfused Lung," *Pharm Res*, vol. 23, no. 2, pp. 270–279, Feb. 2006, https://doi.org/10.1007/s11095-005-9226-0.
108. C. Lonati *et al.*, "Mesenchymal stem cell–derived extracellular vesicles improve the molecular phenotype of isolated rat lungs during ischemia/reperfusion injury," *The Journal of Heart and Lung Transplantation*, vol. 38, no. 12, pp. 1306–1316, Dec. 2019, https://doi.org/10.1016/j.healun.2019.08.016.

# Chapter 4
# Multipotent Embryonic Lung Progenitors: Foundational Units of In Vitro and In Vivo Lung Organogenesis

**Laertis Ikonomou, Maria Yampolskaya, and Pankaj Mehta**

## 4.1 Introduction

Tissue-specific embryonic progenitors are important gateway cell populations at various stages of organ development [49]. Within the endoderm germ layer, a substantial body of work has deciphered the emergence of domain-specific primordial progenitors, including signaling requirements for tissue specification, and has uncovered key transcription factors (TFs), such as *Nkx2-1* for the lung and thyroid domains, *Pdx1* for the pancreatic domain, and *Cdx2* for the intestinal domain, for tissue specification or development or both [30, 52, 55]. Most importantly,

The original version of this chapter was previously published non-open access. A Correction to this chapter is available at https://doi.org/10.1007/978-3-031-26625-6_18

L. Ikonomou (✉)
Department of Oral Biology, University at Buffalo, The State University of New York, Buffalo, NY, USA

Division of Pulmonary, Critical Care and Sleep Medicine, Department of Medicine, University at Buffalo, The State University of New York, Buffalo, NY, USA

Cell, Gene and Tissue Engineering Center, University at Buffalo, The State University of New York, Buffalo, NY, USA
e-mail: laertisi@buffalo.edu

M. Yampolskaya
Department of Physics, Boston University, Boston, MA, USA

P. Mehta
Department of Physics, Boston University, Boston, MA, USA

Faculty of Computing and Data Science, Boston University, Boston, MA, USA

Biological Design Center, Boston University, Boston, MA, USA



C. M. Magin (ed.), *Engineering Translational Models of Lung Homeostasis and Disease*, Advances in Experimental Medicine and Biology 1413,
https://doi.org/10.1007/978-3-031-26625-6_4

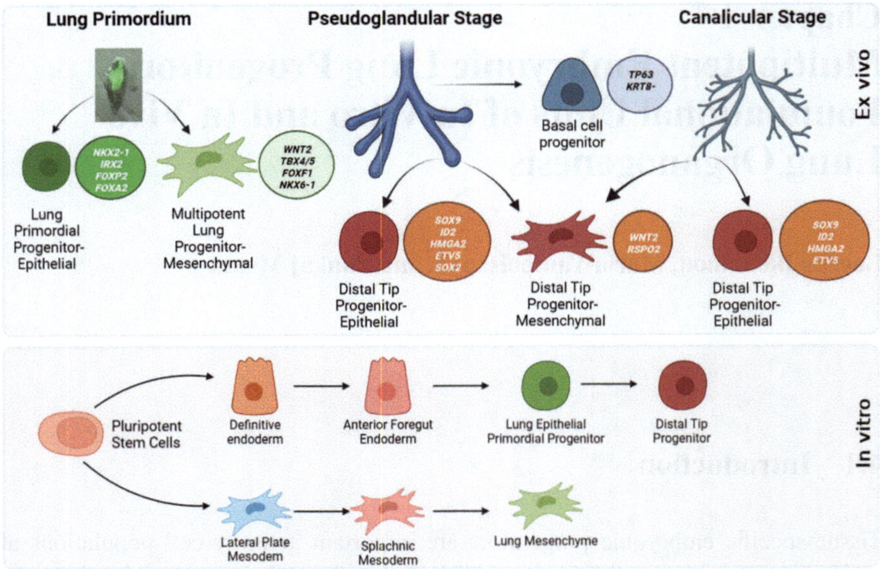


**Fig. 4.1** Major types of progenitor cells in the developing lung and their in vitro engineered counterparts. *SOX2* is expressed only in human distal lung tips

self-renewing, transplantable, multipotent lung progenitors are of value in lung regenerative medicine; mechanistic understanding of their cell fate decisions will be key for their successful clinical application.

In this chapter, we review the biology of multipotent embryonic progenitors of the respiratory system and recent efforts to employ such progenitors for the modeling and study of cell fate decisions in ex vivo organotypic models of human lung development (Fig. 4.1). We also assess efforts to derive de novo lung progenitors of different developmental stages from pluripotent stem cells (PSCs). As the areas of mammalian developmental and PSC biology continuously inform each other [96], there is an increasing need for computational models to evaluate and quantify the similarity between embryonic and PSC-derived progenitors. We critically appraise the nascent literature on cell similarity models, including their limitations and underlying assumptions.

## 4.2 Overview of Embryonic Lung Progenitors

### *4.2.1 Stage-Specific Epithelial Progenitors (Primordial, Distal Tip, Basal)*

The respiratory system is specified within the anterior ventral foregut endoderm (the anterior part of the gut tube) at around embryonic day (E)9.0 (18–22 somites) in mice and 4 weeks of gestation in humans [48, 102]. During subsequent stages (embryonic, pseudoglandular, canalicular, saccular, and alveolar), developmental processes, such as tracheoesophageal septation, formation and elongation of the primary lung buds, branching morphogenesis, and alveologenesis, lead to the anatomically and functionally distinct compartments of the adult lung that perform the basic functions of air conduction, mucociliary clearance, and gas exchange. These processes are concurrent with and inseparable from diversification of cell fates driven by stage-specific embryonic progenitors. Genetic alteration or ablation of such progenitors can lead to profound developmental defects, such as lung agenesis and mispatterning of the lung proximal-distal axis.

## Lung Primordial Progenitors

The lung epithelial primordial progenitors are the first epithelial cells within the developing gut tube to adopt a lung fate at around E9.0 in the mouse. The appearance of this rare, transient progenitor population is marked by expression of the homeodomain TF *Nkx2-1* [48, 54]. *Nkx2-1*$^+$ lung primordial progenitors are multipotent progenitors and give rise to the vast majority of lung epithelial lineages (both airway and alveolar), as shown by a lineage tracing study performed in our laboratory [48]. Although *Nkx2-1* is not necessary for lung specification [55], it is the earliest marker of lung epithelial fate, and its expression has been used extensively to benchmark PSC-derived lung multipotent progenitors, as discussed below. Remarkably, *ventx2*, a *Xenopus* homeodomain TF, was recently shown to be expressed in the prospective respiratory domain at Nieuwkoop and Faber (NF)31/32 stage, earlier than *Nkx2-1* (NF33/34), raising the tantalizing possibility that it is the earliest marker of respiratory (tracheal and lung) fate in *Xenopus* and potentially in humans [90].

Given the inaccessibility of the lung primordial population in humans, most information on the molecular pathways regulating its emergence and potential gene regulatory networks derives from mouse and *Xenopus* studies, both in vivo and ex vivo [15, 48, 89, 102]. *Xenopus (Xenopus laevis, African clawed frog)* has provided an excellent, complementary to mouse, model of foregut development, due to highly conserved endoderm formation, rapid development, and facile acquisition of large numbers of embryos, offering the possibility of medium-to-high-throughput loss-of-function studies [62, 89]. Mesoderm-derived Wnt signaling is indispensable for lung specification [32, 38], and Bmp signaling is necessary for correct spatial restriction of respiratory fated progenitors and tracheoesophageal separation [26]. Retinoic acid (RA) signaling confers lung competence to endodermal cells, which can then respond to Wnt and Bmp signals [87, 88]. The minimal signaling requirements for lung endodermal fate have also been demonstrated by pathway analysis between the transcriptomes of flow sorted lung primordial progenitors and prespecified endoderm, which demonstrated upregulation of Wnt and Tgf-β superfamily pathways in the former population [48].

In stark contrast to other Nkx2-1$^+$ embryonic tissues, such as thyroid and forebrain [24, 97], little else is known regarding the gene regulatory program underlying the singular lung primordial identity. Apart from Nkx2-1, pan-endodermal TFs, such as *Foxa2* [99], and TFs with known roles in later lung development, such as

*Irx2* and *Foxp2* [36, 48], are also expressed in the lung primordium. Future work will certainly decipher TF interactions that uniquely define the lung primordial identity and are involved in early epithelial cell fate decisions in lung development.

## Distal Tip Progenitors

Following respiratory system specification, its anterior-posterior axis is rapidly formed as witnessed by both morphological changes, such as trachea formation and elongation and emergence of primary lung buds, and mutually exclusive expression of the TFs *Sox2* and *Sox9* in the murine tracheal and lung domains [4]. During the ensuing pseudoglandular and canalicular stages, highly proliferating progenitors at the distal tips of the developing lung are essential for branching morphogenesis and formation of airway and alveolar epithelia [4, 81, 91]. Extensive work in mouse genetic models and, most recently, in human fetal organoids has shed light on the molecular profile of distal tip progenitors, their competence, and gene regulatory networks.

In mice, distal tip progenitors possess a distinct transcriptional profile with high expression of a constellation of TFs, such as *Sox9, Id2, Nmyc,* and *Etv5* [4, 44, 61, 67, 77, 80, 91]; matrisomal genes such as *Thbs1, Lama3,* and *Fbn2* [67, 81, 95]; and cell surface markers such as *Clu* [12, 81]. Interestingly, human distal tip progenitors also express *SOX2*, which is considered a proximal airway marker in murine development, during the pseudoglandular stage, but its distal expression is extinguished at the canalicular stage [22, 72, 78]. Lineage tracing studies have demonstrated that tip progenitors have a broad differentiation repertoire during the pseudoglandular stage (as early as E11.5), giving rise to major lung epithelial lineages (club, ciliated, neuroendocrine, and alveolar type I (AT1) and type II (AT2) cells) [91], while at later time points (E16.5) their differentiation potential is restricted to alveolar epithelial types. Nevertheless, late tip progenitors appear to possess some degree of plasticity since they can give rise to airway progeny when grafted in cultured E12.5 lung explants [61]. Further insights in the developmental potential of distal tip progenitors have been derived by combining RNA-sequencing (RNA-seq) with assay for transposase-accessible chromatin using sequencing (ATAC-seq) at two developmental time points, E11.5 and E16.5 [53]. Remarkably, almost 25% of open chromatin regions were specific to each time point, probably reflecting the time-dependent competence of tip progenitors. This integrated analysis identified increased expression and chromatin accessibility of PI3K signaling pathway components at the later point, and inhibition of this pathway at an early time point increased the percentage of $SOX9^+$/NKX2-$1^+$ lung epithelial progenitors.

Integration of several pathways, such as β-catenin, Fgf, and Bmp signaling, ensure the proliferation and maintenance of the distal tip progenitor identity, as has been demonstrated in numerous genetic mouse studies. β-catenin-dependent Wnt signaling, which is indispensable for lung specification, is also required to maintain distal tip progenitors, by controlling *Sox9* and *Nkx2-1* expression and acting downstream of Fgf/Kras signaling [81]. Precise modulation of β-catenin levels is critical for respiratory identity of tip progenitors, as either deletion of hyperactivation of β-catenin leads to the ectopic expression of gastrointestinal or liver genes within the developing respiratory epithelium [79, 81]. Glucocorticoid signaling, which is necessary for differentiation of AT2 cells [31, 105], also dictates, at late stages of respiratory development, the formation of proximal-distal boundary, namely, the bronchoalveolar duct junction (BADJ). Premature activation or deletion of glucocorticoid signaling leads to proximal or distal shift of the BADJ, respectively [4]. Glucocorticoid signaling and STAT3 signaling, activated by LIF and other IL6 family ligands, can act in concert to promote alveolar differentiation of distal tip progenitors from the canalicular stage [61].

Murine distal tip progenitors have a critical role in lung epithelial differentiation and morphogenesis. Information gleaned from developmental studies have propelled the development of lung-directed differentiation protocols from human and mouse PSCs, and ex vivo organoid culture of such human fetal progenitors has provided valuable insights in similarities and differences in respiratory development between the two species, as detailed below.

## Airway Basal Cells

An important stem cell population in the adult respiratory system are P63⁺/KRT5⁺ airway basal cells that can self-renew and differentiate to a wide variety of proximal cell types, such as secretory, ciliated, tuft, ionocyte, and neuroendocrine [6, 93]. There is a clear species difference in basal cell distribution as basal cell-containing pseudostratified epithelium is limited in the trachea and mainstem bronchi in mice but extends more distally spanning several airway generations in humans [94]. Recent work has shed light to the developmental origins of this population [73, 112]. Intriguingly, dispersed P63+ cells are already present in the Nkx2-1$^{GFP+}$ lung primordium (around E9.0–9.5) and can give rise to cells in both the airway and alveolar compartments [112]. Epithelial P63+ cells are rapidly restricted to the tracheal domain by E10.5 with concomitant lineage restriction to tracheal and proximal airways. Lineage tracing studies at later time points indicate that the pool of embryonic progenitors to E18.5 p63+Krt8- prebasal cells is established as early as E13.5–E14.5, as cells labeled at this time point comprise more than 80% of perinatal prebasal cells [112]. ITGB4$^{hi}$ multipotent progenitors have been described in early developing trachea (E14.5), but it is not clear to which extent these progenitors overlap with the P63+ progenitor pool [11]. In human fetal lungs, P63+/KRT5+ cells, considered bona fide basal cells, are present in trachea and mainstem bronchi by 12 weeks [73]. Interestingly, P63+/KRT5- in non-cartilaginous and small-cartilaginous airways have nuclear pSMAD staining, and they may be the human equivalent of basal cell progenitors described in the study of Yang et al. [112].

### 4.2.2 Stage-Specific Mesenchymal Progenitors

Mesenchymal-epithelial interactions are indispensable at every stage of lung development, from lung epithelial specification to branching morphogenesis. For example, different degrees of disruption of endodermally derived Shh signaling to splanchnic mesoderm lead to severe developmental effects, such as lung hypoplasia and tracheoesophageal fistula or even lung, tracheal, and esophageal agenesis [66, 74]. On the other hand, FGF10 expressed in the lung epithelium-adjacent mesoderm is important for bud induction [15], whereas, at later stages, mesenchymal FGF10 expression in the vicinity of distal tips is part of a complex signaling circuit that regulates tip outgrowth and proliferation [10, 63].

As the lung develops, the mesenchyme is also subject to a diversification of fates, leading to the various mesoderm-derived cell populations in the adult lung, such as vascular and airway smooth muscle cells, endothelial cells, pericytes, and pleura. Lineage tracing studies alone or in combination with new high-resolution genomic methods, such as single-cell RNA-seq (scRNA-seq), have allowed for the characterization of various mesenchymal progenitor populations and their developmental potential, as they have been comprehensively reviewed in Riccetti et al. [92]. For example, lineage tracing using different gene drivers has shown that several lung mesenchymal lineages, such as vascular and airway smooth muscle cells, proximal endothelial cells, and Pdgfrβ+ pericyte-like cells, originate from Wnt2+/Gli1+/Isl1+ cardiopulmonary progenitors [82]. Systems that employ *Tbx4* lung-specific enhancers have also been described and used for lineage tracing (as Cre or rtTA drivers) of mesenchymal progenitor progeny at various windows during lung development [59, 115].

Zorn et al. has recently used scRNA-seq to describe signaling networks governing interactions between mesenchymal and epithelial domains in early foregut development (E8.5–E9.5) [36]. This study revealed several mesenchymal cell clusters corresponding to domain-specific mesenchymal populations that engage in signaling crosstalk with the corresponding epithelial populations. In the developing lung mesenchyme, there was expression of previously described lung mesenchymal markers, such as *Wnt2* [32], *Osr1* [86], *Foxf1* [19, 84], and *Tbx5* [13]. *Nkx6-1*, a known pancreatic epithelial TF, was also expressed in the respiratory mesenchymal domain (both by transcript and protein), and it was one of the key TFs on the cell-state trajectory from foregut lateral mesoderm to lung mesenchyme [36].

Similar transcriptomic characterization of the developing human lung mesenchyme was performed using fetal human tissue from various endodermal organs, including the lung, from 7 to 21 weeks postconception [113]. As predicted, human lung mesenchyme was quite different than intestinal and stomach mesenchyme, with *WNT2* being one of the most differentially expressed genes. The TF *PRRX1* marked chondrocyte-like cells and pericytes in the lung. Furthermore, distal tip mesenchymal progenitors were also identified using human lung fetal tissue from pseudoglandular to canalicular stages [43]. These progenitors were RSPO2+ and could be purified based on LIFR expression. The RSPO2+ progenitors were found to maintain the distal tip undifferentiated state by epithelial LGR5-mediated Wnt signaling [43].

## 4.3 Ex Vivo Culture of Multipotent Embryonic Lung Progenitors

As discussed in the previous section, use of mouse models has been instrumental in our understanding of lung embryonic progenitors, including their differentiation repertoire and signaling pathways that control their specification and self-renewal. Embryonic explant culture has provided another powerful tool to study early lung development, due to ease of manipulation of the soluble environment (including activators and inhibitors of developmental pathways) and molecular characterization of the explants [15, 16, 25, 61]. Recently, lung developmental studies have received considerable impetus by the ex vivo expansion and differentiation of fetal respiratory progenitors that are phenotypically and functionally similar to their in vivo counterparts. Ex vivo progenitor organoids combined with state-of-the-art technologies, such as multimodal single-cell profiling and clustered regularly interspaced short palindromic repeats (CRISPR)-mediated gene editing [50], are already enabling high-resolution studies of lung embryonic progenitor biology.

### *4.3.1 Ex Vivo Culture of Mouse Embryonic Progenitors*

As embryonic progenitors are transient populations that exist within defined developmental windows, their capture and ex vivo expansion and multilineage differentiation offer a powerful, tractable system to study developmental questions and develop organoid platforms. While ex vivo culture of endodermal progenitors, such as pancreatic and hepatic progenitors, has been reported for some time now [35, 109], only recently has culture of respiratory progenitors been established. Nichane and coworkers were able to purify mouse *Sox9*+ distal tip progenitors using a *Sox9* fluorescent reporter strain and develop defined culture conditions (lung progenitor medium, LPM) for their ex vivo expansion [77]. 3D culture in Matrigel, activation of Wnt signaling by GSK-3β inhibition (CHIR99021), inhibition of Tgf-β and p38 signaling, and addition of Fgf (FGF9, FGF10) ligands and EGF were necessary to maintain E12.5 tip progenitors in an undifferentiated, multipotent state, with high expression of known tip markers, such as *Id2, Nmyc, Etv5,* and *Sox9*. Importantly, ex vivo expanded progenitors were able to differentiate to airway and alveolar lineages both in vitro and after transplantation to the mouse lungs following naphthalene or bleomycin injury [77]. Interestingly, the independent and contemporaneous work of Spence et al. identified FGF7, CHIR99021, and RA (3F medium) as the minimal signaling requirements for ex vivo maintenance of distal tip progenitor cells from E12.5 mouse lungs [72]. A similar system established earlier in lung development (E11.5) was employed for a genetic screen of *Sox9*+ tip progenitor regulators [2]. This screen identified *Aurkb*, among others, as a regulator of tip progenitor cell cycle, and its relevance was confirmed by in vivo deletion in distal tip progenitors, leading to profound defects in branching morphogenesis.

In our work, we were able to isolate rare E9.0 lung primordial epithelial progenitors using the previously published Nkx2-1$^{GFP}$ reporter mouse [48, 68]. scRNA-seq of in vivo and ex vivo expanded lung primordial progenitors guided modifications of the distal tip expansion medium, thus allowing us to successfully culture primordial progenitors that proliferated and maintained Nkx2-1$^{GFP}$ expression for several passages [64]. This ex vivo system can be used to study cell fate decisions during the embryonic stage of lung development.

### *4.3.2 Ex Vivo Culture of Human Embryonic Progenitors*

The need for ex vivo systems is even more salient with human fetal progenitors due to the relative inaccessibility of human fetal tissue and severe restrictions on related research in certain regions [3]. The possibility of culturing ex vivo human distal tip progenitors was first shown by Rawlins et al. [78]. Use of a cocktail of growth factors (EGF, FGF7, FGF10), Tgf-β (SB431542) and BMP (Noggin) inhibitors, and Wnt activators (CHIR99021) and potentiators (R-spondin 1) resulted in long-term culture of mesenchyme-depleted tip progenitors from 5- to 9-week fetal lungs. There was absolute requirement for activation of FGF and Wnt and inhibition of Tgf-β signaling for the self-renewal of the progenitors. The multilineage differentiation potential of the ex vivo expanded progenitors was convincingly demonstrated in a series of both in vitro (PneumaCult for bronchiolar differentiation, alveolar differentiation medium with expanded human lung fetal fibroblasts) and in vivo (kidney capsule co-transplantation with E13.5 mouse fetal cells) assays. Miller et al. showed that the 3F medium (FGF7, CHIR99021, RA) described for the expansion of the mouse distal tip progenitors, as well as the 4-factor medium (FG7, FGF10, CHR99021, RA), were sufficient for the ex vivo stable culture of 12-week dissected human lung distal tip progenitors [72]. The expanded progenitors had low expression of airway markers, such as *P63, FOXJ1,* and *SCGB1A1*, and high co-expression of tip markers, such as *SOX9* and *SOX2*. Notably, expanded progenitors had significant transcriptional similarity with freshly isolated fetal lung tips. Although BMP4 removal resulted in higher *SOX9* expression in Miller et al. [72] similarly to BMP inhibition in Nikolic et al. [78], the absence of Tgf-β inhibition requirement in the former protocol is intriguing, and it may be explained by subtle differences in the two systems due to different developmental times of the tissue of origin, including endogenous TGF-β levels.

Establishing tractable systems of human fetal lung progenitors opens up the exciting possibility of ex vivo human development studies, as shown by recent work of Rawlins et al. and Spence et al. [18, 65, 100, 101]. Engineering a CRISPRi system in human distal tip progenitors, Sun and coworkers were able to perform loss-of-function studies and systematically screen the role of several TFs in self-renewal of SOX9+ progenitors [100]. For example, *MYBL2* (proliferation-related gene) and *CTNNB1* (Wnt signaling effector) guide RNAs were highly depleted in a 2-week culture of organoids, indicating a role of these genes in progenitor self-renewal.

Knockdown of *SOX9* resulted in downregulation of 455 common genes in two organoid lines, and further gain-of-function and genomic occupancy studies identified *ETV4, ETV5,* and *NMYC* and Wnt-related genes, such as *LGR5* and *CD44*, as direct transcriptional SOX9 targets. Removal of Wnt activators from the tip medium resulted in SOX9 expression loss. Interestingly, genes suppressed by *SOX9* included secretory cell-related genes, as well as non-lung (stomach, liver) endodermal genes, similarly to mouse respiratory development [81].

Human distal tip progenitors from a later (canalicular) stage were used to study alveolar differentiation due to their restricted differentiation potential [65]. CD36 was specifically expressed in these progenitors and along with CD44 marks distal tip progenitors from 16- to 21-week human fetal lungs. CD36/CD44 co-expression has also functional significance as CD36+CD44+ ($Lin^{Pos}$) organoids had greater propensity to differentiate to alveolar cells. In this system, *NKX2-1* and *TFAP2C* were found to regulate opposing fates, with the former TF promoting alveolar and the latter promoting airway (basal) fate, as shown by gain-of-function studies.

Distal tip organoids have been also used to interrogate lineage relationships, focusing on newly identified progenitors [18]. In scRNA-seq of the developing human lung, *SCGB3A2/SFTPB/CFTR* co-expression identified a progenitor cell, named fetal airway secretory (FAS) cell. FAS cells emerged during airway differentiation [72] of distal tip progenitors and were shown, following barcoding-mediated lineage tracing, to give rise to pulmonary neuroendocrine cells and a subset (C6+) of multiciliated cells.

Overall, culture of lung fetal progenitors has offered important insights in lung epithelial development and provided sophisticated systems that are being used to study the intricate relationships between progenitor competence, gene regulatory networks, and signaling pathways.

## 4.4 In Vitro Derivation of Multipotent Embryonic Lung Progenitors

A large body of work reviewed recently [29, 104] has led to the development of directed differentiation protocols for the derivation of a variety of lung lineages from both mouse and human PSCs. It is now well established that derivation of lung-competent anterior foregut endoderm via dual Smad inhibition followed by Wnt and Bmp signaling can generate lung multipotent primordial-like progenitors in vitro [27, 33, 34, 39, 47, 48, 68, 75]. These progenitors can be highly purified by the use of either knock-in fluorescent reporters, such as eGFP and mCherry, or cell surface marker sorting algorithms, such as $CPM^{hi}$ and $CD47^{hi}/CD26^{lo}$ [33, 39, 48, 68, 98, 110]. Interestingly, *Cd26 (Dpp4)* is one of the most downregulated genes upon in vivo lung specification in mice [48]. Wnt signaling is also important post-specification in vitro with timing of Wnt addition or withdrawal leading to different differentiation trajectories (airway or alveolar) [23, 51, 58, 71].

As PSC-based systems have been further refined, one major goal is to derive and stably expand in culture other important lung epithelial and mesenchymal progenitors. Earlier efforts had led to the long-term culture of complex PSC-derived organoids that initially contained tip progenitors and appeared to recapitulate the transcriptional profile of second-trimester human lung [17]. At the same time, there is an imperative need to develop computational methods to reliably benchmark such populations against their presumed in vivo counterparts or ex vivo expanded embryonic progenitors, as discussed in the following section. Creation of human or mouse developmental atlases at single-cell resolution [41, 73, 76, 113] can satisfy this need while also driving the study of developmental trajectories and cell fate decisions.

Hawkins and coworkers reported the derivation and isolation of airway basal-like cells (iBCs) from human PSCs using a bi-fluorescent (NKX2-1$^{GFP}$; TP63$^{tdTomato}$) reporter cell line [40]. These cells were expandable in airway media (PneumaCult-Ex) containing dual Smad inhibitors and rapidly acquired and maintained expression of the airway basal cell marker, NGFR [93]. Furthermore, their airway multilineage potential (ciliated, secretory, goblet) was demonstrated in both in vitro air-liquid interface (ALI) culture and in tracheal xenografts.

The derivation of stable distal tip-like progenitors from PSCs has also been reported [42, 45]. Refinement of a previously published protocol for the derivation of mouse lung primordial progenitors [48] and subsequent use of the LPM medium [77] led to the stable expansion of a progenitor population that maintained *Nkx2-1*$^{mCherry}$ expression and also expressed distal tip markers (*Sox9, Id2*) [45]. Remarkably, these PSC-derived progenitors were able to engraft (up to 15 weeks) in the distal lung, following transplantation into bleomycin-injured, syngeneic, immunocompetent recipients. The transplanted progenitors differentiated into AT1- and AT2-like cells that were highly similar to their endogenous counterparts as evidenced by protein expression of mature markers, such as PDPN and proSFTPC, and high similarity scores using a novel scRNA-seq-based computational method (single-cell type order parameters, scTOP) [45, 111]. Similarly, human PSC-derived lung progenitors adopted a bud tip fate in a previously described medium [72] and were expanded over several weeks as bud tip organoids (iBTOs) [42]. These progenitors express several markers of human pseudoglandular stage distal tip cells (*SOX9, ETV5, ID2, HMGA2*), had high similarity computational scores to culture-expanded primary human distal tip lung progenitors, and were able to differentiate to either alveolar or airway cells in respective media.

The value of PSC-based systems as a discovery/validation tool for hitherto unknown progenitor populations was evident in two publications [8, 70]. McCauley et al. used a human PSC reporter line to derive and isolate *SCGB3A2*+ early secretory progenitors. By using scRNA-seq, a Wnt-dependent subpopulation of these progenitors was found to co-express an AT2-like program, including genes such as *SFTPC, NAPSA*, and *PGC,* and even possess functional lamellar bodies. This in vitro engineered population was further validated in Basil et al. [8] and appears to have high similarity with the newly discovered SCGB3A2+ respiratory airway secretory (RAS) cells in human terminal bronchioles, which are progenitors to AT2 cells and whose alveolar differentiation is regulated by Notch and Wnt signals.

While lung epithelial progenitors are routinely derived from PSCs, derivation of multipotent lung mesenchymal progenitors has lagged behind due to the early developmental broad expression of several mesenchymal TFs and dearth of information on the signals specifying lung mesenchyme. Recently, several papers have reported efforts to derive functional multipotent lung mesenchyme from either human or mouse PSCs [5, 36, 56]. In a proof-of-principle study, Han and coworkers gleaned information from their study of coordinated epithelial-mesenchymal development within the early foregut endoderm to derive distinct subtypes of tissue-specific mesenchyme, including lung mesenchyme, from human PSCs [36]. Following derivation of lateral plate mesoderm, they used BMP4, FGF2, and RA with simultaneous inhibition of Wnt and Tgf-β signaling and activation of Shh signaling by purmorphamine (PMA), a Shh signaling agonist, to induce an anterior foregut splanchnic mesoderm fate. Further activation of BMP, Shh, and RA signaling with activation of Wnt at later stages led to the derivation of lunglike mesenchyme, based on co-expression of *NKX6-1, TBX5, FOXF1*, and *WNT2* [57].

Similar work of Morimoto et al. identified β-catenin-dependent Wnt signaling as an important signal for early lung mesodermal *Tbx4* expression in the tracheal domain [56]. Lateral plate mesoderm (*Foxf1*+) was treated with BMP4 and CHIR99021 to produce tracheal mesoderm from mouse PSCs, and the downstream differentiation to chondrocytes and smooth muscle cells of the latter was demonstrated with immunostaining and qPCR. Induction of tracheal mesoderm with chondrocyte and smooth muscle competence from human PSCs required further addition of PMA.

Alber and coworkers made heavy use of a *Tbx4* enhancer Cre mouse PSC line (Tbx4-rtTA; TetO-Cre; mTmG; Tbx4-LER) to indelibly mark and purify PSC-derived lung-specific mesenchyme (iLM) [5, 115]. KDR+ lateral plate mesoderm (also Foxf1+) was treated with various combinations of Wnt, Bmp4, PMA, and RA, and it was determined that RA and PMA were the minimal signaling requirements for derivation of lung-specific mesenchyme, based on the percentage of Tbx4-LER$^{GFP+}$ cells [5]. The functionality of the engineered iLM was demonstrated by the following: (a) enhanced lung epithelial differentiation in organoid culture with PSC-derived lung progenitors, (b) induction of lung fate (Nkx2-1$^{+}$ cells) in non-lung epithelial progenitors, and (c) maintenance of primary AT2 cells in organoid coculture. Tbx4-LER$^{GFP+}$ iLM was also shown to be multipotential and was able to differentiate to smooth muscle cells, adipocytes, and mesenchymal alveolar niche cells (MANCs) [114].

## 4.5 Progenitor Cell Similarity Models

A major challenge facing the field is to develop tractable computational methods for assessing cell similarity between engineered and natural cells. An ideal method would assess similarity along multiple phenotypic axes: transcriptomic, epigenetic, metabolomic, etc. [28, 37]. However, such multimodal comparisons remain uncommon due to the difficulty and expense of generating the required datasets. For this reason, the vast majority of cell similarity methods restrict themselves to analyzing transcriptomic measurements made using RNA-seq methods. RNA-seq can occur in bulk, where a multitude of cells in a tissue are sequenced simultaneously, or in single-cell form (scRNA-seq), where individual cells are sequenced. The latter, although more expensive and resource intensive, provides granular resolution of the gene expression of individual cells and has become standard for identifying cell types. For this reason, here, we focus on scRNA-seq-based computational methods for understanding cell similarity (compared and reviewed in Abdelaal et al. [1] and Xie et al. [108]).

Any scRNA-seq-based method for quantifying cell similarity must account for several technical challenges [46]. Single-cell data tends to be extremely sparse, with many low-expressing genes being measured as zero expression (a phenomenon often called "dropout" in the computational literature). The noisy nature of scRNA-seq data is exacerbated by the large amount of cell-to-cell technical variation. These problems become especially acute when comparing data across experiments and biological conditions, where batch effects can dominate variation in scRNA-seq data, making it difficult to assess cell similarity, especially in the context of closely related cell types [83]. This presents a major technical hurdle for utilizing the comprehensive cell atlases now being generated by the community [69, 103].

To address the noisy nature of scRNA-seq, numerous computational methods have been proposed (many of which are collected in the now indispensable Bioconductor package) [7]. To deal with the missing zeroes and data sparsity in scRNA-seq datasets, analysis pipelines often include methods for data imputation (see Dai et al. [20] for a recent comparison of imputation methods). Although data imputation methods vary in their technical details (model-dependent or model-free deep learning methods), they all make use of the expression profiles of similar cells to impute missing values. This process tends to homogenize expression profiles across similar cells, making it difficult to assess small differences in gene expression profiles between cells. Most importantly, for our purposes, this homogenization can lead to inflated cell similarity scores, giving rise to an overly optimistic assessments of similarity between engineered and natural cell types. This general phenomenon is also true of commonly used batch correction methods, which decrease variation across conditions or experiments and thus can conflate technical variability with biological variability [69]. For this reason, great care must be taken when using imputation and batch correction methods in analysis pipelines whose goal is comparing cell similarity.

Perhaps the biggest challenge for assessing cell similarity is the high-dimensional nature of scRNA-seq data, with RNA from thousands of genes being measured simultaneously in individual cells. High dimensionality requires the use of specialized methods for visualization and interpretation. A typical data analysis pipeline [7] for discovering and classifying cell types involves applying a dimensional reduction algorithm to summarize the information provided by thousands of genes down to just two dimensions (e.g., using a dimensional reduction method such as SPRING [106] or uniform manifold approximation and projection (UMAP) [9], followed by clustering, and the utilization of marker genes to assign clusters to cell types. Many analysis pipelines also restrict the features being analyzed to a smaller subset, such as highly variable genes. Finally, to assess cell similarity of the new cell, the expression profile of the new cell is compared to expression profiles of the clusters that are used as a proxy for cell type.

A powerful feature of this pipeline is that it allows for easy visualization and outputs biologically interpretable clusters. This allows experimentalists to quickly assess experiments, identify biologically interesting features, and develop novel biological hypothesis that can be tested in new experiments. Given the widespread adoption, it is worth highlighting a number of limitations of this analysis pipeline for assessing cell similarity. First, both the feature selection and the dimensional reduction depend on the datasets that are included in the analysis. The inclusion or exclusion of different datasets generally yields different results even for the same experiment. This dependence on choice of data is especially important when trying to assess cell similarity between closely related cell types or comparing natural and engineered cells. Second, dimensional reduction techniques are primarily useful for visualization and almost always fundamentally distort biological patterns that occur in the full high-dimensional space [14]. In general, axes and distances in a UMAP or SPRING plot have no natural biological interpretation, making it challenging to rigorously interpret biological experiments.

To address these limitations, the authors have developed a novel computational method for computing cell similarity at the single-cell level (single-cell type order parameters, scTOP) [45, 111]. scTOP utilizes insights developed from statistical physics-inspired epigenetic landscape models to assess cell similarity in a holistic manner [21, 48, 60, 85]. The inputs to scTOP are a "cell basis," consisting of the transcriptomic profiles of cell types of interest (e.g., the expression profiles of lung cells in a cell atlas, such as the single-cell lung cell atlas) and the scRNA-seq measurements of the cell being analyzed (e.g., an engineered cell whose cell similarity we wish to compute). The output of scTOP is a set of cell similarity scores that characterize the cell similarity with each cell type in the "cell basis" (e.g., a score to be a club cell, AT1 cell, AT2 cell, etc.). An appealing feature of the scTOP algorithm is that it allows for biologically meaningful visualizations and sensitive and accurate classification of cell similarity using scRNA-seq data [45]. Importantly, the method does not require any dimensional reduction techniques, feature selection, produces visualizations with meaningful axes, and does not depend on what other data is being analyzed. One limitation of scTOP is that it requires experimentalists to specify cell types of interest ahead of time through the choice of cellular basis. For this reason, scTOP cannot be used to assess or discover novel cell types. Nonetheless, we have found scTOP to be an excellent tool for comparing engineered and natural cells, where the cells of interest are often known [111].

Assessing cell similarity of progenitor populations is an especially computationally challenging task [108]. Transcriptome profiles of developmental progenitor cell types are generally more similar to each other than their adult counterparts, highlighting the need to collect scRNA-seq data on multiple progenitor populations. Furthermore, during the course of development, progenitor cell populations often rapidly change their transcriptome profiles as they differentiate, making it difficult to accurately measure transcriptomes in vivo.

This has important implications for thinking about lung progenitors at various developmental stages. Even when in vivo progenitors can be stabilized and expanded ex vivo, their transcriptomes can be altered due to radical differences in in vivo and ex vivo environments [42]. Hence, it is important to decide how to benchmark in vitro engineered cells, and characterization of progenitor-based organoids should extend beyond similarity scores to encompass functional differentiation, global chromatin accessibility, and proteomics.

## 4.6 Conclusion

As our understanding of lung multipotent progenitor biology deepens, the range of potential applications increases (Fig. 4.2). Ex vivo or engineered PSC-derived lung organoids have already been deployed for modeling of cell fate decisions and production of cells of clinical relevance. Tissue interactions and morphogenetic processes have been long-standing areas of interest in developmental biology [107], and one could expect modern-day organoids to be increasingly used to similar ends. Other future applications include respiratory disease modeling and the study of lung-microbiome interactions, thus reducing the need for animal experimentation. Ultimately, lung progenitor-based organoids will become widely and routinely used tools in medicine and developmental biology.

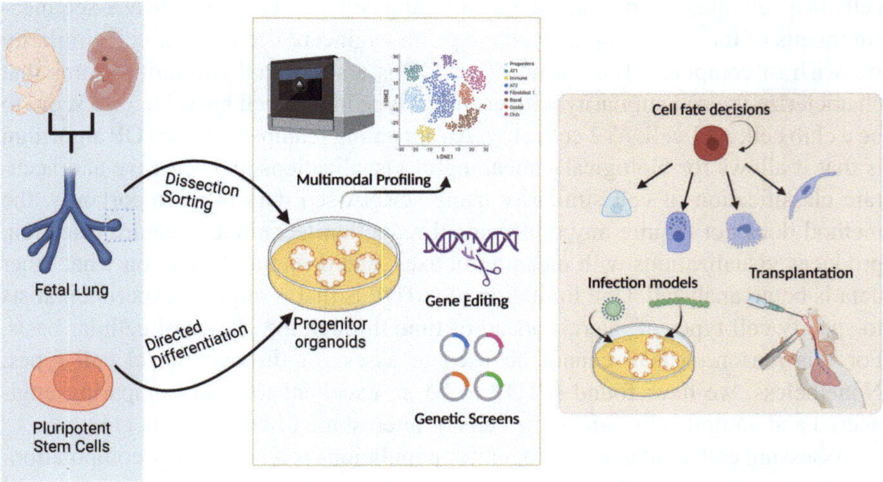


**Fig. 4.2** Lung progenitor-derived tractable organoids, current and future uses

**Acknowledgments** We apologize to those authors whose work we could not cite due to space limitations. L.I. was supported by the National Institutes of Health funding (R56 DE028545-01 and R01 HL158965-01) and the University at Buffalo (UB) Research Foundation Start-up Funds. M.Y. and P.M. were supported by the National Institutes of Health NIGMS funding (2R35 GM119461). Figures were created with Biorender.com.

# References

1. Abdelaal, T., Michielsen, L., Cats, D., Hoogduin, D., Mei, H.L., Reinders, M.J.T., and Mahfouz, A. (2019). A comparison of automatic cell identification methods for single-cell RNA sequencing data. Genome Biol. *20*, 194. https://doi.org/10.1186/s13059-019-1795-z.
2. Ah-Cann, C., Wimmer, V.C., Weeden, C.E., Marceaux, C., Law, C.W., Galvis, L., Filby, C.E., Liu, J., Breslin, K., Willson, T., et al. (2021). A functional genetic screen identifies aurora kinase b as an essential regulator of Sox9-positive mouse embryonic lung progenitor cells. Development *148*, dev199543. https://doi.org/10.1242/dev.199543.
3. Al Alam, D., Ballard, P.L., Le Saux, C.J., Ingram, J.L., Mariani, T.J., Moore, N.S., Weiss, D.J., and Yu, M. (2021). Human Fetal Tissue Regulation Impact on Pediatric and Adult Respiratory-related Research. Ann. Am. Thoracic Society *18*, 204–208. https://doi.org/10.1513/AnnalsATS.202005-460PS.
4. Alanis, D.M., Chang, D.R., Akiyama, H., Krasnow, M.A., and Chen, J.C. (2014). Two nested developmental waves demarcate a compartment boundary in the mouse lung. Nat. Commun. *5*, 3923. https://doi.org/10.1038/ncomms4923.
5. Alber, A.B., Marquez, H.A., Ma, L., Kwong, G., Thapa, B.R., Villacorta-Martin, C., Lindstrom-Vautrin, J., Bawa, P., Luo, Y., Ikonomou, L., et al. (2022). Directed differentiation of mouse pluripotent stem cells into functional lung-specific mesenchyme. bioRxiv, 2022.2008.2012.502651. https://doi.org/10.1101/2022.08.12.502651.
6. Alysandratos, K.D., Herriges, M.J., and Kotton, D.N. (2021). Epithelial Stem and Progenitor Cells in Lung Repair and Regeneration. In Annual Review of Physiology, Vol 83, M.T. Nelson, and K. Walsh, eds. (Annual Reviews), pp. 529–550. https://doi.org/10.1146/annurev-physiol-041520-092904.
7. Amezquita, R.A., Lun, A.T.L., Becht, E., Carey, V.J., Carpp, L.N., Geistlinger, L., Marini, F., Rue-Albrecht, K., Risso, D., Soneson, C., et al. (2020). Orchestrating single-cell analysis with Bioconductor. Nat. Methods *17*, 137–145. https://doi.org/10.1038/s41592-019-0654-x.
8. Basil, M.C., Cardenas-Diaz, F.L., Kathiriya, J.J., Morley, M.P., Carl, J., Brumwell, A.N., Katzen, J., Slovik, K.J., Babu, A., Zhou, S., et al. (2022). Human distal airways contain a multipotent secretory cell that can regenerate alveoli. Nature *604*, 120–126. https://doi.org/10.1038/s41586-022-04552-0.
9. Becht, E., McInnes, L., Healy, J., Dutertre, C.A., Kwok, I.W.H., Ng, L.G., Ginhoux, F., and Newell, E.W. (2019). Dimensionality reduction for visualizing single-cell data using UMAP. Nat. Biotechnol. *37*, 38–44. https://doi.org/10.1038/nbt.4314.
10. Bellusci, S., Grindley, J., Emoto, H., Itoh, N., and Hogan, B.L.M. (1997). Fibroblast growth factor 10(FGF10) and branching morphogenesis in the embryonic mouse lung. Development *124*, 4867–4878.
11. Bilodeau, M., Shojaie, S., Ackerley, C., Post, M., and Rossant, J. (2014). Identification of a Proximal Progenitor Population from Murine Fetal Lungs with Clonogenic and Multilineage Differentiation Potential. Stem Cell Rep. *3*, 634–649. https://doi.org/10.1016/j.stemcr.2014.07.010.
12. Chang, D.R., Alanis, D.M., Miller, R.K., Ji, H., Akiyama, H., McCrea, P.D., and Chen, J.C. (2013). Lung epithelial branching program antagonizes alveolar differentiation. Proc. Natl. Acad. Sci. U. S. A. *110*, 18042–18051. https://doi.org/10.1073/pnas.1311760110.

13. Chapman, D.L., Garvey, N., Hancock, S., Alexiou, M., Agulnik, S.I., GibsonBrown, J.J., CebraThomas, J., Bollag, R.J., Silver, L.M., and Papaioannou, V.E. (1996). Expression of the T-box family genes, Tbx1-Tbx5, during early mouse development. Dev. Dyn. *206*, 379–390.
14. Chari, T., Banerjee, J., and Pachter, L. (2021). The Specious Art of Single-Cell Genomics. bioRxiv, 2021.2008.2025.457696. https://doi.org/10.1101/2021.08.25.457696.
15. Chen, F., Cao, Y.X., Qian, J., Shao, F.Z., Niederreither, K., and Cardoso, W.V. (2010). A retinoic acid-dependent network in the foregut controls formation of the mouse lung primordium. J. Clin. Invest. *120*, 2040–2048. https://doi.org/10.1172/jci40253.
16. Chen, F., and Cardoso, W.V. (2015). Culture of Mouse Embryonic Foregut Explants. In Tissue Morphogenesis: Methods and Protocols, C.M. Nelson, ed. (Humana Press Inc), pp. 163–169.
17. Chen, Y.W., Huang, S.X., de Carvalho, A., Ho, S.H., Islam, M.N., Volpi, S., Notarangelo, L.D., Ciancanelli, M., Casanova, J.L., Bhattacharya, J., et al. (2017). A three-dimensional model of human lung development and disease from pluripotent stem cells. Nat. Cell Biol. *19*, 542–549. https://doi.org/10.1038/ncb3510.
18. Conchola, A.S., Frum, T., Xiao, Z., Hsu, P.P., Hein, R.F.C., Miller, A., Tsai, Y.-H., Wu, A., Kaur, K., Holloway, E.M., et al. (2022). Distinct airway progenitor cells drive epithelial heterogeneity in the developing human lung. bioRxiv, 2022.2006.2013.495813. https://doi.org/10.1101/2022.06.13.495813.
19. Costa, R.H., Kalinichenko, V.V., and Lim, L. (2001). Transcription factors in mouse lung development and function. Am. J. Physiol.-Lung Cell. Mol. Physiol. *280*, L823–L838.
20. Dai, C.C., Jiang, Y., Yin, C.L., Su, R., Zeng, X.X., Zou, Q., Nakai, K., and Wei, L.Y. (2022). scIMC: a platform for benchmarking comparison and visualization analysis of scRNA-seq data imputation methods. Nucleic Acids Res. *50*, 4877–4899. https://doi.org/10.1093/nar/gkac317.
21. Dame, K., Cincotta, S., Lang, A.H., Sanghrajka, R.M., Zhang, L.Y., Choi, J.Y., Kwok, L., Wilson, T., Kandula, M.M., Monti, S., et al. (2017). Thyroid Progenitors Are Robustly Derived from Embryonic Stem Cells through Transient, Developmental Stage-Specific Overexpression of Nkx2-1. Stem Cell Rep. *8*, 216–225. https://doi.org/10.1016/j.stemcr.2016.12.024.
22. Danopoulos, S., Alonso, I., Thornton, M.E., Grubbs, B.H., Bellusci, S., Warburton, D., and Al Alam, D. (2018). Human lung branching morphogenesis is orchestrated by the spatiotemporal distribution of ACTA2, SOX2, and SOX9. Am. J. Physiol.-Lung Cell. Mol. Physiol. *314*, L144-L149. https://doi.org/10.1152/ajplung.00379.2017.
23. de Carvalho, A., Strikoudis, A., Liu, H.Y., Chen, Y.W., Dantas, T.J., Vallee, R.B., Correia-Pinto, J., and Snoeck, H.W. (2019). Glycogen synthase kinase 3 induces multilineage maturation of human pluripotent stem cell-derived lung progenitors in 3D culture. Development *146*, dev171652. https://doi.org/10.1242/dev.171652.
24. De Felice, M., and Di Lauro, R. (2011). Minireview: Intrinsic and Extrinsic Factors in Thyroid Gland Development: An Update. Endocrinology *152*, 2948–2956. https://doi.org/10.1210/en.2011-0204.
25. Desai, T.J., Malpel, S., Flentke, G.R., Smith, S.M., and Cardoso, W.V. (2004). Retinoic acid selectively regulates Fgf10 expression and maintains cell identity in the prospective lung field of the developing foregut. Dev. Biol. *273*, 402–415. https://doi.org/10.1016/j.ydbio.2004.04.039.
26. Domyan, E.T., Ferretti, E., Throckmorton, K., Mishina, Y., Nicolis, S.K., and Sun, X. (2011). Signaling through BMP receptors promotes respiratory identity in the foregut via repression of Sox2. Development *138*, 971–981. https://doi.org/10.1242/dev.053694.
27. Dye, B.R., Hill, D.R., Ferguson, M.A.H., Tsai, Y.H., Nagy, M.S., Dyal, R., Wells, J.M., Mayhew, C.N., Nattiv, R., Klein, O.D., et al. (2015). In vitro generation of human pluripotent stem cell derived lung organoids. eLife *4*, 25. https://doi.org/10.7554/eLife.05098.
28. Efremova, M., and Teichmann, S.A. (2020). Computational methods for single-cell omics across modalities. Nat. Methods *17*, 14–17. https://doi.org/10.1038/s41592-019-0692-4.
29. Frum, T., and Spence, J.R. (2021). hPSC-derived organoids: models of human development and disease. J. Mol. Med. *99*, 463–473. https://doi.org/10.1007/s00109-020-01969-w.

30. Gao, N., White, P., and Kaestner, K.H. (2009). Establishment of Intestinal Identity and Epithelial-Mesenchymal Signaling by Cdx2. Dev. Cell *16*, 588–599. https://doi.org/10.1016/j.devcel.2009.02.010.
31. Gonzales, L.W., Guttentag, S.H., Wade, K.C., Postle, A.D., and Ballard, P.L. (2002). Differentiation of human pulmonary type II cells in vitro by glucocorticoid plus cAMP. Am. J. Physiol.-Lung Cell. Mol. Physiol. *283*, L940-L951. https://doi.org/10.1152/ajplung.00127.2002.
32. Goss, A.M., Tian, Y., Tsukiyama, T., Cohen, E.D., Zhou, D., Lu, M.M., Yamaguchi, T.P., and Morrisey, E.E. (2009). Wnt2/2b and beta-Catenin Signaling Are Necessary and Sufficient to Specify Lung Progenitors in the Foregut. Dev. Cell *17*, 290–298. https://doi.org/10.1016/j.devcel.2009.06.005.
33. Gotoh, S., Ito, I., Nagasaki, T., Yamamoto, Y., Konishi, S., Korogi, Y., Matsumoto, H., Muro, S., Hirai, T., Funato, M., et al. (2014). Generation of Alveolar Epithelial Spheroids via Isolated Progenitor Cells from Human Pluripotent Stem Cells. Stem Cell Rep. *3*, 394–403. https://doi.org/10.1016/j.stemcr.2014.07.005.
34. Green, M.D., Chen, A., Nostro, M.C., d'Souza, S.L., Schaniel, C., Lemischka, I.R., Gouon-Evans, V., Keller, G., and Snoeck, H.W. (2011). Generation of anterior foregut endoderm from human embryonic and induced pluripotent stem cells. Nat. Biotechnol. *29*, 267-U153. https://doi.org/10.1038/nbt.1788.
35. Greggio, C., De Franceschi, F., Figueiredo-Larsen, M., Gobaa, S., Ranga, A., Semb, H., Lutolf, M., and Grapin-Botton, A. (2013). Artificial three-dimensional niches deconstruct pancreas development in vitro. Development *140*, 4452–4462. https://doi.org/10.1242/dev.096628.
36. Han, L., Chaturvedi, P., Kishimoto, K., Koike, H., Nasr, T., Iwasawa, K., Giesbrecht, K., Witcher, P.C., Eicher, A., Haines, L., et al. (2020). Single cell transcriptomics identifies a signaling network coordinating endoderm and mesoderm diversification during foregut organogenesis. Nat. Commun. *11*, 4158. https://doi.org/10.1038/s41467-020-17968-x.
37. Hao, Y., Hao, S., Andersen-Nissen, E., Mauck, W.M., 3rd, Zheng, S., Butler, A., Lee, M.J., Wilk, A.J., Darby, C., Zager, M., et al. (2021). Integrated analysis of multimodal single-cell data. Cell *184*, 3573–3587.e3529. https://doi.org/10.1016/j.cell.2021.04.048.
38. Harris-Johnson, K.S., Domyan, E.T., Vezina, C.M., and Sun, X. (2009). beta-Catenin promotes respiratory progenitor identity in mouse foregut. Proc. Natl. Acad. Sci. U. S. A. *106*, 16287–16292. https://doi.org/10.1073/pnas.0902274106.
39. Hawkins, F., Kramer, P., Jacob, A., Driver, I., Thomas, D.C., McCauley, K.B., Skvir, N., Crane, A.M., Kurmann, A.A., Hollenberg, A.N., et al. (2017). Prospective isolation of NKX2-1-expressing human lung progenitors derived from pluripotent stem cells. J. Clin. Invest. *127*, 2277–2294. https://doi.org/10.1172/jci89950.
40. Hawkins, F.J., Suzuki, S., Beermann, M.L., Barilla, C., Wang, R.B., Villacorta-Martin, C., Berical, A., Jean, J.C., Le Suer, J., Matte, T., et al. (2021). Derivation of Airway Basal Stem Cells from Human Pluripotent Stem Cells. Cell Stem Cell *28*, 79–95. https://doi.org/10.1016/j.stem.2020.09.017.
41. He, P., Lim, K., Sun, D.W., Pett, J.P., Jeng, Q., Polanski, K., Dong, Z.Q., Bolt, L., Richardson, L., Mamanova, L., et al. (2022). A human fetal lung cell atlas uncovers proximal-distal gradients of differentiation and key regulators of epithelial fates. Cell, *185*(25), 4841–4860. https://doi.org/10.1016/j.cell.2022.11.005.
42. Hein, R.F.C., Conchola, A.S., Fine, A.S., Xiao, Z., Frum, T., Brastrom, L.K., Akinwale, M.A., Childs, C.J., Tsai, Y.-H., Holloway, E.M., et al. (2022a). Stable iPSC-derived NKX2-1+ lung bud tip progenitor organoids give rise to airway and alveolar cell types. Development *149*, dev200693. https://doi.org/10.1242/dev.200693.
43. Hein, R.F.C., Wu, J.H., Holloway, E.M., Frum, T., Conchola, A.S., Tsai, Y.-H., Wu, A., Fine, A.S., Miller, A.J., Szenker-Ravi, E., et al. (2022b). R-SPONDIN2+ mesenchymal cells form the bud tip progenitor niche during human lung development. Dev. Cell *57*, 598–1614. https://doi.org/10.1016/j.devcel.2022.05.010.

44. Herriges, J.C., Yi, L., Hines, E.A., Harvey, J.F., Xu, G.L., Gray, P.A., Ma, Q.F., and Sun, X. (2012). Genome-scale study of transcription factor expression in the branching mouse lung. Dev. Dyn. *241*, 1432–1453. https://doi.org/10.1002/dvdy.23823.
45. Herriges, M.J., Yampolskaya, M., Thapa, B.R., Lindstrom-Vautrin, J., Wang, F., Na, C.-L., Ma, L., Montminy, M.M., Huang, J., Villacorta-Martin, C., et al. (2022). Durable alveolar engraftment of PSC-derived lung epithelial cells into immunocompetent mice. bioRxiv, 2022.2007.2026.501591. https://doi.org/10.1101/2022.07.26.501591.
46. Hicks, S.C., Townes, F.W., Teng, M.X., and Irizarry, R.A. (2018). Missing data and technical variability in single-cell RNA-sequencing experiments. Biostatistics *19*, 562–578. https://doi.org/10.1093/biostatistics/kxx053.
47. Huang, S.X.L., Islam, M.N., O'Neill, J., Hu, Z., Yang, Y.G., Chen, Y.W., Mumau, M., Green, M.D., Vunjak-Novakovic, G., Bhattacharya, J., and Snoeck, H.W. (2014). Efficient generation of lung and airway epithelial cells from human pluripotent stem cells. Nat. Biotechnol. *32*, 84–91. https://doi.org/10.1038/nbt.2754.
48. Ikonomou, L., Herriges, M.J., Lewandowski, S.L., Marsland, R., Villacorta-Martin, C., Caballero, I.S., Frank, D.B., Sanghrajka, R.M., Dame, K., Kandula, M.M., et al. (2020a). The in vivo genetic program of murine primordial lung epithelial progenitors. Nat. Commun. *11*, 635. https://doi.org/10.1038/s41467-020-14348-3.
49. Ikonomou, L., and Kotton, D.N. (2015). Derivation of Endodermal Progenitors From Pluripotent Stem Cells. J. Cell. Physiol. *230*, 246–258. https://doi.org/10.1002/jcp.24771.
50. Ikonomou, L., Wagner, D.E., Gilpin, S.E., Weiss, D.J., and Ryan, A.L. (2020b). Technological advances in study of lung regenerative medicine: perspective from the 2019 Vermont lung stem cell conference. Cytotherapy *22*, 519–520. https://doi.org/10.1016/j.jcyt.2020.04.001.
51. Jacob, A., Morley, M., Hawkins, F., McCauley, K.B., Jean, J.C., Heins, H., Na, C.L., Weaver, T.E., Vedaie, M., Hurley, K., et al. (2017). Differentiation of Human Pluripotent Stem Cells into Functional Lung Alveolar Epithelial Cells. Cell Stem Cell *21*, 472–488. https://doi.org/10.1016/j.stem.2017.08.014.
52. Jonsson, J., Carlsson, L., Edlund, T., and Edlund, H. (1994). Insulin-Promoter-Factor-1 Is Required for Pancreas Development in Mice. Nature *371*, 606–609.
53. Khattar, D., Fernandes, S., Snowball, J., Guo, M., Gillen, M.C., Jain, S.S., Sinner, D., Zacharias, W., and Swarr, D.T. (2022). PI3K signaling specifies proximal-distal fate by driving a developmental gene regulatory network in SOX9+ mouse lung progenitors. eLife *11*, e67954. https://doi.org/10.7554/eLife.67954.
54. Kimura, S., Hara, Y., Pineau, T., FernandezSalguero, P., Fox, C.H., Ward, J.M., and Gonzalez, F.J. (1996). The T/ebp null mouse thyroid-specific enhancer-binding protein is essential for the organogenesis of the thyroid, lung, ventral forebrain, and pituitary. Genes Dev. *10*, 60–69.
55. Kimura, S., Ward, J.M., and Minoo, P. (1999). Thyroid-specific enhancer-binding protein thyroid transcription factor 1 is not required for the initial specification of the thyroid and lung primordia. Biochimie *81*, 321–327.
56. Kishimoto, K., Furukawa, K.T., Luz-Madrigal, A., Yamaoka, A., Matsuoka, C., Habu, M., Alev, C., Zorn, A.M., and Morimoto, M. (2020). Bidirectional Wnt signaling between endoderm and mesoderm confers tracheal identity in mouse and human cells. Nat. Commun. *11*, 4159. https://doi.org/10.1038/s41467-020-17969-w.
57. Kishimoto, K., Iwasawa, K., Sorel, A., Ferran-Heredia, C., Han, L., Morimoto, M., Wells, J.M., Takebe, T., and Zorn, A.M. (2022). Directed differentiation of human pluripotent stem cells into diverse organ-specific mesenchyme of the digestive and respiratory systems. Nat. Protoc. *17*, 2699–2719. https://doi.org/10.1038/s41596-022-00733-3.
58. Konishi, S., Gotoh, S., Tateishi, K., Yamamoto, Y., Korogi, Y., Nagasaki, T., Matsumoto, H., Muro, S., Hirai, T., Ito, I., et al. (2016). Directed Induction of Functional Multi-ciliated Cells in Proximal Airway Epithelial Spheroids from Human Pluripotent Stem Cells. Stem Cell Rep. *6*, 18–25. https://doi.org/10.1016/j.stemcr.2015.11.010.
59. Kumar, M.E., Bogard, P.E., Espinoza, F.H., Menke, D.B., Kingsley, D.M., and Krasnow, M.A. (2014). Defining a mesenchymal progenitor niche at single-cell resolution. Science *346*, 1258810, 1258810. https://doi.org/10.1126/science.1258810.

60. Lang, A.H., Li, H., Collins, J.J., and Mehta, P. (2014). Epigenetic Landscapes Explain Partially Reprogrammed Cells and Identify Key Reprogramming Genes. PLoS Comput. Biol. *10*, e1003734. https://doi.org/10.1371/journal.pcbi.1003734.
61. Laresgoiti, U., Nikolic, M.Z., Rao, C., Brady, J.L., Richardson, R.V., Batchen, E.J., Chapman, K.E., and Rawlins, E.L. (2016). Lung epithelial tip progenitors integrate glucocorticoid- and STAT3-mediated signals to control progeny fate. Development *143*, 3686–3699. https://doi.org/10.1242/dev.134023.
62. Lasser, M., Pratt, B., Monahan, C., Kim, S.W., and Lowery, L.A. (2019). The Many Faces of Xenopus: Xenopus laevis as a Model System to Study Wolf Hirschhorn Syndrome. Front. Physiol. *10*, 12, 817. https://doi.org/10.3389/fphys.2019.00817.
63. Lebeche, D., Malpel, S., and Cardoso, W.V. (1999). Fibroblast growth factor interactions in the developing lung. Mech. Dev. *86*, 125–136.
64. Lewandowski, S., Villacorta-Martin, C., Kotton, D.N., and Ikonomou, L. (2020). Single Cell Analysis and Novel Ex Vivo Culture of Lung Primordial Progenitors. Am. J. Respir. Crit. Care Med. *201*, 1.
65. Lim, K., Tang, W., Sun, D., He, P., Teichmann, S.A., Marioni, J.C., Meyer, K.B., and Rawlins, E.L. (2021). Acquisition of alveolar fate and differentiation competence by human fetal lung epithelial progenitor cells. bioRxiv, 2021.2006.2030.450501. https://doi.org/10.1101/2021.06.30.450501.
66. Litingtung, Y., Lei, L., Westphal, H., and Chiang, C. (1998). Sonic hedgehog is essential to foregut development. Nature Genet. *20*, 58–61.
67. Liu, Y., and Hogan, B.L. (2002). Differential gene expression in the distal tip endoderm of the embryonic mouse lung. Gene Expr Patterns *2*, 229–233.
68. Longmire, T.A., Ikonomou, L., Hawkins, F., Christodoulou, C., Cao, Y.X., Jean, J.C., Kwok, L.W., Mou, H.M., Rajagopal, J., Shen, S.S., et al. (2012). Efficient Derivation of Purified Lung and Thyroid Progenitors from Embryonic Stem Cells. Cell Stem Cell *10*, 398–411. https://doi.org/10.1016/j.stem.2012.01.019.
69. Luecken, M.D., Buttner, M., Chaichoompu, K., Danese, A., Interlandi, M., Mueller, M.F., Strobl, D.C., Zappia, L., Dugas, M., Colome-Tatche, M., and Theis, F.J. (2022). Benchmarking atlas-level data integration in single-cell genomics. Nat. Methods *19*, 41–50. https://doi.org/10.1038/s41592-021-01336-8.
70. McCauley, K.B., Alysandratos, K.D., Jacob, A., Hawkins, F., Caballero, I.S., Vedaie, M., Yang, W.L., Slovik, K.J., Morley, M., Carraro, G., et al. (2018). Single-Cell Transcriptomic Profiling of Pluripotent Stem Cell-Derived SCGB3A2+Airway Epithelium. Stem Cell Rep. *10*, 1579–1595. https://doi.org/10.1016/j.stemcr.2018.03.013.
71. McCauley, K.B., Hawkins, F., Serra, M., Thomas, D.C., Jacob, A., and Kotton, D.N. (2017). Efficient Derivation of Functional Human Airway Epithelium from Pluripotent Stem Cells via Temporal Regulation of Wnt Signaling. Cell Stem Cell *20*, 844–857. https://doi.org/10.1016/j.stem.2017.03.001.
72. Miller, A.J., Hill, D.R., Nagy, M.S., Aoki, Y., Dye, B.R., Chin, A.M., Huang, S., Zhu, F., White, E.S., Lama, V., and Spence, J.R. (2018). In Vitro Induction and In Vivo Engraftment of Lung Bud Tip Progenitor Cells Derived from Human Pluripotent Stem Cells. Stem Cell Rep. *10*, 101–119. https://doi.org/10.1016/j.stemcr.2017.11.012.
73. Miller, A.J., Yu, Q.H., Czerwinski, M., Tsai, Y.H., Conway, R.F., Wu, A., Holloway, E.M., Walker, T., Glass, I.A., Treutlein, B., et al. (2020). In Vitro and In Vivo Development of the Human Airway at Single-Cell Resolution. Dev. Cell *53*, 117–128. https://doi.org/10.1016/j.devcel.2020.01.033.
74. Motoyama, J., Liu, J., Mo, R., Ding, Q., Post, M., and Hui, C.C. (1998). Essential function of Gli2 and Gli3 in the formation of lung, trachea and oesophagus. Nature Genet. *20*, 54–57.
75. Mou, H.M., Zhao, R., Sherwood, R., Ahfeldt, T., Lapey, A., Wain, J., Sicilian, L., Izvolsky, K., Musunuru, K., Cowan, C., and Rajagopal, J. (2012). Generation of Multipotent Lung and Airway Progenitors from Mouse ESCs and Patient-Specific Cystic Fibrosis iPSCs. Cell Stem Cell *10*, 385–397. https://doi.org/10.1016/j.stem.2012.01.018.

76. Negretti, N.M., Plosa, E.J., Benjamin, J.T., Schuler, B.A., Habermann, A.C., Jetter, C.S., Gulleman, P., Bunn, C., Hackett, A.N., Ransom, M., et al. (2021). A single-cell atlas of mouse lung development. Development *148*, dev199512. https://doi.org/10.1242/dev.199512.
77. Nichane, M., Javed, A., Sivakamasundari, V., Ganesan, M., Ang, L.T., Kraus, P., Lufkin, T., Loh, K.M., and Lim, B. (2017). Isolation and 3D expansion of multipotent Sox9(+) mouse lung progenitors. Nat. Methods *14*, 1205–1212. https://doi.org/10.1038/nmeth.4498.
78. Nikolic, M.Z., Caritg, O., Jeng, Q., Johnson, J.A., Sun, D.W., Howell, K.J., Brady, J.L., Laresgoiti, U., Allen, G., Butler, R., et al. (2017). Human embryonic lung epithelial tips are multipotent progenitors that can be expanded in vitro as long-term selfrenewing organoids. eLife *6*, e26575. https://doi.org/10.7554/eLife.26575.001.
79. Okubo, T., and Hogan, B.L. (2004). Hyperactive Wnt signaling changes the developmental potential of embryonic lung endoderm. J Biol *3*, 11. https://doi.org/10.1186/jbiol3.
80. Okubo, T., Knoepfler, P.S., Eisenman, R.N., and Hogan, B.L.M. (2005). Nmyc plays an essential role during lung development as a dosage-sensitive regulator of progenitor cell proliferation and differentiation. Development *132*, 1363–1374. https://doi.org/10.1242/dev.01678.
81. Ostrin, E.J., Little, D.R., Gerner-Mauro, K.N., Sumner, E.A., Rios-Corzo, R., Ambrosio, E., Holt, S.E., Forcioli-Conti, N., Akiyama, H., Hanash, S.M., et al. (2018). beta-Catenin maintains lung epithelial progenitors after lung specification. Development *145*, 13, dev160788. https://doi.org/10.1242/dev.160788.
82. Peng, T., Tian, Y., Boogerd, C.J., Lu, M.M., Kadzik, R.S., Stewart, K.M., Evans, S.M., and Morrisey, E.E. (2013). Coordination of heart and lung co-development by a multipotent cardiopulmonary progenitor. Nature *500*, 589–592. https://doi.org/10.1038/nature12358.
83. Peng, Y.R., Shekhar, K., Yan, W.J., Herrmann, D., Sappington, A., Bryman, G.S., van Zyl, T., Do, M.T.H., Regev, A., and Sanes, J.R. (2019). Molecular Classification and Comparative Taxonomics of Foveal and Peripheral Cells in Primate Retina. Cell *176*, 1222–1237. https://doi.org/10.1016/j.cell.2019.01.004.
84. Peterson, R.S., Lim, L., Ye, H.G., Zhou, H.P., Overdier, D.G., and Costa, R.H. (1997). The winged helix transcriptional activator HFH-8 is expressed in the mesoderm of the primitive streak stage of mouse embryos and its cellular derivatives. Mech. Dev. *69*, 53–69. https://doi.org/10.1016/s0925-4773(97)00153-6.
85. Pusuluri, S.T., Lang, A.H., Mehta, P., and Castillo, H.E. (2018). Cellular reprogramming dynamics follow a simple 1D reaction coordinate. Phys. Biol. *15*, 016001. https://doi.org/10.1088/1478-3975/aa90e0.
86. Rankin, S.A., Gallas, A.L., Neto, A., Gomez-Skarmeta, J.L., and Zorn, A.M. (2012). Suppression of Bmp4 signaling by the zinc-finger repressors Osr1 and Osr2 is required for Wnt/beta-catenin-mediated lung specification in Xenopus. Development *139*, 3010–3020. https://doi.org/10.1242/dev.078220.
87. Rankin, S.A., Han, L., McCracken, K.W., Kenny, A.P., Anglin, C.T., Grigg, E.A., Crawford, C.M., Wells, J.M., Shannon, J.M., and Zorn, A.M. (2016). A Retinoic Acid-Hedgehog Cascade Coordinates Mesoderm-Inducing Signals and Endoderm Competence during Lung Specification. Cell Reports *16*, 66–78. https://doi.org/10.1016/j.celrep.2016.05.060.
88. Rankin, S.A., McCracken, K.W., Luedeke, D.M., Han, L., Wells, J.M., Shannon, J.M., and Zorn, A.M. (2018). Timing is everything: Reiterative Wnt, BMP and RA signaling regulate developmental competence during endoderm organogenesis. Dev. Biol. *434*, 121–132. https://doi.org/10.1016/j.ydbio.2017.11.018.
89. Rankin, S.A., and Zorn, A.M. (2014). Gene regulatory networks Governing lung specification. J. Cell. Biochem. *115*, 1343–1350. https://doi.org/10.1002/jcb.24810.
90. Rankin, S.A., and Zorn, A.M. (2022). The homeodomain transcription factor Ventx2 regulates respiratory progenitor cell number and differentiation timing during Xenopus lung development. Dev. Growth Diff. *64*, 347–361. https://doi.org/10.1111/dgd.12807.
91. Rawlins, E.L., Clark, C.P., Xue, Y., and Hogan, B.L.M. (2009). The Id2(+) distal tip lung epithelium contains individual multipotent embryonic progenitor cells. Development *136*, 3741–3745. https://doi.org/10.1242/dev.037317.

92. Riccetti, M., Gokey, J.J., Aronow, B., and Perl, A.K.T. (2020). The elephant in the lung: Integrating lineage-tracing, molecular markers, and single cell sequencing data to identify distinct fibroblast populations during lung development and regeneration. Matrix Biol. *91–92*, 51–74. https://doi.org/10.1016/j.matbio.2020.05.002.
93. Rock, J.R., Onaitis, M.W., Rawlins, E.L., Lu, Y., Clark, C.P., Xue, Y., Randell, S.H., and Hogan, B.L.M. (2009). Basal cells as stem cells of the mouse trachea and human airway epithelium. Proc. Natl. Acad. Sci. U. S. A. *106*, 12771–12775. https://doi.org/10.1073/pnas.0906850106.
94. Rock, J.R., Randell, S.H., and Hogan, B.L.M. (2010). Airway basal stem cells: a perspective on their roles in epithelial homeostasis and remodeling. Dis. Model. Mech. *3*, 545–556. https://doi.org/10.1242/dmm.006031.
95. Rockich, B.E., Hrycaj, S.M., Shih, H.P., Nagy, M.S., Ferguson, M.A.H., Kopp, J.L., Sander, M., Wellik, D.M., and Spence, J.R. (2013). Sox9 plays multiple roles in the lung epithelium during branching morphogenesis. Proc. Natl. Acad. Sci. U. S. A. *110*, E4456-E4464. https://doi.org/10.1073/pnas.1311847110.
96. Rossant, J. (2011). The Impact of Developmental Biology on Pluripotent Stem Cell Research: Successes and Challenges. Dev. Cell *21*, 20–23. https://doi.org/10.1016/j.devcel.2011.06.010.
97. Sandberg, M., Flandin, P., Silberberg, S., Su-Feher, L., Price, J.D., Hu, J.S., Kim, C., Visel, A., Nord, A.S., and Rubenstein, J.L.R. (2016). Transcriptional Networks Controlled by NKX2-1 in the Development of Forebrain GABAergic Neurons. Neuron *91*, 1260–1275. https://doi.org/10.1016/j.neuron.2016.08.020.
98. Serra, M., Alysandratos, K.D., Hawkins, F., McCauley, K.B., Jacob, A., Choi, J., Caballero, I.S., Vedaie, M., Kurmann, A.A., Ikonomou, L., et al. (2017). Pluripotent stem cell differentiation reveals distinct developmental pathways regulating lung- versus thyroid-lineage specification. Development *144*, 3879–3893. https://doi.org/10.1242/dev.150193.
99. Sherwood, R.I., Chen, T.Y.A., and Melton, D.A. (2009). Transcriptional Dynamics of Endodermal Organ Formation. Dev. Dyn. *238*, 29–42. https://doi.org/10.1002/dvdy.21810.
100. Sun, D., Llora Batlle, O., van den Ameele, J., Thomas, J.C., He, P., Lim, K., Tang, W., Xu, C., Meyer, K.B., Teichmann, S.A., et al. (2022). SOX9 maintains human foetal lung tip progenitor state by enhancing WNT and RTK signalling. EMBO Journal, *41*, e111338. https://doi.org/10.15252/embj.2022111338.
101. Sun, D.W., Evans, L., Perrone, F., Sokleva, V., Lim, K., Rezakhani, S., Lutolf, M., Zilbauer, M., and Rawlins, E.L. (2021). A functional genetic toolbox for human tissue-derived organoids. eLife *10*, e67886. https://doi.org/10.7554/eLife.67886; https://doi.org/10.7554/eLife.67886.sa1; https://doi.org/10.7554/eLife.67886.sa2.
102. Swarr, D.T., and Morrisey, E.E. (2015). Lung Endoderm Morphogenesis: Gasping for Form and Function. In Annual Review of Cell and Developmental Biology, Vol 31, R. Schekman, ed. (Annual Reviews), pp. 553–573. https://doi.org/10.1146/annurev-cellbio-100814-125249.
103. Travaglini, K.J., Nabhan, A.N., Penland, L., Sinha, R., Gillich, A., Sit, R.V., Chang, S., Conley, S.D., Mori, Y., Seita, J., et al. (2020). A molecular cell atlas of the human lung from single-cell RNA sequencing. Nature *587*, 619–625. https://doi.org/10.1038/s41586-020-2922-4.
104. Varghese, B., Ling, Z.H., and Ren, X. (2022). Reconstructing the pulmonary niche with stem cells: a lung story. Stem Cell Res. Ther. *13*, 161. https://doi.org/10.1186/s13287-022-02830-2.
105. Wade, K.C., Guttentag, S.H., Gonzales, L.W., Maschhoff, K.L., Gonzales, J., Kolla, V., Singhal, S., and Ballard, P.L. (2006). Gene induction during differentiation of human pulmonary type II cells in vitro. Am. J. Respir. Cell Mol. Biol. *34*, 727–737. https://doi.org/10.1165/rcmb.2004-0389OC.
106. Weinreb, C., Wolock, S., and Klein, A.M. (2018). SPRING: a kinetic interface for visualizing high dimensional single-cell expression data. Bioinformatics *34*, 1246–1248. https://doi.org/10.1093/bioinformatics/btx792.
107. Wessells, N.K. (1973). Tissue interactions in development (Addison-Wesley Pub. Co.).

108. Xie, B.B., Jiang, Q., Mora, A., and Li, X.R. (2021). Automatic cell type identification methods for single-cell RNA sequencing. Comp. Struct. Biotechnol. J.. *19*, 5874–5887. https://doi.org/10.1016/j.csbj.2021.10.027.
109. Xiong, A.M., Austin, T.W., Lagasse, E., Uchida, N., Tamaki, S., Bordier, B.B., Weissman, I.L., Glenn, J.S., and Millan, M.T. (2008). Isolation of human fetal liver progenitors and their enhanced proliferation by three-dimensional coculture with endothelial cells. Tissue Eng. Part A *14*, 995–1006. https://doi.org/10.1089/ten.tea.2007.0087.
110. Yamamoto, Y., Gotoh, S., Korogi, Y., Seki, M., Konishi, S., Ikeo, S., Sone, N., Nagasaki, T., Matsumoto, H., Muro, S., et al. (2017). Long-term expansion of alveolar stem cells derived from human iPS cells in organoids. Nat. Methods *14*, 1097–1106. https://doi.org/10.1038/nmeth.4448.
111. Yampolskaya, M., Herriges, M., Ikonomou, L., Kotton, D., & Mehta, P. (2023). scTOP: physics-inspired order parameters for cellular identification and visualization. bioRxiv, 2023.2001.2025.525581. https://doi.org/10.1101/2023.01.25.525581.
112. Yang, Y., Riccio, P., Schotsaert, M., Mori, M., Lu, J.N., Lee, D.K., Garcia-Sastre, A., Xu, J.M., and Cardoso, W.V. (2018). Spatial-Temporal Lineage Restrictions of Embryonic p63(+) Progenitors Establish Distinct Stem Cell Pools in Adult Airways. Dev. Cell *44*, 752–761. https://doi.org/10.1016/j.devcel.2018.03.001.
113. Yu, Q.H., Kilik, U., Holloway, E.M., Tsai, Y.H., Harmel, C., Wu, A., Wu, J.H., Czerwinski, M., Childs, C.J., He, Z.S., et al. (2021). Charting human development using a multi-endodermal organ atlas and organoid models. Cell *184*, 3281–3298.e3222. https://doi.org/10.1016/j.cell.2021.04.028.
114. Zepp, J.A., Zacharias, W.J., Frank, D.B., Cavanaugh, C.A., Zhou, S., Morley, M.P., and Morrisey, E.E. (2017). Distinct Mesenchymal Lineages and Niches Promote Epithelial Self-Renewal and Myofibrogenesis in the Lung. Cell *170*, 1134–1148.e10. https://doi.org/10.1016/j.cell.2017.07.034.
115. Zhang, W.M., Menke, D.B., Jiang, M.S., Chen, H., Warburton, D., Turcatel, G., Lu, C.H., Xu, W., Luo, Y.F., and Shi, W. (2013). Spatial-temporal targeting of lung-specific mesenchyme by a Tbx4 enhancer. BMC Biol. *11*, 111. https://doi.org/10.1186/1741-7007-11-111.

# Part II
# Engineering and Modeling Large Airways

# Chapter 5
# Basic Science Perspective on Engineering and Modeling the Large Airways

**Lalit K. Gautam, Noa C. Harriott, Adrian M. Caceres, and Amy L. Ryan**

## 5.1 Introduction

The airways are critical physical and biochemical barriers protecting the lungs from a plethora of exogenous environmental insults during air inspiration, which can lead to inflammation and infection. Their structure is finely tuned to provide optimal tissue homeostasis and regulation of innate immunity. The effectiveness of the epithelium as a barrier is reliant upon its capacity for mucociliary clearance, immune surveillance, and regeneration subsequent to injury: a product of the functional cells that comprise the airway epithelium and the niche in which they reside. Cellular diversity during both homeostatic and diseased states in the adult lung is tightly regulated by interactions within their specific anatomical niche during the orchestration of epithelial regeneration. Niche components include the neighboring epithelial cells, infiltrating inflammatory cells, surrounding extracellular matrix, and supporting mesenchymal cells. Changes in the extracellular matrix (ECM) during disease can significantly change the functional dynamics of the airways including the airways' ability to stretch. Functional dynamics including stretch and directional airflows should also be carefully considered when engineering models of human airway tissues. Effective recapitulation of physiological and pathological states of the proximal airways requires the generation of complex structures comprising the surface airway epithelium (SAE), submucosal gland epithelium (SMG), extracellular matrix, and niche cells, including airway smooth muscle cells (ASMCs), fibroblasts (FBs), and immune cells in addition to dynamic regulation of stretch and airflow. The field is moving toward more complex models as the tools

L. K. Gautam · N. C. Harriott · A. M. Caceres · A. L. Ryan (✉)
Department of Anatomy and Cell Biology, Carver College of Medicine, University of Iowa, Iowa City, IA, USA
e-mail: amy-l-ryan@uiowa.edu



C. M. Magin (ed.), *Engineering Translational Models of Lung Homeostasis and Disease*, Advances in Experimental Medicine and Biology 1413,
https://doi.org/10.1007/978-3-031-26625-6_5

and technology evolve to facilitate bioengineering and analytical approaches. This chapter will introduce the basic cellular biology that underlies a functional proximal airway epithelium, focusing on the relationships between structure and function in the airways and the challenges of developing complex models that accurately recapitulate the healthy and diseased human airway.

## 5.2 Proximal Airways: Composition and Function

The proximal airways of the respiratory system begin with the trachea, which bifurcates into the mainstem bronchi and then subsequently smaller branches of cartilaginous bronchi, which enter the pulmonary lobule becoming bronchioles. Bronchioles have a diameter less than 5 mm and branch into an additional 5–7 terminal bronchioles before transitioning to alveolar ducts. For this review, we refer to the proximal airways as branching generations with a diameter greater than 5 mm. In contrast to the distal airways, the proximal branches comprise a mucosal epithelium with cartilaginous rings, SMGs, and ciliated pseudostratified columnar epithelium. As the airways decrease in diameter, they simplify into a columnar epithelium, and cellular composition changes along the proximal-distal axis in accordance with the requirements for regional lung function [116, 177].

The principal role of the proximal airways is to facilitate airflow to and from the alveoli, the site of gas exchange and blood oxygenation [191]. As with other epithelia, the airway epithelium forms a protective barrier against environmental hazards. This barrier comprises two core components: mucociliary clearance and tight junctions. When either of these fail airway inflammation, mucus accumulation and infection all ensue. These are hallmarks of many obstructive lung diseases, including asthma, cystic fibrosis (CF), chronic obstructive pulmonary disease (COPD), and primary ciliary dyskinesia (PCD), among others [242]. Mucociliary clearance relies on the synchronous beating of 100–300 cilia on specialized multiciliated cells, occupying most of the surface area, and creates a mucociliary escalator to carry debris to the pharynx. Here, fluid and mucus can be swallowed [261]. Successful mucociliary clearance relies upon maintenance of airway surface liquid (ASL) homeostasis [31]. ASL comprises a mucous layer to trap inhaled particles overlying a lower viscosity periciliary layer (PCL) that serves to lubricate the airway surface and facilitate ciliary beating [125]. The functional coordination of multiciliated cells can be restricted in situations of excess mucus accumulation in the airways [212, 242]. Mucus is a viscous mixture of large glycoproteins, known as mucins, combined with fluids, electrolytes, and other antimicrobial proteins, and is a critical component of airway homeostasis responsible for capture of particulates and pathogens entering the airways. Polymeric mucins, MUC5B and MUC5AC, are most common in the airway and are primarily secreted from specialized goblet cells [232]. Goblet cells comprise approximately 25% of the epithelial cells in the proximal airways [203]. In addition, the SMGs contribute to mucus secretion through the release of MUC5B strands coated with MUC5AC [142], slowing the transport of fluid up the airway [69, 242].

Increased occurrence of goblet cells, or goblet cell hyperplasia, results in the hypersecretion of mucus, associated with many lung diseases, including asthma, CF, COPD, and bronchitis [68, 86, 120, 122, 123, 133, 144, 146, 216].

As the epithelium transitions from the larger diameter bronchi to the smaller bronchioles, goblet cells become less frequent and club cells predominate [264]. This distribution can be significantly changed in inflammatory airway diseases [23, 173, 184, 206]. Club cells are another secretory cell type, largely responsible for secretion of surfactant proteins, including surfactant proteins A, B, and D (SPA/SPB/SPD) and club cell secretory protein (CCSP), into the airways [247]. The primary function of SPA and SPD are to regulate immune responses, while SPB primarily decreases surface tension at the air-liquid interface [16, 83, 135, 143, 209]. SPA is critical in the response to infection and allergens and for maintaining lung homeostasis, as it is responsible for pathogen opsonization [57]. Additionally, SPA appears to be linked to the modulation of eosinophils, and its expression protects against inflammation in asthmatics [40, 57, 73]. SPD also regulates inflammation and innate immunity in the airways and, thus, can be used as a marker for inflammatory lung damage [118, 262]. SPB has been shown to have anti-inflammatory and antioxidant effects, indicating that it may modulate oxidative stress response [57, 112, 151]. SPB, along with SPA and SPD, are important biomarkers for COPD [139, 250] and can be used to predict the early development of non-small cell lung cancer tumors and the development of lung cancer, in general [67, 219]. Altered ratios of goblet and club cells are characteristic of changes in the airway associated with chronic lung diseases, such as asthma and COPD. Goblet cell metaplasia, occurring at the expense of club cells, increases mucus production and is connected to increased airway inflammation [181, 214]. A reduction in club cells lowers the secretion of CCSP (aka CC10), which has important anti-inflammatory properties and protects against oxidative stress [107, 237, 247].

While secretory and ciliated cells are the main functional cells at the apical surface of the epithelium, basal cells (BCs) are stem cells that line the basement membrane [234]. In the larger bronchi, BCs form a continuous monolayer with decreasing numbers forming smaller clusters and eventually single cells in the terminal bronchioles. BCs are perhaps the best characterized stem cell in the airway epithelium capable of both self-renewal and differentiation into all the functional airway cells (Fig. 5.1). They are typically quiescent during homeostasis and primed to rapidly respond to injury restoring barrier function. Their stemness makes them ideal targets for gene therapy/gene editing and of particular interest in lung regeneration and cellular therapeutics. In addition, they play an important functional role in the anchorage of the epithelium to the basement membrane [164].

There are also several, less common, cell types present in the airway epithelium: pulmonary neuroendocrine cells (PNECs), brush cells, and ionocytes [46, 177]. PNECs are rare, multifunctional epithelial cells, whose main function are as intrapulmonary chemosensors, capable of detecting changes in the composition of inhaled air and adapt to signal accordingly. Recently, three different types of PNECs were identified having differential expression of neuropeptide markers and changes in functional responses to injury and disease [161]. PNECs primarily perform

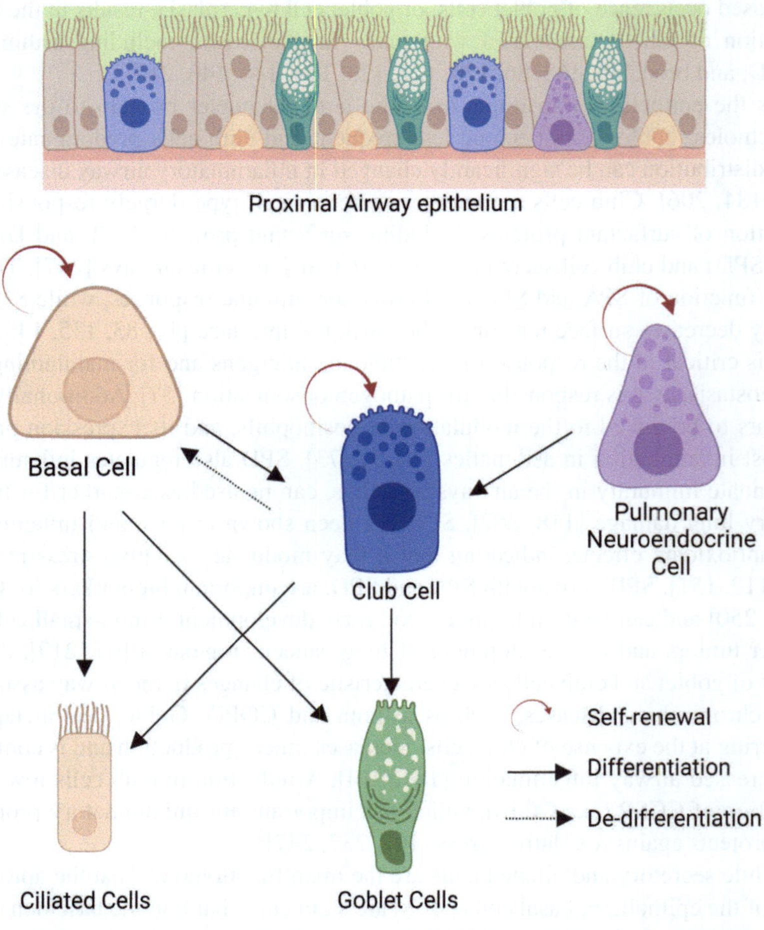


**Fig. 5.1** Stem cells of the proximal airway epithelium. The proximal airway has several cells capable of regeneration. Basal cells can self-renew and give rise to club, goblet, and ciliated cells. Club cells can also give rise to the same progeny and can dedifferentiate to basal cells. Pulmonary neuroendocrine cells have been shown to self-renew and differentiate under specific injury conditions. (This figure was created at Biorender.com)

chemosensory functions in the airway regulating responses to injury and disease [179, 224]. While their specific origins have been debated, a recent study showed that BCs are able to give rise to specific subpopulations of PNECs [161] (Fig. 5.1). Ionocytes became a specific cell of functional interest after the discovery that they highly express the cystic fibrosis transmembrane regulator (CFTR), the defective protein causative for CF disease [51, 156, 211]. The precise functional role of ionocytes in the airways is still an active area of research [95, 211].

## 5.3 Regeneration of the Airways

Engineering a functional airway, either as a model or for transplantation, will rely upon the presence of stem cells to both generate and regenerate the airway epithelium. While generally quiescent, the respiratory epithelium has an immense capacity for rapid regeneration to restore barrier function and physiological homeostasis in response to injury. As mentioned above, the predominant stem cells in the proximal airways are BCs [200, 201, 231]; alveolar type 2 (AT2) cells are stem cells of the alveoli and are reviewed in detail in the chapter focused on the developing lung [4, 5, 38, 43, 136]. These cells are precisely regulated by signals they receive from their immediate niche comprising the surrounding mesenchyme, neighboring epithelium, infiltrating immune cells, and interactions with the ECM. A valuable tissue-level model of the airways should consider all these factors. In this next section we will briefly describe airway BCs and the airway niche.

### *5.3.1 Endogenous Stem Cells*

In the proximal airways, BCs are the main stem cell population that are capable of self-renewal and differentiation to mature airway progeny, including core functional secretory, ciliated, and goblet cells [13, 98, 200]. They are crucial for maintenance and regeneration of cells in the airway after injury and are known to be regulated by changes in Notch signaling [157, 200, 222]. In previous research BCs were considered a homogeneous population, but recent studies suggest there is more complexity [35, 252]. To study early origins in development process, lineage tracing of cytokeratin 5 (KRT5)-expressing BCs indicated that two BC populations existed in the upper airway, one acting as self-renewing stem cell and another population committed to differentiation [240]. More recently, our understanding of the identity of stem cells has been significantly facilitated through the application of single-cell RNA sequencing (scRNAseq) to evaluate cells of the human airways [37, 55, 81, 85, 101, 108, 114, 159, 194, 244, 246, 259, 263]. These studies have provided critical information on stem cell subtypes, transitional cellular states, and the complex pathways implicated in stem cell differentiation [156, 177, 182]. Despite being collectively well studied in their capacity to generate all the differentiated cells of the airway, we are only beginning to understand the functional relevance of the subtypes identified through transcriptomic analysis. Basal stem cells are primarily identified through their expression of core markers, KRT5 and tumor protein 63 (TP63). Small subsets (approx. 20%) also express KRT14, and under homeostatic environments KRT5/TP63/KRT14 populations are considered self-renewing and responsible for maintaining the basal cell stemness. Extensive lineage tracing experiments have been performed in mice, establishing the capacity of KRT14-expressing BCs to generate both secretory and ciliated cells of the airway epithelium [76].

Other putative progenitor cells in the proximal airways include the club cells and PNECs. Club cells, expressing secretoglobin family 1A, member A1 (SCGB1A1), act as facultative progenitor cells and have been shown to harbor the capacity to self-renew and dedifferentiate in cases of severe injury where all basal cells are ablated [12, 119, 128, 149, 195, 196, 231, 257]. While PNECs are less well characterized, they have been shown to act as a more classical unipotent stem cell, self-renewing in response to injury [176] (Fig. 5.1).

### 5.3.2 *The Stem Cell Niche*

Airway function and stem cell behavior are inherently influenced by changes in the cellular microenvironment. Despite this, models of the epithelium are often studied in the absence of niche factors. During lung development the temporal regulation of growth factors and cytokines secreted by the surrounding mesenchyme intricately directs lineage specification, generating contiguous branches of specialized epithelium. This has been extensively reviewed in Riccetti et al. [199], Yuan et al. [256], Volckaert and De Langhe [235], and Morrisey et al. [158]. In the adult lung, signals from the niche remain critical regulators of stem cell function and homeostasis [13]. During homeostasis the airway BCs are relatively quiescent but are rapidly mobilized in response to tissue injury [12, 13]. This epithelial injury is sensed by the adjoining epithelial cells, leading to the production of alarmins, and damage-induced phenotypic alterations of the BCs, together initiating the tissue repair. As a result of such events, immune cells are prompted to generate a suitable response mechanism for tissue homeostasis [204]. The ECM is also active in wound healing through the provision of signals to activate the mesenchymal cells to promote inflammation, cellular migration, and tissue remodeling [213]. The pulmonary mesenchyme consists of FBs, endothelial cells (ECs), ASMCs, and macrophages [39, 148].

Following injury, dynamic changes also occur in the cellular niche to coordinate epithelial recovery. ASMCs secrete fibroblast growth factor 10 (FGF10), stimulating airway regeneration through activation of BCs throughout the airways [256]. Further exploration of the ASMC niche in the airways recently identified a particular subset of ASMC, coined "repair-supportive mesenchymal cells" (RSMCs), which are distinct from conventional ASMCs and highly express platelet-derived growth factor alpha (PDGFR$\alpha$). While RSMCs expressed typical ASMC markers at much lower levels, they were enriched FGF10 [155]. As a niche regulated signal FGF10 has been well studied in the context of regulating both airway development and repair. After a major lung injury, the surviving epithelial cells rapidly spread to sustain barrier function. Through recruitment of integrin-linked kinase signaling, the Hippo pathway is downregulated, increasing the nuclear translocation of YAP and ultimately the secretion of WNT7b from the epithelium. This Wnt signal stimulates cellular cross talk to the niche and induces Fgf10 expression in ASMC [236]. The functional importance of Wnt signaling in the lung was initially demonstrated during lung development, as early as at the stage of lung endoderm specification [109, 145]. In the adult lung, Wnt signaling influences the differentiation of regionally distinct populations of epithelial cells during homeostasis, regeneration, and disease [25, 78, 193, 201]. Wnt is crucial for lineage commitment of glandular stem cells within tracheal niches [140] and can inhibit differentiation of club cells into goblet cells through the RYK receptor [87, 130, 195, 231].

Similarly, interleukin-6 (IL-6) signals from the mesenchymal stromal cells in the niche, alongside neighboring epithelial cells, can regulate cell fate decisions, influencing ciliogenesis through STAT3-dependent pathways [41, 229]. Severe tracheal damage leading to complete loss of luminal epithelial cells differentially regulates p-STAT3, suggestive of influence over two different differentiation states. Firstly, in BCs and early undifferentiated progenitors, STAT3 inhibits differentiation into secretory cells and promotes ciliated cell differentiation through inhibition of Notch signaling. Secondly, in ciliated progenitors, STAT3 promotes differentiation by upregulation of FOXJ1, MCIDAS, and CDC20b/miR-449 [41, 229].

An important cellular niche unique to the proximal airways, and of specific relevance in humans, are the SMGs. The microenvironmental niches provided by the SMG contain progenitor cells capable of restoration of the surface airway epithelium [138, 140, 141, 230]. It still is unclear exactly how the surface airway signals to the niche to recruit the progenitors. Anomalous proliferation and differentiation SMGs have been associated with pathophysiology observed in hypersecretory lung diseases, including asthma, chronic bronchitis, and CF, highlighting their importance in formulating disease models [101, 248].

While resident lung cells provide a significant contribution to the epithelial niche in the proximal airways, inflammation is also a major player in disease pathogenesis and a critical component of lung health in chronic airway disease [63]. Inflammation involves the recruitment and activation of immune cells, including macrophages and neutrophils, to respond to infection and tissue injury. Recently, Puttur and colleagues reviewed how pulmonary macrophages (PMs) act as guardians of tissue repair in the lung environments, discussing how PMs respond to tissue inflammation and contribute to lung remodeling in response to infection and how this inflammatory response is sustained in chronic lung diseases, such as asthma, COPD, and CF [190]. In the mouse trachea three populations of macrophages have been identified: submucosal macrophages, interstitial macrophages, and intraepithelial macrophages. Injury stimulates the recruitment of neutrophils and submucosal macrophages, and successful airway regeneration is facilitated by c-c chemokine receptor type 2 (CCR2)-dependent recruitment of monocytes, which can interact with BCs and impact their regenerative behavior [63]. Macrophages have been more extensively studied for their roles in the alveolar niche, and further understanding of their impact on the proximal airways may shed new light on critical pathways to restore homeostasis.

The influence of the niche on epithelial cell behavior cannot be underestimated and is a critical component to consider in the engineering of new cellular models and the construction of lung tissues. Understanding the intimate regulation of airway homeostasis, disease, and regeneration through the coordination of ECM, mesenchymal cells and BCs will be critical as we strive to model and restore functional airways in disease and aging.

### 5.3.3 *Stem Cell Attrition with Disease and Aging*

Aging and disease both take their toll on the lungs. Stem cell phenotype and function has been shown to decline with age; however, we do not yet have a comprehensive understanding of how BCs are dynamically regulated by their niches, especially in the context of injury and aging. Some recent studies have highlighted the importance of spatiotemporal regulation of Wnt-secreting niches in both airway regeneration and aging [6]. PDGFRα-expressing cells, in the intercartilaginous zone (ICZ) of the tracheal and bronchial niche, transiently secrete Wnt ligand in response to injury to stimulate BC proliferation during regeneration [6]. Aging can also stimulate Wnt-ligand-dependent formation of glandular-like epithelial invaginations (GLEIs) generated from BCs in this ICZ. With aging comes accumulation of airway injury exposures, which can accelerate the "biological aging" of airway BCs, augmenting abnormal repair and disease initiation. Such stem cell aging and/or exhaustion is a pathological feature of many chronic lung diseases, which are typically associated with repeated airway injuries [6, 137, 172, 228]. Stem cell attrition is associated with repeated proliferation, leading to terminal differentiation of BCs [77]. Concomitant with stem cell attrition are changes in the ECM, which has perhaps been most widely studied in the context of aging, fibrosis, and COPD [1, 33, 66, 163, 165, 187, 218, 243]. Since many lung diseases develop over time and have a predominance in the aged population, it is important to consider such changes when constructing phenotypic models of disease. This may require models where matrix stiffness can be tuned over time or where airway models can be subject to long-term repeated insult/injury to understand stem cell attrition in the context of their cellular microenvironment.

## 5.4 Developing Cellular Therapies for Regeneration of Airway Tissues

While cellular therapeutics for airway disease are being actively pursued, how close these treatments are to availability in the clinic remains indeterminate for now. To develop cellular therapies or to engineer for engraftment or for disease modeling, an abundance of autologous cells is likely necessary. Current research efforts are focusing on the identity of the most suitable cells to be delivered to replace the damaged stem cells and the methods to generate sufficient cells, ex vivo, to be able to successfully regenerate the desired lung tissue. Unfortunately, ex vivo expansion of these cells is still plagued by rapid phenotypic changes induced by culture [132, 198]. The advances in induced pluripotent stem cell (iPSC) research have steered an exciting new age of regenerative medicine. Pluripotent stem cells are widely used in vitro by applying combinations of growth factors and cytokines, mimicking the mechanisms of respiratory development. Over the past decade, there had been significant progress in the generation of airway epithelium from iPSC [59, 72, 96, 97, 127, 150, 162]. Figure 5.2 summarizes the timeline for early specification of iPSC-derived lung progenitors, mimicking stages of lung development. While challenges in efficiency and purity of the differentiated progeny persist, RNAseq indicates that there is significant similarity between the derived cells and their endogenous counterparts [96]. The purification of iPSC-derived basal cells (iBCs) that recapitulate the biological and functional properties of basal cells, including mucociliary differentiation in tracheal xenografts, has facilitated a new wave of studies focused on gene editing and functional restoration of the airways [96]. These iBCs can be maintained in 3D spheroid cultures over several passages. Although challenges remain in the functional characterization of iBCs, these recent advances should facilitate the study of inherited or acquired diseases of the human airway and open the door for potential use of iBCs in regenerative medicine [96, 227]. Challenges in regenerating an iBC truly akin to its endogenous counterpart likely lie with the impact of the environment of the epigenetic manipulation of cellular state [249]. Epigenetic comparisons between differentiated iBCs and endogenous BCs may be important in generating an iBC that truly recapitulates the functional phenotype of primary cells. Patient- and/or mutation-specific models can be readily developed to model genetic airway diseases, including chloride channel dysfunction in CF and ciliary defects in PCD [34, 96, 177].

The identity and selection of long-term repopulating BCs and the differentiation of iPSCs to bona fide basal cells are being actively pursued [10, 49, 71, 75, 80, 96, 117, 147, 160, 223, 227]. Once this is achieved, as technologies evolve and progress toward tissue regeneration and cellular therapeutics, it will be critical to first determine the prospect of primary or induced BCs engrafting and integrating in the airway epithelium. While delivery itself perhaps poses the biggest challenge in the success of cellular therapeutics, once this is achieved, consideration of the diseased niche is likely to be equally important in regulating the fate of those delivered cells or therapeutics. Unfortunately, there is a paucity of studies considering the niche in terms of successful stem cell regulation, cellular engraftment, and tissue regeneration.

## 5.5 In Vitro Models of the Human Airways

Over the past decade, there have been several advances in the development of more complex and physiologically/pathophysiologically relevant ex vivo lung models, enhancing our ability to model lung disease. Air-liquid interface culture of primary airway cells, established in Transwell, has long provided a gold standard for evaluation of airway injury and repair. More recently, three-dimensional (3D) lung spheroid and organoid structures have been used to create more spatially relevant structures, and lung-on-a-chip technology adds the complexities of airflow and stretch. Each of these models will be discussed below. Table 5.1 summarizes the core pros and cons of using each of these lung models in scientific research, and Fig. 5.3 summarizes each model system.

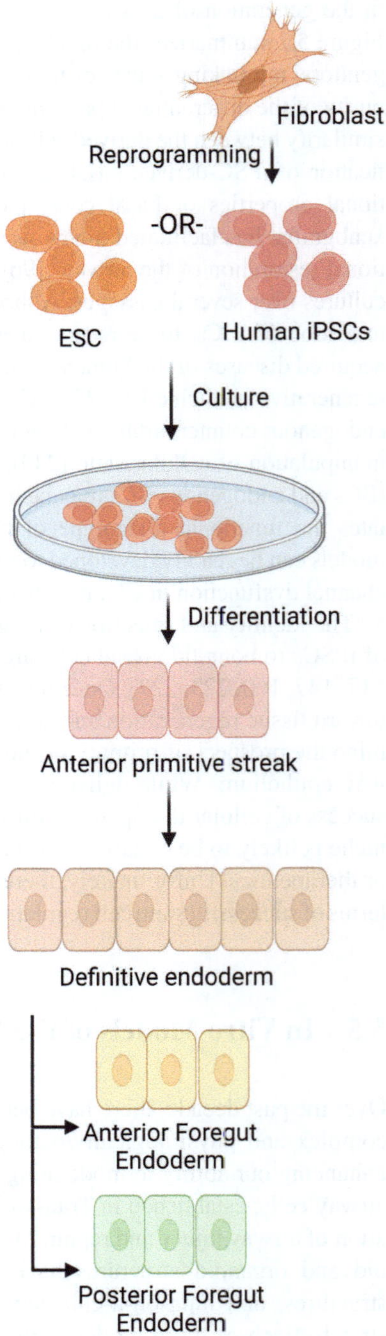


**Fig. 5.2** Early lung development from iPSC. Pluripotent stem cells are isolated and expanded in vitro from the inner cell mass of the blastocyst (embryonic stem cells or ESC) or reprogrammed from somatic cells, such as fibroblasts, from individuals (induced pluripotent stem cells or iPSC). To generate lung endoderm cells are differentiated using a combination of growth factors and small molecules to mimic lung development through the anterior primitive streak and definitive endoderm, where anteriorizing of the definitive endoderm will give rise to the anterior foregut endoderm. This will subsequently have the capacity to generate NKx2–1-expressing primordial lung progenitor cells. (This figure was created at Biorender.com)

**Table 5.1** In vitro models of the airway system

| Model | Application | Advantages | Drawbacks | References |
|---|---|---|---|---|
| Air-liquid interface | Disease models, drug delivery, toxicity, mucociliary clearance, cilia function, barrier function, CFTR activity | Differentiation of a pseudostratified airway epithelium<br>Mucociliary flow measurements<br>Cilia beat frequency measurements<br>Mucus production<br>Dose controlling for inhaled agents<br>Useful for drug transport/delivery<br>Wound-repair investigation<br>TEER measurements and CFTR activity<br>Inflammatory cell interactions<br>Can be long-lived (>3 months) | Not all primary donor cells differentiate well at the ALI<br>No directional flow (confined by Transwell size and shape)<br>No mechanical stretch<br>Lot-to-lot variability in Transwell<br>28-day differentiation | [25, 53, 77, 129, 171, 184, 185, 197, 200, 202, 210, 225, 229] |
| Organoid | Disease models, drug delivery, toxicity, lung development | Mimics lung development and branching morphogenesis<br>Provides the structural architecture and maintains cellular interactions of the lung microenvironment<br>It can be used for personalized medicine studies | Lack of breathing mechanics<br>Lacks structures that can be useful for studying the effect of inhaled therapy on involving mouth-to-airway transit<br>Long culture period and difficulty to reproduce at industrial level<br>Lacks an air-liquid interface (more applicable to development than adult lung)<br>Complexity limits analysis | [39, 54, 58, 93, 99, 115, 130, 152, 208] |
| Spheroids | Disease models, drug delivery, toxicity, CFTR activity | 3D structure allows for more physiological ECM interactions and structure<br>Swelling model for CFTR activity | Apical surface is limited (typically inside the spheroid) with no airflow<br>Variation in spheroid size<br>No niche cells present | [21, 32, 45, 52, 62, 91, 103, 110, 208, 221] |

(continued)

**Table 5.1** (continued)

| Model | Application | Advantages | Drawbacks | References |
|---|---|---|---|---|
| Lung-on-a-chip | Disease models, drug delivery, toxicity, CFTR activity, inflammation, mechanical breathing, airflow | Regulated mechanical stretch<br>Regulated airflow<br>Vascular channel for endothelial/ inflammatory cell interactions<br>Time-dependent drug treatment and monitoring | Challenging to assemble<br>Donor-specific success in differentiation without migration<br>Expensive chips and equipment | [106, 126, 258]<br>[8, 15, 100, 183, 217, 166, 167] |
| Xenografts | Regeneration, gene editing/therapy, drug delivery, toxicity | Regeneration can be evaluated in context of the native niche | Implantation can be technically complicated<br>Damage response generated during tracheal denudation | [9, 44, 48, 56, 82, 88, 89, 94, 121, 124, 239] |

### *5.5.1 Transwell Air-Liquid Interface (ALI) Cultures*

For decades, most pulmonary in vitro studies were performed using cell lines in submerged cell culture conditions, resulting in compromised physiological conditions [74, 192]. The transition to cell growth on Transwell, where epithelial BCs are grown to confluence on ECM-coated synthetic membranes and then polarized by exposing the apical surface to air, created a more physiologically applicable air-liquid interface model (ALI) [74, 102]. Over approximately 28 days, cells polarize and mature into a differentiated mucociliary airway epithelium, comprising BCs, club and goblet secretory cells, and multiciliated cells with scattered ionocytes present [207]. This model rapidly became a gold standard for recapitulating functional properties of the epithelium, including ciliary flow and mucus production. Such ALI cultures are widely used as an effective tool for cell-cell interaction studies, disease models, and drug efficacy/toxicity testing in the lung [3, 14, 19, 22, 26, 42, 192, 233]. They have been instrumental in the validation of small molecules that are now widely used for the treatment of CF [11, 92, 226, 251]. While extremely effective, they do have their limitations. Despite featuring all the major cell types of the proximal airways, these ALI cultures are static; they have no epithelial supporting niche; they are not designed to drive epithelial differentiation toward different branching generations; and choice of media, donor cells, and protocol can significantly impact phenotypic, transcriptomic, and physiological features of the differentiated epithelia [207, 245]. These limitations constrain the capacity of the models to completely recapitulate disease pathophysiology.

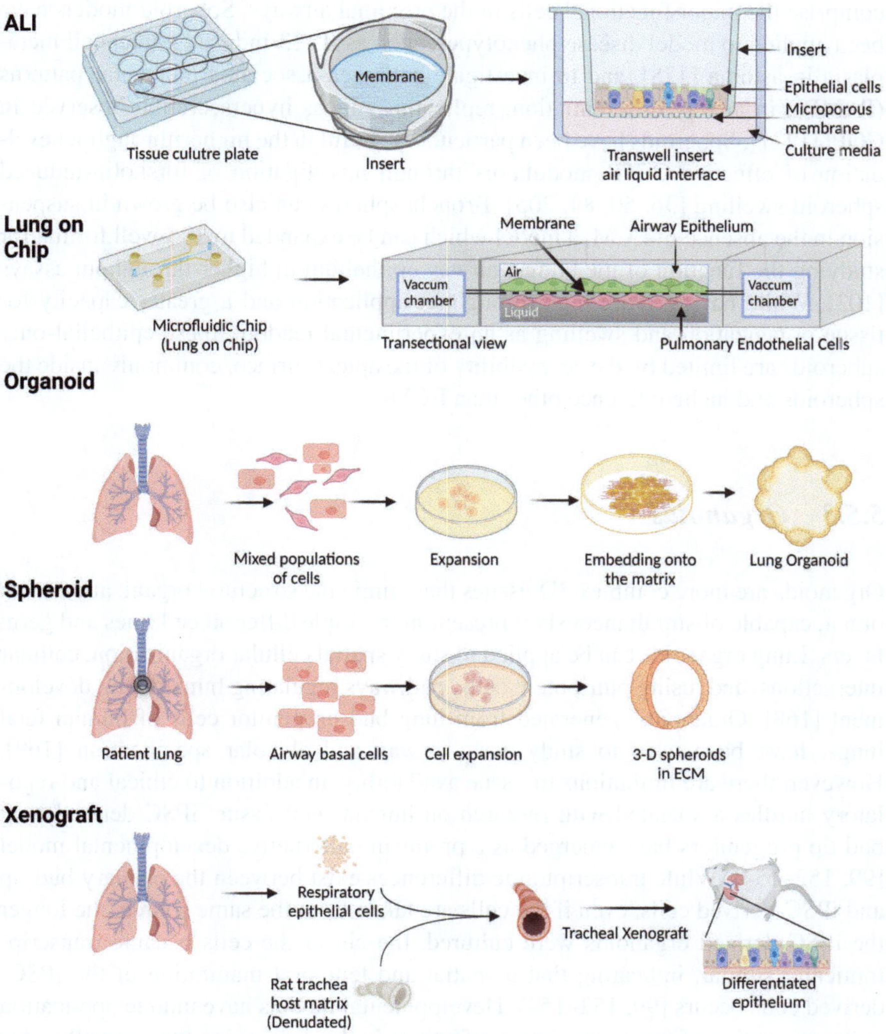


**Fig. 5.3** Cellular models of the proximal lung. This schematic summarizes the currently available in vitro model systems to study the proximal lung epithelium, including air-liquid interface in Transwell, microfluidic lung-on-a-chip, multicellular organoids, spheroids (tracheo/bronchospheres), and xenografts. (This figure was created at Biorender.com)

### 5.5.2 *Airway Spheroids: Tracheo/Bronchospheres*

Moving from a 2D to a 3D culture system allows for a closer replication of cell behaviors during tissue assembly. In vivo cells are exposed to circulating molecules, neighboring cells, and ECM, which are factors more effectively recapitulated in 3D models. The first airway spheroids, known as tracheo- or bronchospheres, were generated from murine basal cells [201], and, akin to ALI, matured bronchospheres comprise the major functional cells in the proximal airways. Spheroid models have been applied to model disease phenotypes, such as IL-13-induced goblet cell metaplasia in asthma [178], and to investigate pathogen-associated molecular patterns (PAMPs) in airway differentiation, replicating mucus hypersecretion observed in COPD [221]. Spheroids have been particularly useful in the higher-throughput evaluation of effective CFTR modulators through investigation of forskolin-induced spheroid swelling [36, 50, 84, 205]. Bronchospheres can also be grown in suspension in the absence of ECM, a model which can be expanded to 384-well format for studying the function of the human airway epithelium in higher-throughput assays [103]. While having more high-throughput application and a greater capacity for tissue organization and swelling as an experimental readout, these epithelial-only spheroids are limited by the accessibility of the apical surface, commonly inside the spheroids and niche influence other than ECM.

### *5.5.3 Organoids*

Organoids are more complex 3D tissues that mimic the structural organization of an organ, capable of simultaneously representing multiple different cell lines and germ layers. Lung organoids can be applied to study spatial cellular organization, cellular interactions, and, using pluripotent cells, pathways regulating human lung development [168]. Organoids, generated from lung bud progenitor cells of human fetal lungs, have been used to study early airway and alveolar specification [169]. However, there are limitations to tissue availability, in addition to ethical and regulatory hurdles associated with research on human fetal tissue. iPSC-derived lung bud tip progenitors have emerged as a promising alternative developmental model [99, 152–154]. While transcriptomic differences exist between the primary bud tip and iPSC-derived cells, even if the cells are taken from the same source, the longer the iPSC-derived organoids were cultured, the closer the cells became transcriptomically similar, indicating that a spatial and temporal maturation of the iPSC-derived cells occurs [99, 152–154]. Developmental models have unique application to the evaluation of respiratory virus infections in the developing lung and allow for study of infection in both proximal and distal lung tissues [186]. There is an enormous success in culturing relevant organoids under desired physiological conditions; however, there are several challenges associated with its real-life application. With the organoid system, variability of self-organizing growth has been observed, and there is limited experimental and analytical feasibility [105]. iPSC-derived lung bud tip organoids have not been documented to mature past an equivalent of the second trimester and thus not reach terminal maturation [186]. Additionally, branching morphogenesis is seemingly random, and how closely the branching morphogenesis and transitional zones mimic those of adult human lungs remains uncertain; similarly, the nature and pattern of the mesenchyme is unclear [39, 59]. Beyond genetic diseases, iPSC-derived airway organoids have been applied to predict drug responses, to study patient-specific host response to respiratory pathogens [238] and viral infection, including SARS-CoV-2 [134, 180, 220].

### *5.5.4 Lung-on-a-Chip*

While ALI, spheroid and organoid models all have advantages for studying functional human lung epithelium, they also have their limitations. Lung-on-a-chip systems have evolved over the past decade that overcome some of the shortcomings of other model systems [8, 15, 17, 166]. These engineered cell-based systems facilitate the evaluation of cells in an environment that mimics an in vivo tissue environment [2, 20, 106]. Most models of the human lung do not account for dynamic cues, such as blood flow and breathing cycles, that can impact airway biology. Recently, second-generation chips have been optimized for studying the effects of physiological airflow and stretch on differentiation, cellular composition, and mucociliary clearance in the proximal and distal airways [167]. Chip perfusion with air in the apical compartment accelerates maturation and polarization of mucociliary clearance, when compared to traditional ALI Transwell cultures, and the additional application of airflow and stretch influences cellular composition and the inflammatory status of the epithelium [167]. Commercially available microfluidic chips (Chip-S1®, Emulate Inc.) feature adjacent perfusable microchannels, where BCs can be cultured in the apical channel, which is separated from a vascular channel by a flexible and highly porous poly(dimethylsiloxane) (PDMS) membrane [167]. This specific membrane can be dynamically stretched in the presence of airflow, compared to previously developed protocols for airway chips using cell-culture optimized, but rigid, membranes [15, 113, 166].

### *5.5.5 Xenografts*

Tracheal xenograft models were first developed over two decades ago for the analysis of airway regeneration and the identification of progenitor cell subpopulations mediating the process [61, 215, 260]. In these models, human epithelial cells were seeded into an epithelium-denuded rat trachea, which was subsequently grafted into a xenon-tolerant/athymic mouse. These models have the benefit of having the entire cellular niche present, including tracheal cartilaginous rings [32, 64, 70, 131, 189]. Some of the earliest studies in xenografts enabled the dynamics of cell turnover, lineage, and differentiation to be evaluated. Patterns of clonal expansion of transgene-expressing cells allowed for the identification of airway progenitor populations, giving rise to clones of different sizes, and were instrumental in the initial identification of stem cells in the human airway epithelium capable of self-renewal. Such studies led to a conclusion that several stem cell subpopulations of cells are involved in the regeneration process of airway epithelium [60, 189, 260]. In an approach to establish intrapulmonary genetic reconstruction, CF xenografts were infected with CFTR recombinant adenovirus, reconstituting cAMP-driven Cl transport [82]. Though this correction was partial and variable, it established the applicability of this approach for disease correction. The presence of an intact epithelial niche in xenografts has been capitalized upon to evaluate epithelial-mesenchymal trophic units generated through the seeding of control and asthmatic epithelial cells [89]. Interestingly, this system demonstrated the importance of cellular cross talk between the niche and epithelium; in these studies the presence of an asthmatic epithelium alone was sufficient to drive mesenchymal remodeling [89].Advances in bioengineering have enabled the evolution of airway models beyond the static, 2D ALI airway models, adding physiological and mechanical cues to mimic tissue dynamics. Each model system has its own qualities and limitations and should be carefully considered for their applicability to specific research questions. For the successful preclinical evaluation of a novel inhaled therapy, for example, a human model capable of mimicking the physiological parameters that facilitate the in vivo transport of drug molecules is critical. Engineered complete lung tissue models with 3D lung architecture, cellular composition, and mechanical and physiological cues would be ideal, enabling evaluation of drug efficacy, transport kinetics, toxicity, and therapeutic potential [42]. Continued collaboration between bioengineers and basic cell biologists will be instrumental to bring new innovations to this space.

## 5.6 Cell-Matrix Interactions

An important component of developing more advanced models relies on the precise understanding of the role of the ECM in regulating airway stem cell function during both homeostasis and disease. The ECM forms a major component of the multicellular 3D architecture of the human lungs, providing structural, mechanical, and biochemical support. The importance of the lung ECM in regulating BC stemness and function was recently reviewed in Busch, Lorenzana, and Ryan [30]. Lung ECM is composed of fibrous proteins (elastin and collagen), glycoproteins (fibronectin, laminin), glycosaminoglycans, and proteoglycans, which are all important for normal lung development. All the major components of the lung (conducting airways, respiratory airways, lymphatics, and the pulmonary vasculature) require their unique ECM environments for specialized cell survival, proliferation, and differentiation [27]. Apart from providing structural integrity, ECM regulates cell behavior due to changes in its molecular composition and stiffness [79, 253] and acts as a reservoir for growth factors and cytokines. Several growth factors (FGF-10, TGF-β1, etc.) and cytokines are reported to regulate the breakdown of the ECM, in addition to regulating lung development and basal cell phenotype [188, 204]. However, precise implications of compartment-specific ECM composition, substrate stiffness, and other mechanical properties, as well as biochemical signaling, remain an active area of research [7, 29, 79, 254].

ECM remodeling is a core feature of many chronic lung diseases. In idiopathic pulmonary fibrosis (IPF), changes in the interstitial matrix within airways (conducting and respiratory) are indicative of disease progression [90]. Activated fibroblasts and myofibroblasts within fibroblast foci are considered as the key producers of ECM. In IPF tissues, type I procollagen, versican, hyaluronan, and tenascin C are highly expressed within the fibroblastic foci, while decorin and biglycan are present at relatively low level of expression, indicative of enhanced synthesis of ECM and remodeling [18, 65]. In COPD, ECM remodeling changes the composition of the basement membranes and the interstitial matrix at different anatomical niches of the lung and is also considered to be associated with progression of disease [90].

Modeling the lung ECM can provide insight into the interactions between the ECM and functional elements of lung airway. While 2D models of cells grown on tissue culture plastic are relatively simple by design, they are typically significantly stiffer than normal lung tissue [24, 104, 241]. In vitro 3D tissue-engineered model systems (lung-on-a-chip models, organoids, lung slices, ECM gels, and coculture systems) have been used in recent years for modeling the unique spatial geometry of the lung and understanding the complexity of cell-cell-ECM interactions [174, 175]. Generation of complex ECM scaffolds, which can facilitate cell growth and differentiation, is required to continue development of these complex models. Recently, a protocol was devised for generation of decellularized human lung bioink for downstream process of 2D and 3D lung cell culture [47]. While the importance of the ECM in organizing niche architecture and regulating cellular function has been extensively studied, there still is a significant knowledge gap in terms of the ECM-mediated regulation of cell behavior and phenotype. The complexity, including cross-linking, and insoluble nature of the ECM has posed a persistent obstacle to determining the generation and function of the ECM [28, 111]. Now, engineered biomaterials are being applied to mimic the in vivo characteristics of stem cell niches, which facilitate a desired in vitro tool for investigating the different roles exercised by the ECM on the stem cell phenotype. Recently, Yu et al. developed a natural ECM biomaterial enriched with miR-29-loaded exosomes for the treatment of pulmonary fibrosis, which promises gene therapy options for treatment of different diseases using ECM [255]. In the lung physiology, there is a strong relationship between ECM proteins and leukocyte migration to the locations of injury [170]. A detailed understanding on the complex ECM-cell interactions will advance the design and development of functional models and engineering tissues for lung regeneration.

## 5.7 Conclusion

The ECM is not simply a scaffold. It is an integral component of lung physiology and pathophysiology that changes with disease progression. Abnormal ECM, niche cells, and BC phenotype are closely related in the context of chronic lung disease. Increasing our understanding of the functional interactions between the ECM, niche cells, and stem cells in the epithelium will have a substantial impact of the recapitulation of lung homeostasis and disease in model systems. Perhaps the most challenging application of ECM-BC research is the potential to repopulate a decellularized ECM scaffold to create functional lung tissues. Techniques for tissue regeneration are reliant upon the creation of a receptive cellular microenvironment that can be applied to both ex vivo expansion and in vivo engraftment. Tunable models that can reflect disease progression and be integrated with niche cells would allow for evaluation of mechanical and biochemical factors in disease pathogenesis. The subsequent chapters will focus on bioengineering approaches for the generation of models for lung homeostasis and disease.

**Acknowledgments** ALR is funded by the NIH: NHLBI (R01HL139328 and R01HL153622-A1), the Cystic Fibrosis Foundation (CFFT17XX0 and CFFT21XX0), and the Daniel Tyler Health and Education Trust.

**Conflict of Interest Statement** The authors have declared that no conflict-of-interest exists.

## References

1. Aghali, A., L. Khalfaoui, A. B. Lagnado, L. Drake, J. J. Teske, C. M. Pabelick, J. F. Passos, and Y. S. Prakash. 2022. 'Cellular Senescence is Increased in Airway Smooth Muscle Cells of Elderly Asthmatics', *Am J Physiol Lung Cell Mol Physiol.*
2. Ahadian, S., R. Civitarese, D. Bannerman, M. H. Mohammadi, R. Lu, E. Wang, L. Davenport-Huyer, B. Lai, B. Zhang, Y. Zhao, S. Mandla, A. Korolj, and M. Radisic. 2018. 'Organ-On-A-Chip Platforms: A Convergence of Advanced Materials, Cells, and Microscale Technologies', *Adv Healthc Mater*, 7.
3. Ahmad, S., D. P. Nichols, M. Strand, R. C. Rancourt, S. H. Randell, C. W. White, and A. Ahmad. 2011. 'SERCA2 regulates non-CF and CF airway epithelial cell response to ozone', *PLoS One*, 6: e27451.
4. Ahmadvand, N., F. Khosravi, A. Lingampally, R. Wasnick, A. I. Vazquez-Armendariz, G. Carraro, M. Heiner, S. Rivetti, Y. Lv, J. Wilhelm, A. Gunther, S. Herold, D. Al Alam, C. Chen, P. Minoo, J. S. Zhang, and S. Bellusci. 2021. 'Identification of a novel subset of alveolar type 2 cells enriched in PD-L1 and expanded following pneumonectomy', *Eur Respir J*, 58.
5. Ali, M., X. Zhang, R. LaCanna, D. Tomar, J. W. Elrod, and Y. Tian. 2022. 'MICU1-dependent mitochondrial calcium uptake regulates lung alveolar type 2 cell plasticity and lung regeneration', *JCI Insight*, 7.
6. Aros, C. J., P. Vijayaraj, C. J. Pantoja, B. Bisht, L. K. Meneses, J. M. Sandlin, J. A. Tse, M. W. Chen, A. Purkayastha, D. W. Shia, J. M. S. Sucre, T. M. Rickabaugh, E. K. Vladar, M. K. Paul, and B. N. Gomperts. 2020. 'Distinct Spatiotemporally Dynamic Wnt-Secreting Niches Regulate Proximal Airway Regeneration and Aging', *Cell Stem Cell*, 27: 413–29 e4.
7. Arteaga-Solis, E., C. Settembre, A. Ballabio, and G. Karsenty. 2012. 'Sulfatases are determinants of alveolar formation', *Matrix Biol*, 31: 253–60.
8. Artzy-Schnirman, A., N. Hobi, N. Schneider-Daum, O. T. Guenat, C. M. Lehr, and J. Sznitman. 2019. 'Advanced in vitro lung-on-chip platforms for inhalation assays: From prospect to pipeline', *Eur J Pharm Biopharm*, 144: 11–17.
9. Baconnais, S., R. Tirouvanziam, J. M. Zahm, S. de Bentzmann, B. Peault, G. Balossier, and E. Puchelle. 1999. 'Ion composition and rheology of airway liquid from cystic fibrosis fetal tracheal xenografts', *Am J Respir Cell Mol Biol*, 20: 605–11.

10. Balasooriya, G. I., M. Goschorska, E. Piddini, and E. L. Rawlins. 2017. 'FGFR2 is required for airway basal cell self-renewal and terminal differentiation', *Development*, 144: 1600–06.
11. Barry, P. J., M. A. Mall, A. Alvarez, C. Colombo, K. M. de Winter-de Groot, I. Fajac, K. A. McBennett, E. F. McKone, B. W. Ramsey, S. Sutharsan, J. L. Taylor-Cousar, E. Tullis, N. Ahluwalia, L. S. Jun, S. M. Moskowitz, V. Prieto-Centurion, S. Tian, D. Waltz, F. Xuan, Y. Zhang, S. M. Rowe, D. Polineni, and V. X. Study Group. 2021. 'Triple Therapy for Cystic Fibrosis Phe508del-Gating and -Residual Function Genotypes', *N Engl J Med*, 385: 815–25.
12. Basil, M. C., F. L. Cardenas-Diaz, J. J. Kathiriya, M. P. Morley, J. Carl, A. N. Brumwell, J. Katzen, K. J. Slovik, A. Babu, S. Zhou, M. M. Kremp, K. B. McCauley, S. Li, J. D. Planer, S. S. Hussain, X. Liu, R. Windmueller, Y. Ying, K. M. Stewart, M. Oyster, J. D. Christie, J. M. Diamond, J. F. Engelhardt, E. Cantu, S. M. Rowe, D. N. Kotton, H. A. Chapman, and E. E. Morrisey. 2022. 'Human distal airways contain a multipotent secretory cell that can regenerate alveoli', *Nature*, 604: 120–26.
13. Basil, M. C., and E. E. Morrisey. 2020. 'Lung regeneration: a tale of mice and men', *Semin Cell Dev Biol*, 100: 88–100.
14. Becker, M. N., M. S. Sauer, M. S. Muhlebach, A. J. Hirsh, Q. Wu, M. W. Verghese, and S. H. Randell. 2004. 'Cytokine secretion by cystic fibrosis airway epithelial cells', *Am J Respir Crit Care Med*, 169: 645–53.
15. Benam, K. H., R. Novak, J. Nawroth, M. Hirano-Kobayashi, T. C. Ferrante, Y. Choe, R. Prantil-Baun, J. C. Weaver, A. Bahinski, K. K. Parker, and D. E. Ingber. 2016. 'Matched-Comparative Modeling of Normal and Diseased Human Airway Responses Using a Microengineered Breathing Lung Chip', *Cell Syst*, 3: 456–66 e4.
16. Benfante, A., S. Battaglia, S. Principe, C. Di Mitri, A. Paterno, M. Spatafora, and N. Scichilone. 2016. 'Asthmatics with high levels of serum surfactant protein D have more severe disease', *Eur Respir J*, 47: 1864–7.
17. Bennet, T. J., A. Randhawa, J. Hua, and K. C. Cheung. 2021. 'Airway-On-A-Chip: Designs and Applications for Lung Repair and Disease', *Cells*, 10.
18. Bensadoun, E. S., A. K. Burke, J. C. Hogg, and C. R. Roberts. 1996. 'Proteoglycan deposition in pulmonary fibrosis', *Am J Respir Crit Care Med*, 154: 1819–28.
19. Bernacki, S. H., A. L. Nelson, L. Abdullah, J. K. Sheehan, A. Harris, C. W. Davis, and S. H. Randell. 1999. 'Mucin gene expression during differentiation of human airway epithelia in vitro. Muc4 and muc5b are strongly induced', *Am J Respir Cell Mol Biol*, 20: 595–604.
20. Bhatia, S. N., and D. E. Ingber. 2014. 'Microfluidic organs-on-chips', *Nat Biotechnol*, 32: 760–72.
21. Białkowska, K., P. Komorowski, M. Bryszewska, and K. Miłowska. 2020. 'Spheroids as a Type of Three-Dimensional Cell Cultures-Examples of Methods of Preparation and the Most Important Application', *Int J Mol Sci*, 21.
22. Bluhmki, T., S. Traub, A. K. Muller, S. Bitzer, E. Schruf, M. T. Bammert, M. Leist, F. Gantner, J. P. Garnett, and R. Heilker. 2021. 'Functional human iPSC-derived alveolar-like cells cultured in a miniaturized 96Transwell air-liquid interface model', *Sci Rep*, 11: 17028.
23. Bodas, Manish, Andrew R. Moore, Bharathiraja Subramaniyan, Constantin Georgescu, Jonathan D. Wren, Willard M. Freeman, Brent R. Brown, Jordan P. Metcalf, and Matthew S. Walters. 2021. 'Cigarette Smoke Activates NOTCH3 to Promote Goblet Cell Differentiation in Human Airway Epithelial Cells', *American Journal of Respiratory Cell and Molecular Biology*, 64: 426–40.
24. Booth, A. J., R. Hadley, A. M. Cornett, A. A. Dreffs, S. A. Matthes, J. L. Tsui, K. Weiss, J. C. Horowitz, V. F. Fiore, T. H. Barker, B. B. Moore, F. J. Martinez, L. E. Niklason, and E. S. White. 2012. 'Acellular normal and fibrotic human lung matrices as a culture system for in vitro investigation', *Am J Respir Crit Care Med*, 186: 866–76.
25. Brechbuhl, H. M., M. Ghosh, M. K. Smith, R. W. Smith, B. Li, D. A. Hicks, B. B. Cole, P. R. Reynolds, and S. D. Reynolds. 2011. 'beta-catenin dosage is a critical determinant of tracheal basal cell fate determination', *Am J Pathol*, 179: 367–79.

26. Burgess, J. K., M. R. Jonker, M. Berg, N. T. H. Ten Hacken, K. B. Meyer, M. van den Berge, M. C. Nawijn, and I. H. Heijink. 2021. 'Periostin: contributor to abnormal airway epithelial function in asthma?', *Eur Respir J*, 57.
27. Burgess, J. K., T. Mauad, G. Tjin, J. C. Karlsson, and G. Westergren-Thorsson. 2016. 'The extracellular matrix – the under-recognized element in lung disease?', *J Pathol*, 240: 397–409.
28. Burgess, J. K., and G. Tjin. 2015. 'Extracellular Matrix Specification of Regenerative Cells in the Adult Lung', *Stem Cells in the Lung: Development, Repair and Regeneration*: 169–89.
29. Burgstaller, G., B. Oehrle, M. Gerckens, E. S. White, H. B. Schiller, and O. Eickelberg. 2017. 'The instructive extracellular matrix of the lung: basic composition and alterations in chronic lung disease', *Eur Respir J*, 50.
30. Busch, S. M., Z. Lorenzana, and A. L. Ryan. 2021. 'Implications for Extracellular Matrix Interactions With Human Lung Basal Stem Cells in Lung Development, Disease, and Airway Modeling', *Front Pharmacol*, 12: 645858.
31. Bustamante-Marin, X. M., and L. E. Ostrowski. 2017. 'Cilia and Mucociliary Clearance', *Cold Spring Harb Perspect Biol*, 9.
32. Butler, C. R., R. E. Hynds, K. H. Gowers, H. Lee Ddo, J. M. Brown, C. Crowley, V. H. Teixeira, C. M. Smith, L. Urbani, N. J. Hamilton, R. M. Thakrar, H. L. Booth, M. A. Birchall, P. De Coppi, A. Giangreco, C. O'Callaghan, and S. M. Janes. 2016. 'Rapid Expansion of Human Epithelial Stem Cells Suitable for Airway Tissue Engineering', *Am J Respir Crit Care Med*, 194: 156–68.
33. Calhoun, C., P. Shivshankar, M. Saker, L. B. Sloane, C. B. Livi, Z. D. Sharp, C. J. Orihuela, S. Adnot, E. S. White, A. Richardson, and C. J. Le Saux. 2016. 'Senescent Cells Contribute to the Physiological Remodeling of Aged Lungs', *J Gerontol A Biol Sci Med Sci*, 71: 153–60.
34. Calvert, B. A., and A. L. Ryan Firth. 2020. 'Application of iPSC to Modelling of Respiratory Diseases', *Adv Exp Med Biol*, 1237: 1–16.
35. Carraro, G., J. Langerman, S. Sabri, Z. Lorenzana, A. Purkayastha, G. Zhang, B. Konda, C. J. Aros, B. A. Calvert, A. Szymaniak, E. Wilson, M. Mulligan, P. Bhatt, J. Lu, P. Vijayaraj, C. Yao, D. W. Shia, A. J. Lund, E. Israely, T. M. Rickabaugh, J. Ernst, M. Mense, S. H. Randell, E. K. Vladar, A. L. Ryan, K. Plath, J. E. Mahoney, B. R. Stripp, and B. N. Gomperts. 2021. 'Transcriptional analysis of cystic fibrosis airways at single-cell resolution reveals altered epithelial cell states and composition', *Nat Med*, 27: 806–14.
36. Castillon, N., J. Hinnrasky, J. M. Zahm, H. Kaplan, N. Bonnet, P. Corlieu, J. M. Klossek, K. Taouil, A. Avril-Delplanque, B. Péault, and E. Puchelle. 2002. 'Polarized expression of cystic fibrosis transmembrane conductance regulator and associated epithelial proteins during the regeneration of human airway surface epithelium in three-dimensional culture', *Lab Invest*, 82: 989–98.
37. Chaudhary, N., A. Jayaraman, C. Reinhardt, J. D. Campbell, and M. Bosmann. 2022. 'A single-cell lung atlas of complement genes identifies the mesothelium and epithelium as prominent sources of extrahepatic complement proteins', *Mucosal Immunol*, 15: 927–39.
38. Chen, Q., and Y. Liu. 2020. 'Heterogeneous groups of alveolar type II cells in lung homeostasis and repair', *Am J Physiol Cell Physiol*, 319: C991–C96.
39. Chen, Y.-W., S. X. Huang, A. L. R. T. de Carvalho, S.-H. Ho, M. N. Islam, S. Volpi, L. D. Notarangelo, M. Ciancanelli, J.-L. Casanova, J. Bhattacharya, A. F. Liang, L. M. Palermo, M. Porotto, A. Moscona, and H.-W. Snoeck. 2017. 'A three-dimensional model of human lung development and disease from pluripotent stem cells', *Nature Cell Biology*, 19: 542–49.
40. Cheng, G., T. Ueda, T. Numao, Y. Kuroki, H. Nakajima, Y. Fukushima, S. Motojima, and T. Fukuda. 2000. 'Increased levels of surfactant protein A and D in bronchoalveolar lavage fluids in patients with bronchial asthma', *Eur Respir J*, 16: 831–5.
41. Choksi, S. P., G. Lauter, P. Swoboda, and S. Roy. 2014. 'Switching on cilia: transcriptional networks regulating ciliogenesis', *Development*, 141: 1427–41.
42. Cidem, A., P. Bradbury, D. Traini, and H. X. Ong. 2020. 'Modifying and Integrating in vitro and ex vivo Respiratory Models for Inhalation Drug Screening', *Front Bioeng Biotechnol*, 8: 581995.

43. Ciechanowicz, A. 2019. 'Stem Cells in Lungs', *Adv Exp Med Biol*, 1201: 261–74.
44. Copreni, E., S. Castellani, A. Bagnacani, A. Colombo, T. Rizzuti, S. Di Gioia, C. Colombo, and M. Conese. 2010. 'Bacterial internalization is not sufficient to clear Pseudomonas aeruginosa infection in human fetal airway xenografts', *Med Sci Monit*, 16: BR361–6.
45. Cores, J., M. T. Hensley, K. Kinlaw, S. M. Rikard, P. U. Dinh, D. Paudel, J. Tang, A. C. Vandergriff, T. A. Allen, Y. Li, J. Liu, B. Niu, Y. Chi, T. Caranasos, L. J. Lobo, and K. Cheng. 2017. 'Safety and Efficacy of Allogeneic Lung Spheroid Cells in a Mismatched Rat Model of Pulmonary Fibrosis', *Stem Cells Transl Med*, 6: 1905–16.
46. Crystal, R. G., S. H. Randell, J. F. Engelhardt, J. Voynow, and M. E. Sunday. 2008. 'Airway epithelial cells: current concepts and challenges', *Proc Am Thorac Soc*, 5: 772–7.
47. Dabaghi, M., N. Saraei, M. B. Carpio, V. Nanduri, J. Ungureanu, M. Babi, A. Chandiramohan, A. Noble, S. D. Revill, B. Zhang, K. Ask, M. Kolb, Y. Shargall, J. Moran-Mirabal, and J. A. Hirota. 2021. 'A Robust Protocol for Decellularized Human Lung Bioink Generation Amenable to 2D and 3D Lung Cell Culture', *Cells*, 10: 1538.
48. Dajani, R., Y. Zhang, P. J. Taft, S. M. Travis, T. D. Starner, A. Olsen, J. Zabner, M. J. Welsh, and J. F. Engelhardt. 2005. 'Lysozyme secretion by submucosal glands protects the airway from bacterial infection', *Am J Respir Cell Mol Biol*, 32: 548–52.
49. de la Grange, P., A. Jolly, C. Courageux, C. Ben Brahim, and P. Leroy. 2021. 'Genes coding for transcription factors involved in stem cell maintenance are repressed by TGF-beta and downstream of Slug/Snail2 in COPD bronchial epithelial progenitors', *Mol Biol Rep*, 48: 6729–38.
50. Dekkers, J. F., R. A. Gogorza Gondra, E. Kruisselbrink, A. M. Vonk, H. M. Janssens, K. M. de Winter-de Groot, C. K. van der Ent, and J. M. Beekman. 2016. 'Optimal correction of distinct CFTR folding mutants in rectal cystic fibrosis organoids', *Eur Respir J*, 48: 451–8.
51. Deprez, M., L. E. Zaragosi, M. Truchi, C. Becavin, S. Ruiz Garcia, M. J. Arguel, M. Plaisant, V. Magnone, K. Lebrigand, S. Abelanet, F. Brau, A. Paquet, D. Pe'er, C. H. Marquette, S. Leroy, and P. Barbry. 2020. 'A Single-Cell Atlas of the Human Healthy Airways', *Am J Respir Crit Care Med*, 202: 1636–45.
52. Dernowsek, J. A., R. A. Rezende, V. E. Passamai, P. Y. Noritomi, D. T. Kemmoku, J. A. Nogueira, V. F. Lara, V. Mironov, and J. V. L. da Silva. 2016. 'Tissue Spheroids Encaged Into Microscaffolds with Internal Structure to Increase Cell Viability', *Procedia CIRP*, 49: 174–77.
53. Diabaté, S., L. Armand, S. Murugadoss, M. Dilger, S. Fritsch-Decker, C. Schlager, D. Béal, M. E. Arnal, M. Biola-Clier, S. Ambrose, S. Mülhopt, H. R. Paur, I. Lynch, E. Valsami-Jones, M. Carriere, and C. Weiss. 2020. 'Air-Liquid Interface Exposure of Lung Epithelial Cells to Low Doses of Nanoparticles to Assess Pulmonary Adverse Effects', *Nanomaterials (Basel)*, 11.
54. Duan, X., X. Tang, M. S. Nair, T. Zhang, Y. Qiu, W. Zhang, P. Wang, Y. Huang, J. Xiang, H. Wang, R. E. Schwartz, D. D. Ho, T. Evans, and S. Chen. 2021. 'An airway organoid-based screen identifies a role for the HIF1alpha-glycolysis axis in SARS-CoV-2 infection', *Cell Rep*, 37: 109920.
55. Duclos, G. E., V. H. Teixeira, P. Autissier, Y. B. Gesthalter, M. A. Reinders-Luinge, R. Terrano, Y. M. Dumas, G. Liu, S. A. Mazzilli, C. A. Brandsma, M. van den Berge, S. M. Janes, W. Timens, M. E. Lenburg, A. Spira, J. D. Campbell, and J. Beane. 2019. 'Characterizing smoking-induced transcriptional heterogeneity in the human bronchial epithelium at single-cell resolution', *Sci Adv*, 5: eaaw3413.
56. Dupuit, F., D. Gaillard, J. Hinnrasky, E. Mongodin, S. de Bentzmann, E. Copreni, and E. Puchelle. 2000. 'Differentiated and functional human airway epithelium regeneration in tracheal xenografts', *Am J Physiol Lung Cell Mol Physiol*, 278: L165–76.
57. Dy, A. B. C., S. Tanyaratsrisakul, D. R. Voelker, and J. G. Ledford. 2018. 'The Emerging Roles of Surfactant Protein-A in Asthma', *J Clin Cell Immunol*, 9.

58. Dye, B. R., J. T. Decker, R. F. C. Hein, A. J. Miller, S. Huang, J. R. Spence, and L. D. Shea. 2022. 'Human Lung Organoid Culture in Alginate With and Without Matrigel to Model Development and Disease', *Tissue Eng Part A*.
59. Dye, B. R., D. R. Hill, M. A. Ferguson, Y. H. Tsai, M. S. Nagy, R. Dyal, J. M. Wells, C. N. Mayhew, R. Nattiv, O. D. Klein, E. S. White, G. H. Deutsch, and J. R. Spence. 2015. 'In vitro generation of human pluripotent stem cell derived lung organoids', *Elife*, 4.
60. Engelhardt, J. F., H. Schlossberg, J. R. Yankaskas, and L. Dudus. 1995. 'Progenitor cells of the adult human airway involved in submucosal gland development', *Development*, 121: 2031–46.
61. Engelhardt, J. F., J. R. Yankaskas, and J. M. Wilson. 1992. 'In vivo retroviral gene transfer into human bronchial epithelia of xenografts', *J Clin Invest*, 90: 2598–607.
62. Engler, A. E., G. Mostoslavsky, L. Miller, and J. R. Rock. 2019. 'Isolation, Maintenance and Differentiation of Primary Tracheal Basal Cells from Adult Rhesus Macaque', *Methods Protoc*, 2.
63. Engler, A. E., A. B. Ysasi, R. M. F. Pihl, C. Villacorta-Martin, H. M. Heston, H. M. K. Richardson, B. R. Thapa, N. R. Moniz, A. C. Belkina, S. A. Mazzilli, and J. R. Rock. 2020. 'Airway-Associated Macrophages in Homeostasis and Repair', *Cell Rep*, 33: 108553.
64. Escotte, S., C. Catusse, C. Coraux, and E. Puchelle. 2004. 'Reconstitution of human airway tissue in the humanized xenograft model', *J Cyst Fibros*, 3 Suppl 2: 63–5.
65. Estany, S., V. Vicens-Zygmunt, R. Llatjos, A. Montes, R. Penin, I. Escobar, A. Xaubet, S. Santos, F. Manresa, J. Dorca, and M. Molina-Molina. 2014. 'Lung fibrotic tenascin-C upregulation is associated with other extracellular matrix proteins and induced by TGFbeta1', *BMC Pulm Med*, 14: 120.
66. Eurlings, I. M., M. A. Dentener, E. M. Mercken, R. de Cabo, K. R. Bracke, J. H. Vernooy, E. F. Wouters, and N. L. Reynaert. 2014. 'A comparative study of matrix remodeling in chronic models for COPD; mechanistic insights into the role of TNF-alpha', *Am J Physiol Lung Cell Mol Physiol*, 307: L557–65.
67. Ewis, A. A., K. Kondo, F. Dang, Y. Nakahori, Y. Shinohara, M. Ishikawa, and Y. Baba. 2006. 'Surfactant protein B gene variations and susceptibility to lung cancer in chromate workers', *Am J Ind Med*, 49: 367–73.
68. Fahy, J. V. 2002. 'Goblet cell and mucin gene abnormalities in asthma', *Chest*, 122: 320S–26S.
69. Fahy, J. V., and B. F. Dickey. 2010. 'Airway mucus function and dysfunction', *N Engl J Med*, 363: 2233–47.
70. Filali, M., Y. Zhang, T. C. Ritchie, and J. F. Engelhardt. 2002. 'Xenograft model of the CF airway', *Methods Mol Med*, 70: 537–50.
71. Firth, A. L., C. T. Dargitz, S. J. Qualls, T. Menon, R. Wright, O. Singer, F. H. Gage, A. Khanna, and I. M. Verma. 2014. 'Generation of multiciliated cells in functional airway epithelia from human induced pluripotent stem cells', *Proc Natl Acad Sci U S A*, 111: E1723–30.
72. Firth, A. L., T. Menon, G. S. Parker, S. J. Qualls, B. M. Lewis, E. Ke, C. T. Dargitz, R. Wright, A. Khanna, F. H. Gage, and I. M. Verma. 2015. 'Functional Gene Correction for Cystic Fibrosis in Lung Epithelial Cells Generated from Patient iPSCs', *Cell Rep*, 12: 1385–90.
73. Francisco, D., Y. Wang, M. Conway, A. N. Hurbon, A. B. C. Dy, K. J. Addison, H. W. Chu, D. R. Voelker, J. G. Ledford, and M. Kraft. 2020. 'Surfactant Protein-A Protects against IL-13-Induced Inflammation in Asthma', *J Immunol*, 204: 2829–39.
74. Fulcher, M. L., and S. H. Randell. 2013. 'Human nasal and tracheo-bronchial respiratory epithelial cell culture', *Methods Mol Biol*, 945: 109–21.
75. Garrido-Jimenez, S., J. F. Barrera-Lopez, S. Diaz-Chamorro, C. M. Mateos-Quiros, I. Rodriguez-Blanco, F. L. Marquez-Perez, M. J. Lorenzo, F. Centeno, A. C. Roman, and J. M. Carvajal-Gonzalez. 2021. 'p53 regulation by MDM2 contributes to self-renewal and differentiation of basal stem cells in mouse and human airway epithelium', *FASEB J*, 35: e21816.

76. Ghosh, M., H. M. Brechbuhl, R. W. Smith, B. Li, D. A. Hicks, T. Titchner, C. M. Runkle, and S. D. Reynolds. 2011. 'Context-dependent differentiation of multipotential keratin 14-expressing tracheal basal cells', *Am J Respir Cell Mol Biol*, 45: 403–10.
77. Ghosh, M., C. L. Hill, A. Alsudayri, S. W. Lallier, D. Hayes, Jr., S. Wijeratne, Z. H. Tan, T. Chiang, J. E. Mahoney, G. Carraro, B. R. Stripp, and S. D. Reynolds. 2021. 'Repeated injury promotes tracheobronchial tissue stem cell attrition', *Stem Cells Transl Med*, 10: 1696–713.
78. Giangreco, A., L. Lu, C. Vickers, V. H. Teixeira, K. R. Groot, C. R. Butler, E. V. Ilieva, P. J. George, A. G. Nicholson, E. K. Sage, F. M. Watt, and S. M. Janes. 2012. 'β-Catenin determines upper airway progenitor cell fate and preinvasive squamous lung cancer progression by modulating epithelial-mesenchymal transition', *J Pathol*, 226: 575–87.
79. Gilpin, S. E., Q. Li, D. Evangelista-Leite, X. Ren, D. P. Reinhardt, B. L. Frey, and H. C. Ott. 2017. 'Fibrillin-2 and Tenascin-C bridge the age gap in lung epithelial regeneration', *Biomaterials*, 140: 212–19.
80. Giuranno, L., C. Wansleeben, R. Iannone, L. Arathoon, J. Hounjet, A. J. Groot, and M. Vooijs. 2019. 'NOTCH signaling promotes the survival of irradiated basal airway stem cells', *Am J Physiol Lung Cell Mol Physiol*, 317: L414–L23.
81. Goldfarbmuren, K. C., N. D. Jackson, S. P. Sajuthi, N. Dyjack, K. S. Li, C. L. Rios, E. G. Plender, M. T. Montgomery, J. L. Everman, P. E. Bratcher, E. K. Vladar, and M. A. Seibold. 2020. 'Dissecting the cellular specificity of smoking effects and reconstructing lineages in the human airway epithelium', *Nat Commun*, 11: 2485.
82. Goldman, M. J., Y. Yang, and J. M. Wilson. 1995. 'Gene therapy in a xenograft model of cystic fibrosis lung corrects chloride transport more effectively than the sodium defect', *Nat Genet*, 9: 126–31.
83. Goto, H., A. Mitsuhashi, and Y. Nishioka. 2014. 'Role of surfactant protein A in non-infectious lung diseases', *J Med Invest*, 61: 1–6.
84. Graeber, Simon Y., Peter van Mourik, Annelotte M. Vonk, Evelien Kruisselbrink, Stephanie Hirtz, Cornelis K. van der Ent, Marcus A. Mall, and Jeffrey M. Beekman. 2020. 'Comparison of Organoid Swelling and In Vivo Biomarkers of CFTR Function to Determine Effects of Lumacaftor–Ivacaftor in Patients with Cystic Fibrosis Homozygous for the F508del Mutation', *American Journal of Respiratory and Critical Care Medicine*, 202: 1589–92.
85. Greaney, A. M., T. S. Adams, M. S. Brickman Raredon, E. Gubbins, J. C. Schupp, A. J. Engler, M. Ghaedi, Y. Yuan, N. Kaminski, and L. E. Niklason. 2020. 'Platform Effects on Regeneration by Pulmonary Basal Cells as Evaluated by Single-Cell RNA Sequencing', *Cell Rep*, 30: 4250–65 e6.
86. Groneberg, D. A., P. R. Eynott, T. Oates, S. Lim, R. Wu, I. Carlstedt, A. G. Nicholson, and K. F. Chung. 2002. 'Expression of MUC5AC and MUC5B mucins in normal and cystic fibrosis lung', *Respir Med*, 96: 81–6.
87. Guha, A., A. Deshpande, A. Jain, P. Sebastiani, and W. V. Cardoso. 2017. 'Uroplakin 3a(+) Cells Are a Distinctive Population of Epithelial Progenitors that Contribute to Airway Maintenance and Post-injury Repair', *Cell Rep*, 19: 246–54.
88. Guihaire, J., R. Itagaki, M. Stubbendorff, X. Hua, T. Deuse, S. Ullrich, E. Fadel, P. Dorfmuller, R. C. Robbins, H. Reichenspurner, U. Schumacher, and S. Schrepfer. 2016. 'Orthotopic tracheal transplantation using human bronchus: an original xenotransplant model of obliterative airway disorder', *Transpl Int*, 29: 1337–48.
89. Hackett, T. L., S. C. Ferrante, C. E. Hoptay, J. F. Engelhardt, J. L. Ingram, Y. Zhang, S. E. Alcala, F. Shaheen, E. Matz, D. K. Pillai, and R. J. Freishtat. 2017. 'A Heterotopic Xenograft Model of Human Airways for Investigating Fibrosis in Asthma', *Am J Respir Cell Mol Biol*, 56: 291–99.
90. Hackett, T. L., and E. T. Osei. 2021. 'Modeling Extracellular Matrix-Cell Interactions in Lung Repair and Chronic Disease', *Cells*, 10: 2145.
91. Han, Kyuho, Sarah E. Pierce, Amy Li, Kaitlyn Spees, Gray R. Anderson, Jose A. Seoane, Yuan-Hung Lo, Michael Dubreuil, Micah Olivas, Roarke A. Kamber, Michael Wainberg,

Kaja Kostyrko, Marcus R. Kelly, Maryam Yousefi, Scott W. Simpkins, David Yao, Keonil Lee, Calvin J. Kuo, Peter K. Jackson, Alejandro Sweet-Cordero, Anshul Kundaje, Andrew J. Gentles, Christina Curtis, Monte M. Winslow, and Michael C. Bassik. 2020. 'CRISPR screens in cancer spheroids identify 3D growth-specific vulnerabilities', *Nature*, 580: 136–41.

92. Haq, I., M. Almulhem, S. Soars, D. Poulton, and M. Brodlie. 2022. 'Precision Medicine Based on CFTR Genotype for People with Cystic Fibrosis', *Pharmgenomics Pers Med*, 15: 91–104.
93. Harford, T. J., F. Rezaee, B. R. Dye, J. Fan, J. R. Spence, and G. Piedimonte. 2022. 'RSV-induced changes in a 3-dimensional organoid model of human fetal lungs', *PLoS One*, 17: e0265094.
94. Harris, J. E., Jr., J. Shin, B. Lee, K. Pelosky, C. M. Hooker, K. Harbom, A. Hulbert, C. Zahnow, S. C. Yang, S. Baylin, C. Brayton, and M. V. Brock. 2011. 'A murine xenograft model of spontaneous metastases of human lung adenocarcinoma', *J Surg Res*, 171: e75–9.
95. Hawkins, F. J., and D. N. Kotton. 2018. 'Pulmonary Ionocytes Challenge the Paradigm in Cystic Fibrosis', *Trends Pharmacol Sci*, 39: 852–54.
96. Hawkins, F. J., S. Suzuki, M. L. Beermann, C. Barilla, R. Wang, C. Villacorta-Martin, A. Berical, J. C. Jean, J. Le Suer, T. Matte, C. Simone-Roach, Y. Tang, T. M. Schlaeger, A. M. Crane, N. Matthias, S. X. L. Huang, S. H. Randell, J. Wu, J. R. Spence, G. Carraro, B. R. Stripp, A. Rab, E. J. Sorsher, A. Horani, S. L. Brody, B. R. Davis, and D. N. Kotton. 2021. 'Derivation of Airway Basal Stem Cells from Human Pluripotent Stem Cells', *Cell Stem Cell*, 28: 79–95 e8.
97. Hawkins, F., and D. N. Kotton. 2015. 'Embryonic and induced pluripotent stem cells for lung regeneration', *Ann Am Thorac Soc*, 12 Suppl 1: S50–3.
98. Hegab, A. E., V. L. Ha, D. O. Darmawan, J. L. Gilbert, A. T. Ooi, Y. S. Attiga, B. Bisht, D. W. Nickerson, and B. N. Gomperts. 2012. 'Isolation and in vitro characterization of basal and submucosal gland duct stem/progenitor cells from human proximal airways', *Stem Cells Transl Med*, 1: 719–24.
99. Hein, Renee F. C., Ansley S. Conchola, Alexis S. Fine, Zhiwei Xiao, Tristan Frum, Lindy K. Brastrom, Mayowa A. Akinwale, Charlie J. Childs, Yu-Hwai Tsai, Emily M. Holloway, Sha Huang, John Mahoney, Idse Heemskerk, and Jason R. Spence. 2022. 'Stable iPSC-derived NKX2-1+ lung bud tip progenitor organoids give rise to airway and alveolar cell types', *Development*, 149.
100. Henry, O. Y. F., R. Villenave, M. J. Cronce, W. D. Leineweber, M. A. Benz, and D. E. Ingber. 2017. 'Organs-on-chips with integrated electrodes for trans-epithelial electrical resistance (TEER) measurements of human epithelial barrier function', *Lab Chip*, 17: 2264–71.
101. Hewitt, R. J., and C. M. Lloyd. 2021. 'Regulation of immune responses by the airway epithelial cell landscape', *Nat Rev Immunol*, 21: 347–62.
102. Hiemstra, P. S., T. D. Tetley, and S. M. Janes. 2019. 'Airway and alveolar epithelial cells in culture', *Eur Respir J*, 54.
103. Hild, M., and A. B. Jaffe. 2016. 'Production of 3-D Airway Organoids From Primary Human Airway Basal Cells and Their Use in High-Throughput Screening', *Curr Protoc Stem Cell Biol*, 37: IE 9 1–IE 9 15.
104. Hinz, B. 2012. 'Mechanical aspects of lung fibrosis: a spotlight on the myofibroblast', *Proc Am Thorac Soc*, 9: 137–47.
105. Hofer, M., and M. P. Lutolf. 2021. 'Engineering organoids', *Nat Rev Mater*, 6: 402–20.
106. Huh, D. D. 2015. 'A human breathing lung-on-a-chip', *Ann Am Thorac Soc*, 12 Suppl 1: S42–4.
107. Hung, C. H., L. C. Chen, Z. Zhang, B. Chowdhury, W. L. Lee, B. Plunkett, C. H. Chen, A. C. Myers, and S. K. Huang. 2004. 'Regulation of TH2 responses by the pulmonary Clara cell secretory 10-kd protein', *J Allergy Clin Immunol*, 114: 664–70.
108. Hurskainen, M., I. Mizikova, D. P. Cook, N. Andersson, C. Cyr-Depauw, F. Lesage, E. Helle, L. Renesme, R. P. Jankov, M. Heikinheimo, B. C. Vanderhyden, and B. Thebaud. 2021.

'Single cell transcriptomic analysis of murine lung development on hyperoxia-induced damage', *Nat Commun*, 12: 1565.
109. Hussain, M., C. Xu, M. Lu, X. Wu, L. Tang, and X. Wu. 2017. 'Wnt/β-catenin signaling links embryonic lung development and asthmatic airway remodeling', *Biochim Biophys Acta Mol Basis Dis*, 1863: 3226–42.
110. Hynds, R. E., C. R. Butler, S. M. Janes, and A. Giangreco. 2019. 'Expansion of Human Airway Basal Stem Cells and Their Differentiation as 3D Tracheospheres', *Methods Mol Biol*, 1576: 43–53.
111. Hynes, R. O., and A. Naba. 2012. 'Overview of the matrisome – an inventory of extracellular matrix constituents and functions', *Cold Spring Harb Perspect Biol*, 4: a004903.
112. Ikegami, M., J. A. Whitsett, P. C. Martis, and T. E. Weaver. 2005. 'Reversibility of lung inflammation caused by SP-B deficiency', *Am J Physiol Lung Cell Mol Physiol*, 289: L962–70.
113. Izadifar, Z., A. Sontheimer-Phelps, B. A. Lubamba, H. Bai, C. Fadel, A. Stejskalova, A. Ozkan, Q. Dasgupta, A. Bein, A. Junaid, A. Gulati, G. Mahajan, S. Kim, N. T. LoGrande, A. Naziripour, and D. E. Ingber. 2022. 'Modeling mucus physiology and pathophysiology in human organs-on-chips', *Adv Drug Deliv Rev*, 191: 114542.
114. Jackson, N. D., J. L. Everman, M. Chioccioli, L. Feriani, K. C. Goldfarbmuren, S. P. Sajuthi, C. L. Rios, R. Powell, M. Armstrong, J. Gomez, C. Michel, C. Eng, S. S. Oh, J. Rodriguez-Santana, P. Cicuta, N. Reisdorph, E. G. Burchard, and M. A. Seibold. 2020. 'Single-Cell and Population Transcriptomics Reveal Pan-epithelial Remodeling in Type 2-High Asthma', *Cell Rep*, 32: 107872.
115. Jacob, A., M. Morley, F. Hawkins, K. B. McCauley, J. C. Jean, H. Heins, C. L. Na, T. E. Weaver, M. Vedaie, K. Hurley, A. Hinds, S. J. Russo, S. Kook, W. Zacharias, M. Ochs, K. Traber, L. J. Quinton, A. Crane, B. R. Davis, F. V. White, J. Wambach, J. A. Whitsett, F. S. Cole, E. E. Morrisey, S. H. Guttentag, M. F. Beers, and D. N. Kotton. 2017. 'Differentiation of Human Pluripotent Stem Cells into Functional Lung Alveolar Epithelial Cells', *Cell Stem Cell*, 21: 472–88 e10.
116. Jain, Manu, and J. Iasha Sznajder. 2007. 'Bench-to-bedside review: Distal airways in acute respiratory distress syndrome', *Critical Care*, 11: 206.
117. Jensen-Cody, C. W., A. K. Crooke, P. G. Rotti, V. Ievlev, W. Shahin, S. Y. Park, T. J. Lynch, and J. F. Engelhardt. 2021. 'Lef-1 controls cell cycle progression in airway basal cells to regulate proliferation and differentiation', *Stem Cells*, 39: 1221–35.
118. Johansson, S. L., Q. Tan, R. Holst, L. Christiansen, N. C. Hansen, A. T. Hojland, H. Wulf-Johansson, A. Schlosser, I. L. Titlestad, J. Vestbo, U. Holmskov, K. O. Kyvik, and G. L. Sorensen. 2014. 'Surfactant protein D is a candidate biomarker for subclinical tobacco smoke-induced lung damage', *Am J Physiol Lung Cell Mol Physiol*, 306: L887–95.
119. Kathiriya, J. J., A. N. Brumwell, J. R. Jackson, X. Tang, and H. A. Chapman. 2020. 'Distinct Airway Epithelial Stem Cells Hide among Club Cells but Mobilize to Promote Alveolar Regeneration', *Cell Stem Cell*, 26: 346–58 e4.
120. Kato, K., E. H. Chang, Y. Chen, W. Lu, M. M. Kim, M. Niihori, L. Hecker, and K. C. Kim. 2020. 'MUC1 contributes to goblet cell metaplasia and MUC5AC expression in response to cigarette smoke in vivo', *Am J Physiol Lung Cell Mol Physiol*, 319: L82–L90.
121. Keswani, S. G., S. Balaji, L. Le, A. Leung, A. B. Katz, F. Y. Lim, M. Habli, H. N. Jones, J. M. Wilson, and T. M. Crombleholme. 2012. 'Pseudotyped AAV vector-mediated gene transfer in a human fetal trachea xenograft model: implications for in utero gene therapy for cystic fibrosis', *PLoS One*, 7: e43633.
122. Kibe, A., H. Inoue, S. Fukuyama, K. Machida, K. Matsumoto, H. Koto, T. Ikegami, H. Aizawa, and N. Hara. 2003. 'Differential regulation by glucocorticoid of interleukin-13-induced eosinophilia, hyperresponsiveness, and goblet cell hyperplasia in mouse airways', *Am J Respir Crit Care Med*, 167: 50–6.
123. Kim, V., and G. J. Criner. 2013. 'Chronic bronchitis and chronic obstructive pulmonary disease', *Am J Respir Crit Care Med*, 187: 228–37.

124. Kita, K., K. Fukuda, H. Takahashi, A. Tanimoto, A. Nishiyama, S. Arai, S. Takeuchi, K. Yamashita, K. Ohtsubo, S. Otani, N. Yanagimura, C. Suzuki, H. Ikeda, M. Tamura, I. Matsumoto, and S. Yano. 2019. 'Patient-derived xenograft models of non-small cell lung cancer for evaluating targeted drug sensitivity and resistance', *Cancer Sci*, 110: 3215–24.
125. Knowles, M. R., and R. C. Boucher. 2002. 'Mucus clearance as a primary innate defense mechanism for mammalian airways', *J Clin Invest*, 109: 571–7.
126. Konar, D., M. Devarasetty, D. V. Yildiz, A. Atala, and S. V. Murphy. 2016. 'Lung-On-A-Chip Technologies for Disease Modeling and Drug Development', *Biomed Eng Comput Biol*, 7: 17–27.
127. Konishi, S., S. Gotoh, K. Tateishi, Y. Yamamoto, Y. Korogi, T. Nagasaki, H. Matsumoto, S. Muro, T. Hirai, I. Ito, S. Tsukita, and M. Mishima. 2016. 'Directed Induction of Functional Multi-ciliated Cells in Proximal Airway Epithelial Spheroids from Human Pluripotent Stem Cells', *Stem Cell Reports*, 6: 18–25.
128. Kotton, D. N., and E. E. Morrisey. 2014. 'Lung regeneration: mechanisms, applications and emerging stem cell populations', *Nat Med*, 20: 822–32.
129. Kumar, V., J. Bariwal, A. S. Narang, J. Tso, J. Cheong, and R. I. Mahato. 2020. 'Functional similarity of modified cascade impactor to deposit drug particles on cells', *Int J Pharm*, 583: 119404.
130. Lee, J. H., T. Tammela, M. Hofree, J. Choi, N. D. Marjanovic, S. Han, D. Canner, K. Wu, M. Paschini, D. H. Bhang, T. Jacks, A. Regev, and C. F. Kim. 2017. 'Anatomically and Functionally Distinct Lung Mesenchymal Populations Marked by Lgr5 and Lgr6', *Cell*, 170: 1149–63 e12.
131. Lee, J. Y., J. H. Park, and D. W. Cho. 2018. 'Comparison of tracheal reconstruction with allograft, fresh xenograft and artificial trachea scaffold in a rabbit model', *J Artif Organs*, 21: 325–31.
132. Lee, R. E., S. M. Miller, T. M. Mascenik, C. A. Lewis, H. Dang, Z. H. Boggs, R. Tarran, and S. H. Randell. 2020. 'Assessing Human Airway Epithelial Progenitor Cells for Cystic Fibrosis Cell Therapy', *Am J Respir Cell Mol Biol*, 63: 374–85.
133. Lee, S. N., S. J. Kim, S. A. Yoon, J. M. Song, J. S. Ahn, H. C. Kim, A. M. K. Choi, and J. H. Yoon. 2021. 'CD44v3-Positive Intermediate Progenitor Cells Contribute to Airway Goblet Cell Hyperplasia', *Am J Respir Cell Mol Biol*, 64: 247–59.
134. Leibel, S. L., R. N. McVicar, A. M. Winquist, W. D. Niles, and E. Y. Snyder. 2020. 'Generation of Complete Multi-Cell Type Lung Organoids From Human Embryonic and Patient-Specific Induced Pluripotent Stem Cells for Infectious Disease Modeling and Therapeutics Validation', *Curr Protoc Stem Cell Biol*, 54: e118.
135. Leung, J. M., J. Mayo, W. Tan, C. M. Tammemagi, G. Liu, S. Peacock, F. A. Shepherd, J. Goffin, G. Goss, G. Nicholas, A. Tremblay, M. Johnston, S. Martel, F. Laberge, R. Bhatia, H. Roberts, P. Burrowes, D. Manos, L. Stewart, J. M. Seely, M. Gingras, S. Pasian, M. S. Tsao, S. Lam, D. D. Sin, and Group Pan-Canadian Early Lung Cancer Study. 2015. 'Plasma pro-surfactant protein B and lung function decline in smokers', *Eur Respir J*, 45: 1037–45.
136. Li, C., N. Peinado, S. M. Smith, J. Zhou, F. Gao, G. Kohbodi, B. Zhou, M. E. Thornton, B. H. Grubbs, M. K. Lee, S. Bellusci, Z. Borok, Y. W. Chen, and P. Minoo. 2022. 'Wnt5a Promotes AT1 and Represses AT2 Lineage-Specific Gene Expression in a Cell-Context-Dependent Manner', *Stem Cells*, 40: 691–703.
137. Liu, Q., H. Li, Q. Wang, Y. Zhang, W. Wang, S. Dou, and W. Xiao. 2016. 'Increased expression of TROP2 in airway basal cells potentially contributes to airway remodeling in chronic obstructive pulmonary disease', *Respir Res*, 17: 159.
138. Liu, X., and J. F. Engelhardt. 2008. 'The glandular stem/progenitor cell niche in airway development and repair', *Proc Am Thorac Soc*, 5: 682–8.
139. Lv, M. Y., L. X. Qiang, B. C. Wang, Y. P. Zhang, Z. H. Li, X. S. Li, L. L. Jin, and S. D. Jin. 2022. 'Complex Evaluation of Surfactant Protein A and D as Biomarkers for the Severity of COPD', *Int J Chron Obstruct Pulmon Dis*, 17: 1537–52.

140. Lynch, T. J., P. J. Anderson, W. Xie, A. K. Crooke, X. Liu, S. R. Tyler, M. Luo, D. M. Kusner, Y. Zhang, T. Neff, D. C. Burnette, K. S. Walters, M. J. Goodheart, K. R. Parekh, and J. F. Engelhardt. 2016. 'Wnt Signaling Regulates Airway Epithelial Stem Cells in Adult Murine Submucosal Glands', *Stem Cells*, 34: 2758–71.
141. Lynch, T. J., and J. F. Engelhardt. 2014. 'Progenitor cells in proximal airway epithelial development and regeneration', *J Cell Biochem*, 115: 1637–45.
142. Ma, J., B. K. Rubin, and J. A. Voynow. 2018. 'Mucins, Mucus, and Goblet Cells', *Chest*, 154: 169–76.
143. Mackay, R. M., C. L. Grainge, L. C. Lau, C. Barber, H. W. Clark, and P. H. Howarth. 2016. 'Airway Surfactant Protein D Deficiency in Adults With Severe Asthma', *Chest*, 149: 1165–72.
144. Maestrelli, P., M. Saetta, C. E. Mapp, and L. M. Fabbri. 2001. 'Remodeling in response to infection and injury. Airway inflammation and hypersecretion of mucus in smoking subjects with chronic obstructive pulmonary disease', *Am J Respir Crit Care Med*, 164: S76–80.
145. Majidinia, M., J. Aghazadeh, R. Jahanban-Esfahlani, and B. Yousefi. 2018. 'The roles of Wnt/β-catenin pathway in tissue development and regenerative medicine', *J Cell Physiol*, 233: 5598–612.
146. Manevski, M., D. Devadoss, C. Long, S. P. Singh, M. W. Nasser, G. M. Borchert, M. N. Nair, I. Rahman, M. Sopori, and H. S. Chand. 2022. 'Increased Expression of LASI lncRNA Regulates the Cigarette Smoke and COPD Associated Airway Inflammation and Mucous Cell Hyperplasia', *Front Immunol*, 13: 803362.
147. McConnell, A. M., C. Yao, A. R. Yeckes, Y. Wang, A. S. Selvaggio, J. Tang, D. G. Kirsch, and B. R. Stripp. 2016. 'p53 Regulates Progenitor Cell Quiescence and Differentiation in the Airway', *Cell Rep*, 17: 2173–82.
148. McCulley, D., M. Wienhold, and X. Sun. 2015. 'The pulmonary mesenchyme directs lung development', *Curr Opin Genet Dev*, 32: 98–105.
149. McQualter, J. L. 2019. 'Endogenous lung stem cells for lung regeneration', *Expert Opin Biol Ther*, 19: 539–46.
150. Menon, T., A. L. Firth, D. D. Scripture-Adams, Z. Galic, S. J. Qualls, W. B. Gilmore, E. Ke, O. Singer, L. S. Anderson, A. R. Bornzin, I. E. Alexander, J. A. Zack, and I. M. Verma. 2015. 'Lymphoid regeneration from gene-corrected SCID-X1 subject-derived iPSCs', *Cell Stem Cell*, 16: 367–72.
151. Miles, P. R., L. Bowman, K. M. Rao, J. E. Baatz, and L. Huffman. 1999. 'Pulmonary surfactant inhibits LPS-induced nitric oxide production by alveolar macrophages', *Am J Physiol*, 276: L186–96.
152. Miller, A. J., B. R. Dye, D. Ferrer-Torres, D. R. Hill, A. W. Overeem, L. D. Shea, and J. R. Spence. 2019. 'Generation of lung organoids from human pluripotent stem cells in vitro', *Nat Protoc*, 14: 518–40.
153. Miller, A. J., D. R. Hill, M. S. Nagy, Y. Aoki, B. R. Dye, A. M. Chin, S. Huang, F. Zhu, E. S. White, V. Lama, and J. R. Spence. 2018. 'In Vitro Induction and In Vivo Engraftment of Lung Bud Tip Progenitor Cells Derived from Human Pluripotent Stem Cells', *Stem Cell Reports*, 10: 101–19.
154. Miller, A. J., Q. Yu, M. Czerwinski, Y. H. Tsai, R. F. Conway, A. Wu, E. M. Holloway, T. Walker, I. A. Glass, B. Treutlein, J. G. Camp, and J. R. Spence. 2020. 'In Vitro and In Vivo Development of the Human Airway at Single-Cell Resolution', *Dev Cell*, 53: 117–28 e6.
155. Moiseenko, A., A. I. Vazquez-Armendariz, V. Kheirollahi, X. Chu, A. Tata, S. Rivetti, S. Gunther, K. Lebrigand, S. Herold, T. Braun, B. Mari, S. De Langhe, G. Kwapiszewska, A. Gunther, C. Chen, W. Seeger, P. R. Tata, J. S. Zhang, S. Bellusci, and E. El Agha. 2020. 'Identification of a Repair-Supportive Mesenchymal Cell Population during Airway Epithelial Regeneration', *Cell Rep*, 33: 108549.
156. Montoro, D. T., A. L. Haber, M. Biton, V. Vinarsky, B. Lin, S. E. Birket, F. Yuan, S. Chen, H. M. Leung, J. Villoria, N. Rogel, G. Burgin, A. M. Tsankov, A. Waghray, M. Slyper, J. Waldman, L. Nguyen, D. Dionne, O. Rozenblatt-Rosen, P. R. Tata, H. Mou, M. Shivaraju,

H. Bihler, M. Mense, G. J. Tearney, S. M. Rowe, J. F. Engelhardt, A. Regev, and J. Rajagopal. 2018. 'A revised airway epithelial hierarchy includes CFTR-expressing ionocytes', *Nature*, 560: 319–24.

157. Mori, M., J. E. Mahoney, M. R. Stupnikov, J. R. Paez-Cortez, A. D. Szymaniak, X. Varelas, D. B. Herrick, J. Schwob, H. Zhang, and W. V. Cardoso. 2015. 'Notch3-Jagged signaling controls the pool of undifferentiated airway progenitors', *Development*, 142: 258–67.
158. Morrisey, E. E., W. V. Cardoso, R. H. Lane, M. Rabinovitch, S. H. Abman, X. Ai, K. H. Albertine, R. D. Bland, H. A. Chapman, W. Checkley, J. A. Epstein, C. R. Kintner, M. Kumar, P. Minoo, T. J. Mariani, D. M. McDonald, Y. S. Mukouyama, L. S. Prince, J. Reese, J. Rossant, W. Shi, X. Sun, Z. Werb, J. A. Whitsett, D. Gail, C. J. Blaisdell, and Q. S. Lin. 2013. 'Molecular determinants of lung development', *Ann Am Thorac Soc*, 10: S12–6.
159. Morrow, J. D., R. P. Chase, M. M. Parker, K. Glass, M. Seo, M. Divo, C. A. Owen, P. Castaldi, D. L. DeMeo, E. K. Silverman, and C. P. Hersh. 2019. 'RNA-sequencing across three matched tissues reveals shared and tissue-specific gene expression and pathway signatures of COPD', *Respir Res*, 20: 65.
160. Mou, H., V. Vinarsky, P. R. Tata, K. Brazauskas, S. H. Choi, A. K. Crooke, B. Zhang, G. M. Solomon, B. Turner, H. Bihler, J. Harrington, A. Lapey, C. Channick, C. Keyes, A. Freund, S. Artandi, M. Mense, S. Rowe, J. F. Engelhardt, Y. C. Hsu, and J. Rajagopal. 2016. 'Dual SMAD Signaling Inhibition Enables Long-Term Expansion of Diverse Epithelial Basal Cells', *Cell Stem Cell*, 19: 217–31.
161. Mou, H., Y. Yang, M. A. Riehs, J. Barrios, M. Shivaraju, A. L. Haber, D. T. Montoro, K. Gilmore, E. A. Haas, B. Paunovic, J. Rajagopal, S. O. Vargas, R. L. Haynes, A. Fine, W. V. Cardoso, and X. Ai. 2021. 'Airway basal stem cells generate distinct subpopulations of PNECs', *Cell Rep*, 35: 109011.
162. Mucci, A., E. Lopez-Rodriguez, M. Hetzel, S. Liu, T. Suzuki, C. Happle, M. Ackermann, H. Kempf, R. Hillje, J. Kunkiel, E. Janosz, S. Brennig, S. Glage, J. P. Bankstahl, S. Dettmer, T. Rodt, G. Gohring, B. Trapnell, G. Hansen, C. Trapnell, L. Knudsen, N. Lachmann, and T. Moritz. 2018. 'iPSC-Derived Macrophages Effectively Treat Pulmonary Alveolar Proteinosis in Csf2rb-Deficient Mice', *Stem Cell Reports*, 11: 696–710.
163. Murray, L. A. 2012. 'Commonalities between the pro-fibrotic mechanisms in COPD and IPF', *Pulm Pharmacol Ther*, 25: 276–80.
164. Nakajima, M., O. Kawanami, E. Jin, M. Ghazizadeh, M. Honda, G. Asano, K. Horiba, and V. J. Ferrans. 1998. 'Immunohistochemical and ultrastructural studies of basal cells, Clara cells and bronchiolar cuboidal cells in normal human airways', *Pathol Int*, 48: 944–53.
165. Nava, V. E., R. Khosla, S. Shin, F. E. Mordini, and B. C. Bandyopadhyay. 2022. 'Enhanced carbonic anhydrase expression with calcification and fibrosis in bronchial cartilage during COPD', *Acta Histochem*, 124: 151834.
166. Nawroth, J. C., C. Lucchesi, D. Cheng, A. Shukla, J. Ngyuen, T. Shroff, A. Varone, K. Karalis, H. H. Lee, S. Alves, G. A. Hamilton, M. Salmon, and R. Villenave. 2020. 'A Microengineered Airway Lung Chip Models Key Features of Viral-induced Exacerbation of Asthma', *Am J Respir Cell Mol Biol*, 63: 591–600.
167. Nawroth, Janna C., Doris Roth, Annemarie van Schadewijk, Abilash Ravi, Tengku Ibrahim Maulana, Christiana N. Senger, Sander van Riet, Dennis K. Ninaber, Amy M. de Waal, Dorothea Kraft, Pieter S. Hiemstra, Amy L Ryan, and Anne M. van der Does. 2022. 'Breathing on Chip: Dynamic flow and stretch tune cellular composition and accelerate mucociliary maturation of airway epithelium *in vitro*', *bioRxiv*: 2021.05.07.443164.
168. Nikolic, M. Z., O. Caritg, Q. Jeng, J. A. Johnson, D. Sun, K. J. Howell, J. L. Brady, U. Laresgoiti, G. Allen, R. Butler, M. Zilbauer, A. Giangreco, and E. L. Rawlins. 2017. 'Human embryonic lung epithelial tips are multipotent progenitors that can be expanded in vitro as long-term self-renewing organoids', *Elife*, 6.
169. Nyeng, P., G. A. Norgaard, S. Kobberup, and J. Jensen. 2008. 'FGF10 maintains distal lung bud epithelium and excessive signaling leads to progenitor state arrest, distalization, and goblet cell metaplasia', *BMC Dev Biol*, 8: 2.

170. O'Dwyer, D. N., S. J. Gurczynski, and B. B. Moore. 2018. 'Pulmonary immunity and extracellular matrix interactions', *Matrix Biol*, 73: 122–34.
171. Ong, H. X., C. L. Jackson, J. L. Cole, P. M. Lackie, D. Traini, P. M. Young, J. Lucas, and J. Conway. 2016. 'Primary Air-Liquid Interface Culture of Nasal Epithelium for Nasal Drug Delivery', *Mol Pharm*, 13: 2242–52.
172. Ortiz-Zapater, Elena, Dustin C. Bagley, Virginia Llopis Hernandez, Luke B. Roberts, Thomas J. A. Maguire, Felizia Voss, Philipp Mertins, Marieluise Kirchner, Isabel Peset-Martin, Grzegorz Woszczek, Jody Rosenblatt, Michael Gotthardt, George Santis, and Maddy Parsons. 2022. 'Epithelial coxsackievirus adenovirus receptor promotes house dust mite-induced lung inflammation', *Nature Communications*, 13: 6407.
173. Osan, J. K., S. N. Talukdar, F. Feldmann, B. Ann DeMontigny, K. Jerome, K. L. Bailey, H. Feldmann, and M. Mehedi. 2020. 'Goblet Cell Hyperplasia Increases SARS-CoV-2 Infection in COPD', *bioRxiv*.
174. Osei, E. T., S. Booth, and T. L. Hackett. 2020. 'What Have In Vitro Co-Culture Models Taught Us about the Contribution of Epithelial-Mesenchymal Interactions to Airway Inflammation and Remodeling in Asthma?', *Cells*, 9: 1694.
175. Osei, E. T., and T. L. Hackett. 2020. 'Epithelial-mesenchymal crosstalk in COPD: An update from in vitro model studies', *Int J Biochem Cell Biol*, 125: 105775.
176. Ouadah, Y., E. R. Rojas, D. P. Riordan, S. Capostagno, C. S. Kuo, and M. A. Krasnow. 2019. 'Rare Pulmonary Neuroendocrine Cells Are Stem Cells Regulated by Rb, p53, and Notch', *Cell*, 179: 403–16.e23.
177. Parekh, K. R., J. Nawroth, A. Pai, S. M. Busch, C. N. Senger, and A. L. Ryan. 2020. 'Stem cells and lung regeneration', *Am J Physiol Cell Physiol*, 319: C675–C93.
178. Pat, Yagiz, Beate Rückert, Ismail Ogulur, Duygu Yazici, Mario Pérez-Diego, Ozan C. Küçükkase, Manru Li, and Cezmi A. Akdis. 2022. 'Differentiation of bronchial epithelial spheroids in the presence of IL-13 recapitulates characteristic features of asthmatic airway epithelia', *Allergy*, 77: 2229–33.
179. Peake, J. L., S. D. Reynolds, B. R. Stripp, K. E. Stephens, and K. E. Pinkerton. 2000. 'Alteration of pulmonary neuroendocrine cells during epithelial repair of naphthalene-induced airway injury', *Am J Pathol*, 156: 279–86.
180. Peng, L., L. Gao, X. Wu, Y. Fan, M. Liu, J. Chen, J. Song, J. Kong, Y. Dong, B. Li, A. Liu, and F. Bao. 2022. 'Lung Organoids as Model to Study SARS-CoV-2 Infection', *Cells*, 11.
181. Pilette, C., V. Godding, R. Kiss, M. Delos, E. Verbeken, C. Decaestecker, K. De Paepe, J. P. Vaerman, M. Decramer, and Y. Sibille. 2001. 'Reduced epithelial expression of secretory component in small airways correlates with airflow obstruction in chronic obstructive pulmonary disease', *Am J Respir Crit Care Med*, 163: 185–94.
182. Plasschaert, L. W., R. Zilionis, R. Choo-Wing, V. Savova, J. Knehr, G. Roma, A. M. Klein, and A. B. Jaffe. 2018. 'A single-cell atlas of the airway epithelium reveals the CFTR-rich pulmonary ionocyte', *Nature*, 560: 377–81.
183. Plebani, R., R. Potla, M. Soong, H. Bai, Z. Izadifar, A. Jiang, R. N. Travis, C. Belgur, A. Dinis, M. J. Cartwright, R. Prantil-Baun, P. Jolly, S. E. Gilpin, M. Romano, and D. E. Ingber. 2022. 'Modeling pulmonary cystic fibrosis in a human lung airway-on-a-chip', *J Cyst Fibros*, 21: 606–15.
184. Polosukhin, V. V., J. M. Cates, W. E. Lawson, R. Zaynagetdinov, A. P. Milstone, P. P. Massion, S. Ocak, L. B. Ware, J. W. Lee, R. P. Bowler, A. V. Kononov, S. H. Randell, and T. S. Blackwell. 2011a. 'Bronchial secretory immunoglobulin a deficiency correlates with airway inflammation and progression of chronic obstructive pulmonary disease', *Am J Respir Crit Care Med*, 184: 317–27.
185. Polosukhin, Vasiliy V, Justin M Cates, William E Lawson, Aaron P Milstone, Anton G Matafonov, Pierre P Massion, Jae Woo Lee, Scott H Randell, and Timothy S Blackwell. 2011b. 'Hypoxia-inducible factor-1 signalling promotes goblet cell hyperplasia in airway epithelium', *The Journal of Pathology*, 224: 203–11.

186. Porotto, M., M. Ferren, Y.-W. Chen, Y. Siu, N. Makhsous, B. Rima, T. Briese, A. L. Greninger, H.-W. Snoeck, and A. Moscona. 2019. 'Authentic Modeling of Human Respiratory Virus Infection in Human Pluripotent Stem Cell-Derived Lung Organoids', *mBio*, 10: e00723–19.
187. Prakash, Y. S. 2016. 'Emerging concepts in smooth muscle contributions to airway structure and function: implications for health and disease', *Am J Physiol Lung Cell Mol Physiol*, 311: L1113–L40.
188. Ptasinski, V. A., J. Stegmayr, M. G. Belvisi, D. E. Wagner, and L. A. Murray. 2021. 'Targeting Alveolar Repair in Idiopathic Pulmonary Fibrosis', *Am J Respir Cell Mol Biol*, 65: 347–65.
189. Puchelle, E., and B. Peault. 2000. 'Human airway xenograft models of epithelial cell regeneration', *Respir Res*, 1: 125–8.
190. Puttur, F., L. G. Gregory, and C. M. Lloyd. 2019. 'Airway macrophages as the guardians of tissue repair in the lung', *Immunol Cell Biol*, 97: 246–57.
191. Rackley, C. R., and B. R. Stripp. 2012. 'Building and maintaining the epithelium of the lung', *J Clin Invest*, 122: 2724–30.
192. Randell, S. H., M. L. Fulcher, W. O'Neal, and J. C. Olsen. 2011. 'Primary epithelial cell models for cystic fibrosis research', *Methods Mol Biol*, 742: 285–310.
193. Raslan, A. A., and J. K. Yoon. 2020. 'WNT Signaling in Lung Repair and Regeneration', *Mol Cells*, 43: 774–83.
194. Ravindra, N. G., M. M. Alfajaro, V. Gasque, N. C. Huston, H. Wan, K. Szigeti-Buck, Y. Yasumoto, A. M. Greaney, V. Habet, R. D. Chow, J. S. Chen, J. Wei, R. B. Filler, B. Wang, G. Wang, L. E. Niklason, R. R. Montgomery, S. C. Eisenbarth, S. Chen, A. Williams, A. Iwasaki, T. L. Horvath, E. F. Foxman, R. W. Pierce, A. M. Pyle, D. van Dijk, and C. B. Wilen. 2021. 'Single-cell longitudinal analysis of SARS-CoV-2 infection in human airway epithelium identifies target cells, alterations in gene expression, and cell state changes', *PLoS Biol*, 19: e3001143.
195. Rawlins, E. L., T. Okubo, Y. Xue, D. M. Brass, R. L. Auten, H. Hasegawa, F. Wang, and B. L. Hogan. 2009. 'The role of Scgb1a1+ Clara cells in the long-term maintenance and repair of lung airway, but not alveolar, epithelium', *Cell Stem Cell*, 4: 525–34.
196. Reynolds, S. D., K. U. Hong, A. Giangreco, G. W. Mango, C. Guron, Y. Morimoto, and B. R. Stripp. 2000. 'Conditional clara cell ablation reveals a self-renewing progenitor function of pulmonary neuroendocrine cells', *Am J Physiol Lung Cell Mol Physiol*, 278: L1256–63.
197. Reynolds, S. D., P. R. Reynolds, J. C. Snyder, F. Whyte, K. J. Paavola, and B. R. Stripp. 2007. 'CCSP regulates cross talk between secretory cells and both ciliated cells and macrophages of the conducting airway', *Am J Physiol Lung Cell Mol Physiol*, 293: L114–23.
198. Reynolds, S. D., C. Rios, A. Wesolowska-Andersen, Y. Zhuang, M. Pinter, C. Happoldt, C. L. Hill, S. W. Lallier, G. P. Cosgrove, G. M. Solomon, D. P. Nichols, and M. A. Seibold. 2016. 'Airway Progenitor Clone Formation Is Enhanced by Y-27632-Dependent Changes in the Transcriptome', *Am J Respir Cell Mol Biol*, 55: 323–36.
199. Riccetti, M., J. J. Gokey, B. Aronow, and A. T. Perl. 2020. 'The elephant in the lung: Integrating lineage-tracing, molecular markers, and single cell sequencing data to identify distinct fibroblast populations during lung development and regeneration', *Matrix Biol*, 91–92: 51–74.
200. Rock, J. R., X. Gao, Y. Xue, S. H. Randell, Y. Y. Kong, and B. L. Hogan. 2011. 'Notch-dependent differentiation of adult airway basal stem cells', *Cell Stem Cell*, 8: 639–48.
201. Rock, J. R., M. W. Onaitis, E. L. Rawlins, Y. Lu, C. P. Clark, Y. Xue, S. H. Randell, and B. L. Hogan. 2009. 'Basal cells as stem cells of the mouse trachea and human airway epithelium', *Proc Natl Acad Sci U S A*, 106: 12771–5.
202. Rock, J. R., S. H. Randell, and B. L. Hogan. 2010. 'Airway basal stem cells: a perspective on their roles in epithelial homeostasis and remodeling', *Dis Model Mech*, 3: 545–56.
203. Rogers, Duncan F. 2003. 'The airway goblet cell', *The International Journal of Biochemistry & Cell Biology*, 35: 1–6.
204. Ruysseveldt, E., K. Martens, and B. Steelant. 2021. 'Airway Basal Cells, Protectors of Epithelial Walls in Health and Respiratory Diseases', *Front Allergy*, 2: 787128.

205. Sachs, N., A. Papaspyropoulos, D. D. Zomer-van Ommen, I. Heo, L. Böttinger, D. Klay, F. Weeber, G. Huelsz-Prince, N. Iakobachvili, G. D. Amatngalim, J. de Ligt, A. van Hoeck, N. Proost, M. C. Viveen, A. Lyubimova, L. Teeven, S. Derakhshan, J. Korving, H. Begthel, J. F. Dekkers, K. Kumawat, E. Ramos, M. F. van Oosterhout, G. J. Offerhaus, D. J. Wiener, E. P. Olimpio, K. K. Dijkstra, E. F. Smit, M. van der Linden, S. Jaksani, M. van de Ven, J. Jonkers, A. C. Rios, E. E. Voest, C. H. van Moorsel, C. K. van der Ent, E. Cuppen, A. van Oudenaarden, F. E. Coenjaerts, L. Meyaard, L. J. Bont, P. J. Peters, S. J. Tans, J. S. van Zon, S. F. Boj, R. G. Vries, J. M. Beekman, and H. Clevers. 2019. 'Long-term expanding human airway organoids for disease modeling', *EMBO J*, 38.
206. Saetta, M., G. Turato, S. Baraldo, A. Zanin, F. Braccioni, C. E. Mapp, P. Maestrelli, G. Cavallesco, A. Papi, and L. M. Fabbri. 2000. 'Goblet Cell Hyperplasia and Epithelial Inflammation in Peripheral Airways of Smokers with Both Symptoms of Chronic Bronchitis and Chronic Airflow Limitation', *American Journal of Respiratory and Critical Care Medicine*, 161: 1016–21.
207. Saint-Criq, V., L. Delpiano, J. Casement, J. C. Onuora, J. Lin, and M. A. Gray. 2020. 'Choice of Differentiation Media Significantly Impacts Cell Lineage and Response to CFTR Modulators in Fully Differentiated Primary Cultures of Cystic Fibrosis Human Airway Epithelial Cells', *Cells*, 9.
208. Saygili, E., U. Devamoglu, B. Goker-Bagca, O. Goksel, C. Biray-Avci, T. Goksel, and O. Yesil-Celiktas. 2022. 'A drug-responsive multicellular human spheroid model to recapitulate drug-induced pulmonary fibrosis', *Biomed Mater*, 17.
209. Schleh, C., B. M. Rothen-Rutishauser, F. Blank, H. D. Lauenstein, M. Nassimi, N. Krug, A. Braun, V. J. Erpenbeck, P. Gehr, and J. M. Hohlfeld. 2012. 'Surfactant Protein D modulates allergen particle uptake and inflammatory response in a human epithelial airway model', *Respir Res*, 13: 8.
210. Schwab, U. E., M. L. Fulcher, S. H. Randell, M. J. Flaminio, and D. G. Russell. 2010. 'Equine bronchial epithelial cells differentiate into ciliated and mucus producing cells in vitro', *In Vitro Cell Dev Biol Anim*, 46: 102–6.
211. Scudieri, P., I. Musante, A. Venturini, D. Guidone, M. Genovese, F. Cresta, E. Caci, A. Palleschi, M. Poeta, F. Santamaria, F. Ciciriello, V. Lucidi, and L. J. V. Galietta. 2020. 'Ionocytes and CFTR Chloride Channel Expression in Normal and Cystic Fibrosis Nasal and Bronchial Epithelial Cells', *Cells*, 9.
212. Sears, P. R., C. W. Davis, M. Chua, and J. K. Sheehan. 2011. 'Mucociliary interactions and mucus dynamics in ciliated human bronchial epithelial cell cultures', *Am J Physiol Lung Cell Mol Physiol*, 301: L181–6.
213. Shi, Xiaowei, Jingqi Liu, Deying Chen, Minglei Zhu, Jiong Yu, Haiyang Xie, Lin Zhou, Liang Li, and Shusen Zheng. 2019. 'MSC-triggered metabolomic alterations in liver-resident immune cells isolated from CCl4-induced mouse ALI model', *Experimental Cell Research*, 383: 111511.
214. Shijubo, N., Y. Itoh, T. Yamaguchi, A. Imada, M. Hirasawa, T. Yamada, T. Kawai, and S. Abe. 1999. 'Clara cell protein-positive epithelial cells are reduced in small airways of asthmatics', *Am J Respir Crit Care Med*, 160: 930–3.
215. Shimizu, T., P. Nettesheim, F. C. Ramaekers, and S. H. Randell. 1992. 'Expression of "cell-type-specific" markers during rat tracheal epithelial regeneration', *Am J Respir Cell Mol Biol*, 7: 30–41.
216. Shimizu, T., Y. Takahashi, S. Kawaguchi, and Y. Sakakura. 1996. 'Hypertrophic and metaplastic changes of goblet cells in rat nasal epithelium induced by endotoxin', *Am J Respir Crit Care Med*, 153: 1412–8.
217. Si, L., H. Bai, C. Y. Oh, L. Jin, R. Prantil-Baun, and D. E. Ingber. 2021. 'Clinically Relevant Influenza Virus Evolution Reconstituted in a Human Lung Airway-on-a-Chip', *Microbiol Spectr*, 9: e0025721.

218. Sicard, D., A. J. Haak, K. M. Choi, A. R. Craig, L. E. Fredenburgh, and D. J. Tschumperlin. 2018. 'Aging and anatomical variations in lung tissue stiffness', *Am J Physiol Lung Cell Mol Physiol*, 314: L946–L55.
219. Sin, D. D., C. M. Tammemagi, S. Lam, M. J. Barnett, X. Duan, A. Tam, H. Auman, Z. Feng, G. E. Goodman, S. Hanash, and A. Taguchi. 2013. 'Pro-surfactant protein B as a biomarker for lung cancer prediction', *J Clin Oncol*, 31: 4536–43.
220. Spitalieri, P., F. Centofanti, M. Murdocca, M. G. Scioli, A. Latini, S. Di Cesare, G. Citro, A. Rossi, A. Orlandi, S. Miersch, S. S. Sidhu, P. P. Pandolfi, A. Botta, F. Sangiuolo, and G. Novelli. 2022. 'Two Different Therapeutic Approaches for SARS-CoV-2 in hiPSCs-Derived Lung Organoids', *Cells*, 11.
221. Sprott, R. F., F. Ritzmann, F. Langer, Y. Yao, C. Herr, Y. Kohl, T. Tschernig, R. Bals, and C. Beisswenger. 2020. 'Flagellin shifts 3D bronchospheres towards mucus hyperproduction', *Respir Res*, 21: 222.
222. Stupnikov, M. R., Y. Yang, M. Mori, J. Lu, and W. V. Cardoso. 2019. 'Jagged and Delta-like ligands control distinct events during airway progenitor cell differentiation', *Elife*, 8.
223. Succony, L., S. Gomez-Lopez, A. Pennycuick, A. S. N. Alhendi, D. Davies, S. E. Clarke, K. H. C. Gowers, N. A. Wright, K. B. Jensen, and S. M. Janes. 2022. 'Lrig1 expression identifies airway basal cells with high proliferative capacity and restricts lung squamous cell carcinoma growth', *Eur Respir J*, 59.
224. Sui, P., D. L. Wiesner, J. Xu, Y. Zhang, J. Lee, S. Van Dyken, A. Lashua, C. Yu, B. S. Klein, R. M. Locksley, G. Deutsch, and X. Sun. 2018. 'Pulmonary neuroendocrine cells amplify allergic asthma responses', *Science*, 360.
225. Suprynowicz, F. A., G. Upadhyay, E. Krawczyk, S. C. Kramer, J. D. Hebert, X. Liu, H. Yuan, C. Cheluvaraju, P. W. Clapp, R. C. Boucher, Jr., C. M. Kamonjoh, S. H. Randell, and R. Schlegel. 2012. 'Conditionally reprogrammed cells represent a stem-like state of adult epithelial cells', *Proc Natl Acad Sci U S A*, 109: 20035–40.
226. Sutanto, E. N., A. Scaffidi, L. W. Garratt, K. Looi, C. J. Foo, M. A. Tessari, R. A. Janssen, D. F. Fischer, S. M. Stick, A. Kicic, and C. F. Arest. 2018. 'Assessment of p.Phe508del-CFTR functional restoration in pediatric primary cystic fibrosis airway epithelial cells', *PLoS One*, 13: e0191618.
227. Suzuki, S., F. J. Hawkins, C. Barilla, M. L. Beermann, D. N. Kotton, and B. R. Davis. 2021. 'Differentiation of human pluripotent stem cells into functional airway basal stem cells', *STAR Protoc*, 2: 100683.
228. Swatek, A. M., T. J. Lynch, A. K. Crooke, P. J. Anderson, S. R. Tyler, L. Brooks, M. Ivanovic, J. A. Klesney-Tait, M. Eberlein, T. Pena, D. K. Meyerholz, J. F. Engelhardt, and K. R. Parekh. 2018. 'Depletion of Airway Submucosal Glands and TP63(+)KRT5(+) Basal Cells in Obliterative Bronchiolitis', *Am J Respir Crit Care Med*, 197: 1045–57.
229. Tadokoro, T., Y. Wang, L. S. Barak, Y. Bai, S. H. Randell, and B. L. Hogan. 2014. 'IL-6/STAT3 promotes regeneration of airway ciliated cells from basal stem cells', *Proc Natl Acad Sci U S A*, 111: E3641–9.
230. Tata, A., Y. Kobayashi, R. D. Chow, J. Tran, A. Desai, A. J. Massri, T. J. McCord, M. D. Gunn, and P. R. Tata. 2018. 'Myoepithelial Cells of Submucosal Glands Can Function as Reserve Stem Cells to Regenerate Airways after Injury', *Cell Stem Cell*, 22: 668–83.e6.
231. Tata, P. R., H. Mou, A. Pardo-Saganta, R. Zhao, M. Prabhu, B. M. Law, V. Vinarsky, J. L. Cho, S. Breton, A. Sahay, B. D. Medoff, and J. Rajagopal. 2013. 'Dedifferentiation of committed epithelial cells into stem cells in vivo', *Nature*, 503: 218–23.
232. Thornton, D. J., K. Rousseau, and M. A. McGuckin. 2008. 'Structure and function of the polymeric mucins in airways mucus', *Annu Rev Physiol*, 70: 459–86.
233. Upadhyay, S., and L. Palmberg. 2018. 'Air-Liquid Interface: Relevant In Vitro Models for Investigating Air Pollutant-Induced Pulmonary Toxicity', *Toxicol Sci*, 164: 21–30.
234. Volckaert, T., and S. De Langhe. 2014. 'Lung epithelial stem cells and their niches: Fgf10 takes center stage', *Fibrogenesis Tissue Repair*, 7: 8.

235. Volckaert, T., and S. P. De Langhe. 2015. 'Wnt and FGF mediated epithelial-mesenchymal crosstalk during lung development', *Dev Dyn*, 244: 342–66.
236. Volckaert, T., T. Yuan, C. M. Chao, H. Bell, A. Sitaula, L. Szimmtenings, E. El Agha, D. Chanda, S. Majka, S. Bellusci, V. J. Thannickal, R. Fassler, and S. P. De Langhe. 2017. 'Fgf10-Hippo Epithelial-Mesenchymal Crosstalk Maintains and Recruits Lung Basal Stem Cells', *Dev Cell*, 43: 48–59 e5.
237. Wang, H., Y. Liu, and Z. Liu. 2013. 'Clara cell 10-kD protein in inflammatory upper airway diseases', *Curr Opin Allergy Clin Immunol*, 13: 25–30.
238. Wang, Ruobing, Katie B. McCauley, Darrell N. Kotton, and Finn Hawkins. 2020. 'Chapter 5 – Differentiation of human airway-organoids from induced pluripotent stem cells (iPSCs).' in Jason R. Spence (ed.), *Methods in Cell Biology* (Academic Press).
239. Wang, X., Y. Zhang, A. Amberson, and J. F. Engelhardt. 2001. 'New models of the tracheal airway define the glandular contribution to airway surface fluid and electrolyte composition', *Am J Respir Cell Mol Biol*, 24: 195–202.
240. Watson, J. K., S. Rulands, A. C. Wilkinson, A. Wuidart, M. Ousset, A. Van Keymeulen, B. Gottgens, C. Blanpain, B. D. Simons, and E. L. Rawlins. 2015. 'Clonal Dynamics Reveal Two Distinct Populations of Basal Cells in Slow-Turnover Airway Epithelium', *Cell Rep*, 12: 90–101.
241. White, E. S. 2015. 'Lung extracellular matrix and fibroblast function', *Ann Am Thorac Soc*, 12 Suppl 1: S30–3.
242. Whitsett, J. A. 2018. 'Airway Epithelial Differentiation and Mucociliary Clearance', *Ann Am Thorac Soc*, 15: S143–S148.
243. Wicher, S. A., B. B. Roos, J. J. Teske, Y. H. Fang, C. Pabelick, and Y. S. Prakash. 2021. 'Aging increases senescence, calcium signaling, and extracellular matrix deposition in human airway smooth muscle', *PLoS One*, 16: e0254710.
244. Wijk, S. C., P. Prabhala, B. Michalikova, M. Sommarin, A. Doyle, S. Lang, K. Kanzenbach, E. Tufvesson, S. Lindstedt, N. D. Leigh, G. Karlsson, L. Bjermer, G. Westergren-Thorsson, and M. Magnusson. 2021. 'Human Primary Airway Basal Cells Display a Continuum of Molecular Phases from Health to Disease in Chronic Obstructive Pulmonary Disease', *Am J Respir Cell Mol Biol*, 65: 103–13.
245. Wirtz, H. R., and L. G. Dobbs. 2000. 'The effects of mechanical forces on lung functions', *Respir Physiol*, 119: 1–17.
246. Wohnhaas, C. T., J. A. Gindele, T. Kiechle, Y. Shen, G. G. Leparc, B. Stierstorfer, H. Stahl, F. Gantner, C. Viollet, J. Schymeinsky, and P. Baum. 2021. 'Cigarette Smoke Specifically Affects Small Airway Epithelial Cell Populations and Triggers the Expansion of Inflammatory and Squamous Differentiation Associated Basal Cells', *Int J Mol Sci*, 22.
247. Wong, A. P., A. Keating, and T. K. Waddell. 2009. 'Airway regeneration: the role of the Clara cell secretory protein and the cells that express it', *Cytotherapy*, 11: 676–87.
248. Xie, W., J. T. Fisher, T. J. Lynch, M. Luo, T. I. Evans, T. L. Neff, W. Zhou, Y. Zhang, Y. Ou, N. W. Bunnett, A. F. Russo, M. J. Goodheart, K. R. Parekh, X. Liu, and J. F. Engelhardt. 2011. 'CGRP induction in cystic fibrosis airways alters the submucosal gland progenitor cell niche in mice', *J Clin Invest*, 121: 3144–58.
249. Yang, I. V., and D. A. Schwartz. 2011. 'Epigenetic control of gene expression in the lung', *Am J Respir Crit Care Med*, 183: 1295–301.
250. Yang, J., B. Wang, H. X. Zhou, B. M. Liang, H. Chen, C. L. Ma, J. Xiao, J. Deng, L. Yan, Y. P. Chen, C. L. Chen, F. Chen, X. M. Ou, and Y. L. Feng. 2014. 'Association of surfactant protein B gene with chronic obstructive pulmonary disease susceptibility', *Int J Tuberc Lung Dis*, 18: 1378–84.
251. Yang, Q., A. R. Soltis, G. Sukumar, X. Zhang, H. Caohuy, J. Freedy, C. L. Dalgard, M. D. Wilkerson, H. B. Pollard, and B. S. Pollard. 2019. 'Gene therapy-emulating small molecule treatments in cystic fibrosis airway epithelial cells and patients', *Respir Res*, 20: 290.

252. Yang, Y., P. Riccio, M. Schotsaert, M. Mori, J. Lu, D. K. Lee, A. Garcia-Sastre, J. Xu, and W. V. Cardoso. 2018. 'Spatial-Temporal Lineage Restrictions of Embryonic p63(+) Progenitors Establish Distinct Stem Cell Pools in Adult Airways', *Dev Cell*, 44: 752–61 e4.
253. Yeung, T., P. C. Georges, L. A. Flanagan, B. Marg, M. Ortiz, M. Funaki, N. Zahir, W. Ming, V. Weaver, and P. A. Janmey. 2005. 'Effects of substrate stiffness on cell morphology, cytoskeletal structure, and adhesion', *Cell Motil Cytoskeleton*, 60: 24–34.
254. Young, B. M., K. Shankar, C. K. Tho, A. R. Pellegrino, and R. L. Heise. 2019. 'Laminin-driven Epac/Rap1 regulation of epithelial barriers on decellularized matrix', *Acta Biomater*, 100: 223–34.
255. Yu, Y., X. Liu, Z. Zhao, Z. Xu, Y. Qiao, Y. Zhou, H. Qiao, J. Zhong, J. Dai, and G. Suo. 2021. 'The Extracellular Matrix Enriched With Exosomes for the Treatment on Pulmonary Fibrosis in Mice', *Front Pharmacol*, 12: 747223.
256. Yuan, T., T. Volckaert, D. Chanda, V. J. Thannickal, and S. P. De Langhe. 2018. 'Fgf10 Signaling in Lung Development, Homeostasis, Disease, and Repair After Injury', *Front Genet*, 9: 418.
257. Zacharias, W. J., D. B. Frank, J. A. Zepp, M. P. Morley, F. A. Alkhaleel, J. Kong, S. Zhou, E. Cantu, and E. E. Morrisey. 2018. 'Regeneration of the lung alveolus by an evolutionarily conserved epithelial progenitor', *Nature*, 555: 251–55.
258. Zamprogno, P., G. Thoma, V. Cencen, D. Ferrari, B. Putz, J. Michler, G. E. Fantner, and O. T. Guenat. 2021. 'Mechanical Properties of Soft Biological Membranes for Organ-on-a-Chip Assessed by Bulge Test and AFM', *ACS Biomater Sci Eng*, 7: 2990–97.
259. Zaragosi, L. E., M. Deprez, and P. Barbry. 2020. 'Using single-cell RNA sequencing to unravel cell lineage relationships in the respiratory tract', *Biochem Soc Trans*, 48: 327–36.
260. Zepeda, M. L., M. R. Chinoy, and J. M. Wilson. 1995. 'Characterization of stem cells in human airway capable of reconstituting a fully differentiated bronchial epithelium', *Somat Cell Mol Genet*, 21: 61–73.
261. Zepp, J. A., and E. E. Morrisey. 2019. 'Cellular crosstalk in the development and regeneration of the respiratory system', *Nat Rev Mol Cell Biol*, 20: 551–66.
262. Zhou, Y., J. Ming, M. Deng, Y. Li, B. Li, J. Li, Y. Ma, Z. Chen, and S. Liu. 2020. 'Berberine-mediated up-regulation of surfactant protein D facilitates cartilage repair by modulating immune responses via the inhibition of TLR4/NF-kB signaling', *Pharmacol Res*, 155: 104690.
263. Zhou, Y., Y. Yang, L. Guo, J. Qian, J. Ge, D. Sinner, H. Ding, A. Califano, and W. V. Cardoso. 2022. 'Airway basal cells show regionally distinct potential to undergo metaplastic differentiation', *Elife*, 11.
264. Zuo, Wu-Lin, Sushila A. Shenoy, Sheng Li, Sarah L. O'Beirne, Yael Strulovici-Barel, Philip L. Leopold, Guoqing Wang, Michelle R. Staudt, Matthew S. Walters, Christopher Mason, Robert J. Kaner, Jason G. Mezey, and Ronald G. Crystal. 2018. 'Ontogeny and Biology of Human Small Airway Epithelial Club Cells', *American Journal of Respiratory and Critical Care Medicine*, 198: 1375–88.

# Chapter 6
# Computational, Ex Vivo, and Tissue Engineering Techniques for Modeling Large Airways

**Rebecca L. Heise**

## 6.1 Large Airways: Structure-Function Relationship

The large airways may be characterized as the upper airways, consisting of the nasal cavity to the cervical trachea, and the lower airways, consisting of the conducting airspace below the thoracic inlet down through the trachea and bronchi. For the purposes of this chapter, we will focus on the lower airways, beginning with the trachea in our classification and description. The lower large airways are formed by a series of cartilage rings joined together by ligaments. In the human lung, the right bronchus divides into three secondary lobar bronchi, leading to each right lung lobe (upper, middle, lower). The left bronchus divides into two lobar bronchi, leading to each left lung lobe (upper, lower) [1].

The airways have anisotropic mechanical properties due to their structure. In a study by Eskandari et al., the anisotropic properties of the pseudoelastic linear modulus are compiled for sheep and pig airways. Figure 6.1 shows that the axial properties across the trachea, large bronchi, and small bronchi are similar; however, the circumferential moduli increase with decreasing airway diameter (Fig. 6.1). These changing moduli are due to the alterations observed in the layer structure.

Within the airways, there are layers of tissues that perform ventilation of air from the outside to the respiratory surfaces of gas exchange. The four main layers consist of the respiratory mucosa, submucosa, cartilage and/or muscular layer, and adventitia [3].

The composition of the airway layers may be observed in Fig. 6.2. The mucosa layer in particular contains differing amounts of the structural extracellular matrix

R. L. Heise (✉)
Department of Biomedical Engineering, Virginia Commonwealth University, Richmond, VA, USA
e-mail: rlheise@vcu.edu



C. M. Magin (ed.), *Engineering Translational Models of Lung Homeostasis and Disease*, Advances in Experimental Medicine and Biology 1413,
https://doi.org/10.1007/978-3-031-26625-6_6

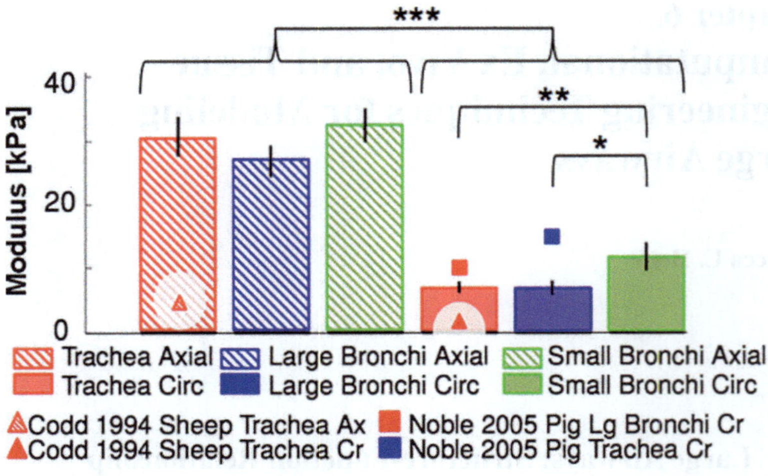


**Fig. 6.1** Reproduced under the CC BY 4.0 from [2]. The pseudoelastic linear modulus for the trachea, large bronchi, and small bronchi samples from axial and circumferential orientations. Anisotropy was evident as greater pseudoelastic linear moduli for axial samples than for corresponding circumferential samples. Heterogeneity was found within circumferential samples, as the modulus of circumferential samples from the small bronchi were significantly larger than that of trachea or large bronchi counterparts. No significant differences in pseudoelastic linear modulus were found among axially oriented samples. *, **, and *** denote *P* values <0.05, 0.01, and 0.001, respectively

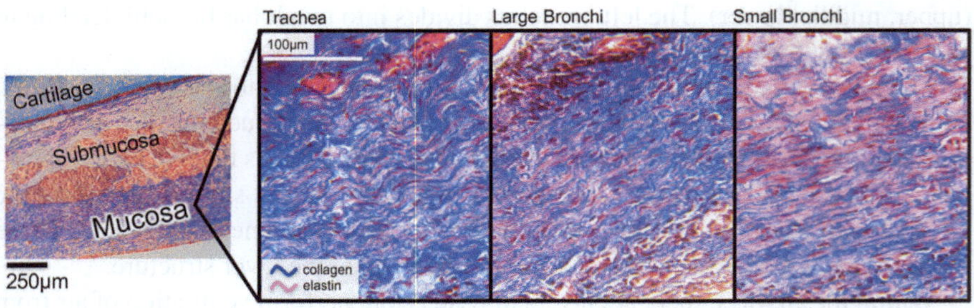


**Fig. 6.2** Reproduced with permission from [6]. Representative histological samples from porcine lungs stained with Masson's trichrome, where collagen fibers are blue and elastin fibers appear red. Fibers in the trachea were crimped, while fibers in the small bronchi were taut and straightened

proteins, collagen and elastin. The Masson's trichrome staining in Fig. 6.2 shows the regional differences between collagen and elastin architecture with crimping of the fibers observed in the tracheal section. The thickness and amount of smooth muscle found in the submucosa and the presence or absence of cartilage rings will also alter the mechanical properties of the airways. Additional extracellular matrix components such as the number of glycosaminoglycans also vary, with increasing amounts found in the smaller bronchi compared with the trachea and large bronchi.

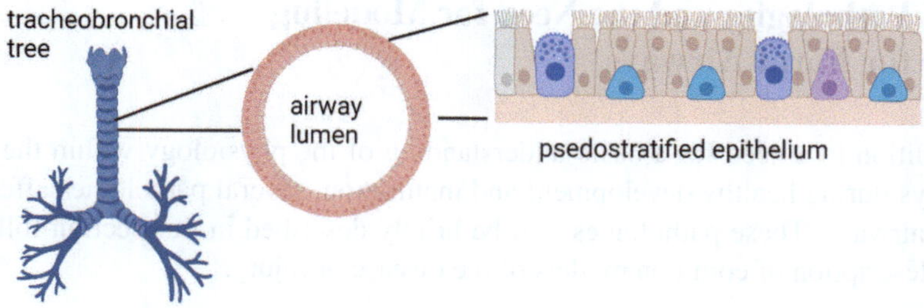


**Fig. 6.3** Structure of the airway. (Created with Biorender.com)

This highly specific arrangement of the fibers and the surrounding extracellular matrix make the airways able to withstand compression, bending, and elongation. These mechanical motions are necessary for the proper function of the large airways [4, 5].

The airways perform important barrier functions that protect from the external environment. The barriers provided are the mucous lining the airway, which prevents loss of moisture and trapping of unwanted particulate, and temperature control, which is performed mostly by warming the air in the upper airways [7].

Within the respiratory mucosa, the epithelium and supporting lamina propria provide the lining that contacts the air. The epithelium are pseudostratified and columnar containing goblet cells and ciliated cells (Fig. 6.3). These cells are critical to the barrier and mucociliary clearance.

Mucociliary clearance begins in the nose and is functional down through the conducting airways. Mucociliary clearance occurs throughout the respiratory tree until reaching the gas exchange regions, where there are no ciliated cells. Mucociliary clearance is an innate mechanism that rids the airways of biological, chemical, or physical particulates. The mucociliary clearance is driven by the ciliated epithelium, rhythmically beating to move the mucus and trapped particles out of the airways to the pharynx [8]. This movement is sometimes known as the mucociliary escalator due to its coordinated movement by the ciliated cells. The physical and chemical properties of the mucus play an important role in this mechanism that are further described in the disease states in Sect. 6.2.

In examining the structure/function relationship in the large airways, the scale is an important factor to consider. While the majority of studies cited above examine bulk tissue properties, the cellular-level scale is also important for functional large airways. The use of cellular-level models [9] and ex vivo tissue slices [10], described in more detail later in this chapter, is useful for determining specific structure/function relationships at the microscale.

## 6.2 Pathologies and the Need for Modeling the Large Airways

In addition to a need for a basic understanding of the physiology within the large airways during healthy development and maturation, several pathologies affect the large airways. These pathologies will be briefly described in this section followed by a description of common models of the disease or injury.

### *6.2.1 Conditions That Cause Large Airway Dysfunction*

Congenital defects, though rare, can affect the trachea and bronchi and cause lung underdevelopment [11]. Some congenital defects are repaired with reconstructive approaches, yet there are no routine gold standards for reconstruction in tissue engineering or non-tissue engineering replacement. Additionally, injuries to the chest or directly to the trachea or pharynx can result in the need for replacement. Beyond the structural need for replacement as in the case of congenital defect or injury, many airway diseases cause dysfunction of the large airways.

Cystic fibrosis is a genetic disease caused by dysfunction of the cystic fibrosis transmembrane receptor (CFTR) [12]. The CFTR dysfunction disrupts the mucociliary clearance making cystic fibrosis patients unable to clear mucus properly causing chronic inflammation and infection. Mucociliary transport is disrupted in other lung diseases as well. Chronic obstructive pulmonary disease (COPD) causes tissue remodeling, or the reorganization of tissue components, and chronic production of mucus and infection. Though commonly associated with smaller airway obstruction, the large airways are also remodeled and inflamed in patients with COPD [13]. As COPD has no known cure, further understanding of the large airway remodeling and obstruction is needed. Airway remodeling is also observed in asthma and can be considered a mechanical feedback process through integrin signaling [14]. Airway constriction and obstruction that occurs in asthma are typically able to be reversed; however, in aging patients, significant airway remodeling and inflammation lead to an obstruction that can be fatal [15]. Additionally, bronchiectasis, or prevention of mucus clearing, can be caused by any of the aforementioned conditions or by repeated infection or injury.

### *6.2.2 Need for Computational and Physiological Models of the Large Airways*

Due to the complicated nature of the airway remodeling and mucociliary clearance disruption in the conditions described in Sect. 6.2.1, there is a significant need for models of the large airway. There are limited animal models for each condition, and none of these animal models fully recapitulate all aspects of the disease, injury, and remodeling observed in the pathologies. For cystic fibrosis, there are several animal models [16] with the most common being genetically modified mouse strains. More recent animal models for cystic fibrosis include genetically modified ferrets [17] and pigs [18], which come closer to the infections and airway structure observed in humans. For COPD, there are animal models that can recapitulate the distal airspace destruction or airway obstruction based on cigarette smoke exposure in rodents [19]. However, the COPD models are costly, take a lot of time to induce, and do not mimic the progressive disease. There are many models of asthma that focus on animal exposure to allergens that cause airway hyperresponsiveness, yet these models do not have the irreversible long-term airway remodeling observed in the human pathology.

With the shortcomings of these animal models, cell culture models provide an opportunity to utilize human cells and potentially allow for precision therapy approaches and personalized medicine. In order to be useful, the cell culture models must recapitulate the aspects of the large airway as much as possible. This includes growing the airway epithelium at an air/liquid interface in coculture with mesenchymal cell types, such as fibroblasts and smooth muscle cells. For conditions that impact the cartilage-ringed airways, chondrocytes may also be added to these in vitro cell culture modes. Inflammatory cells may also be cultured to properly investigate the cell-cell signaling events observed in airway disease. To mimic the complexity of multiple cell types and extracellular matrix structure, tissue engineering techniques are one option that may be employed for models of the large airways. Additional useful models include computational modeling and ex vivo testing. We will now discuss the state of the art for each of these modeling approaches for the large airways. These approaches seek to understand the diseases of the large airway as well as address the growing need for large airway replacement [20].

## 6.3 Computational Modeling

Computational modeling of the large airway has several advantages. Computational models can examine a wide range of conditions, may be personalized based on quantitative imaging of human airways, and can take into account the complicated fluid/structure nature of the large airways. A significant amount of research has been performed to create and utilize computational models of the airway. We will briefly give an overview of some of the current state of the art.

Successful computational models employ multiple approaches, including data from imaging, statistical modeling, constitutive biomechanical models, and computational fluid dynamics models. One example of this multifaceted approach is observed in Eskandari et al. wherein the team linked computational models, histology, and mechanical testing to thoroughly understand airway tissue structure and function [6]. Purely computational or in silico approaches toward normal breathing have also been successful [21]. The most recent computational models utilize deep learning and patient data from spirometry [22] or imaging in asthma [23], COPD [24], or cystic fibrosis [25]. These models may be used for detection of airway stenosis, predicting the severity of disease, and in future studies assess treatment options.

Beyond using models to understand physiology and pathophysiology, a large portion of the computational modeling field has been devoted to models for inhaled drug delivery. Many of these approaches in computational fluid dynamics are reviewed here [26, 27]. Additional approaches include statistical shape modeling and computational respiratory dynamics [28] and fluid/structure airway models [29]. This field of computational modeling employs simulations of the aerosol distribution and deposition of inhaled therapeutics based on parameters, such as the type of inhaler or nebulizer, drug particle size, drug particle charge, and patient airway geometry. The models take into account gravity, lung movement due to breathing, and the nature of the mucus and surfactant linings of the lung. The best computational models are those that can be validated with data so that they can be useful in the diagnosis and design of experimental approaches. These computational models can be combined with physiological models for the large airway to design and simulate new therapies.

## 6.4 Ex Vivo Testing

As observed as a component of the successful computational model validation in the prior section, ex vivo testing of airway tissues can be a useful tool in modeling the large airways. Typically, as employed in computational models, mechanical testing of strips or sections of airways is a common ex vivo testing setup. Mechanical testing may be performed in an ex vivo lung with whole lung mechanical ventilation techniques taking airway pressure/volume relationships [30]. Single airways may also be dissected and tested with tensile testing in either the axial or circumferential directions [2]. While these techniques are useful for bulk mechanical properties, they do not capture complex mechanics at the cellular level.

In developmental studies, ex vivo large airway embryonic branching may be studied by taking animal (usually mouse) explants at early embryonic stages and allowing them to grow and branch in Matrigel cultures [31, 32]. A large amount of information can be gained from these ex vivo branching morphogenesis studies, including the complex epithelial-mesenchymal cell interactions, mechanical stress and strain fields, and spatio-genetic profiles of development.

In order to study cell-cell interaction and response to therapies in adult tissues, precision-cut lung slices are another robust approach. Precision-cut lung slices are prepared from human or animal lung tissue that is filled with agarose and thinly sliced for laboratory study. These lung slices are a valuable tool for studying airway contraction, but they are most useful for the smaller airways [33]. Precision-cut lung slices have offered useful knowledge in the role of extracellular matrix in contractility [34] and provide an excellent drug or toxicity screening platforms [35, 36]. As with all tissue testing, there are significant challenges in procuring and utilizing human tissues and maintaining tissue integrity and cell viability ex vivo [37]. However, utilizing biomaterial approaches can extend the culture of precision-cut lung slices out to 3 weeks [38].

## 6.5 Tissue Engineering Techniques for Modeling the Large Airways

To overcome tissue testing challenges, tissue engineering techniques may be used to model the large airways. Significant advances have been made in tissue engineering the trachea. Tracheal or mainstem bronchi reconstruction or replacement in patients is at times necessary due to resection from cancer or injury. The many techniques available for engineering large airway replacements are comprehensively reviewed here [39]. These techniques include synthetic prosthesis, allograft, or tissue engineering strategies, including decellularization and recellularization and graft engineering in a bioreactor (Fig. 6.4). Many of these same techniques can be used in vitro to model the large airways for future replacement or as stand-alone models to study airway pathophysiology and repair.

One interesting feature of the large airways that is not present in the small airway or lung parenchyma is the presence of cartilage rings. A multidisciplinary tissue engineering approach must be taken to engineer all of the tissue components in a large airway. Most often, naturally derived scaffolds made from either decellularized animal cartilage or purified collagen hydrogels are used as the scaffolding material for the airway's cartilage rings grown in a bioreactor [40] or orthotopically [41]. A modular approach of growing individual tracheal rings and then combining them showed some success [42]. While cartilage is one of the longest studied engineered tissues, there are still challenges remaining with cell viability [43] and reproducing the mechanical strength of cartilage tissues. The rest of this section will focus on the varied approaches for engineering models of the large airway, focusing on repair of the epithelium and smooth muscle cell components.

### 6.5.1 *Biomaterial Scaffolds*

**Decellularized Scaffolds**

Decellularized tissues are made from the removal of cellular components from an explanted tissue. Decellularized tissues provide the appropriate structure and scaffolding components for the large airway. Many approaches toward the construction of decellularized tissue replacements are utilized in the literature. In a minipig model, a decellularized allograft tracheal replacement without recellularization showed success, growing with the animal subjects out to 1 year following

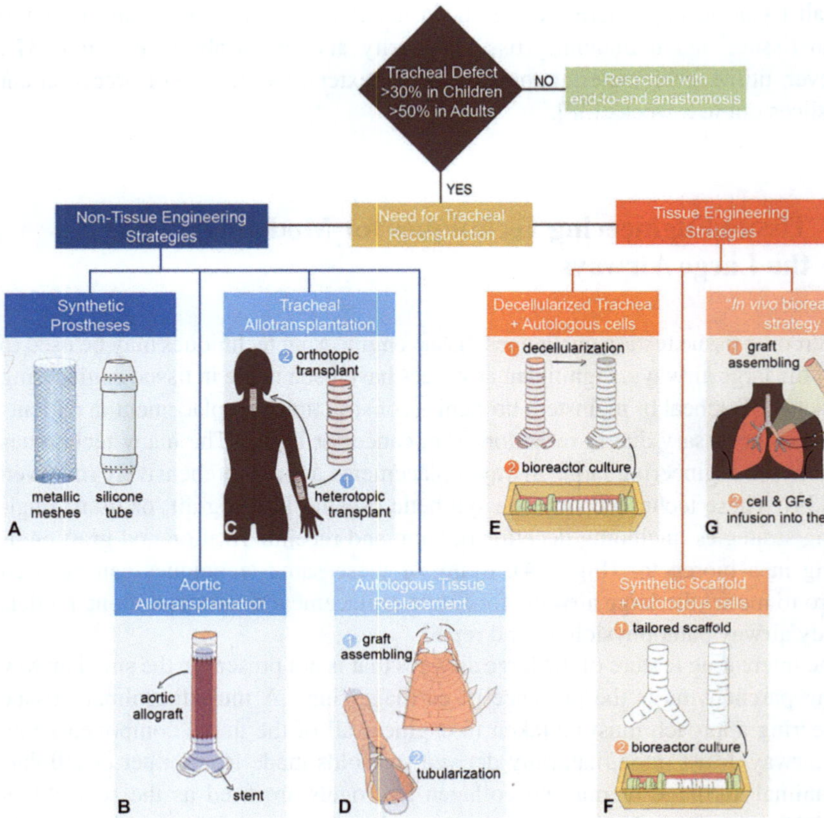


**Fig. 6.4** Overview of tracheal or mainstem bronchi replacement strategies reproduced with permission from the Creative Commons license CCY 4.0 from [20]. (**a**) Synthetic prostheses: metallic meshes and silicone tube. (**b**) Aortic allotransplantation with a supporting stent. (**c**) Tracheal allotransplantation: two-step procedure with pre-vascularization in heterotopic position (forearm) followed by orthotopic transplantation. (**d**) Autologous tissue replacement: stripes of rib cartilage are inserted in a skin forearm free flap (graft assembling phase). Then, the graft is tubularized to reproduce the tracheal lumen. (**e**) An allogeneic trachea is decellularized and then recellularized in a rotating bioreactor with autologous cells. (**f**) Synthetic tailored tracheal grafts seeded with autologous cells in a rotating bioreactor. (**g**) A graft composed of a nitinol stent inserted between two layers of porcine acellularized dermis matrix is seeded with autologous skin keratinocytes (graft assembling phase). Once transplanted, the graft is alternately perfused, through pumps and cannulas, with antibiotics, autologous cells, and growth factors (GFs)

replacement [44]. Similar techniques can use decellularized grafts without reseeding as long as a prelamination technique is used with the implant [45]. Prelamination is a method of layering materials and compressing them to form one graft. For in vitro construction of the large airways, recellularization may be used.

Recellularization, meaning seeding the decellularized scaffold with cells, allows the inclusion of human cells in engineered models of the large airways. Hybrid materials that mix both naturally derived materials with synthetic polymers and multiple cells may be used in these models. One challenge in recellularizing scaffolds is getting a large number of cells to properly adhere and proliferate in the correct spatial arrangement. Recellularization can be enhanced with using rotating bioreactors where mechanical stimulation is used to disperse the cells across the scaffold [46, 47]. Alternatively, vacuum or centrifugation methods may be used to coax cell penetration into decellularized scaffolds [48].

In addition to the bioreactors mentioned above for total recellularization of large airway scaffolds, there are several innovations in bioreactor approaches to partially decellularizing and growing small airways. One novel approach used imaging to guide the removal and new culture of epithelium in the trachea [49]. Other approaches use the shape of the airway smooth muscle tube as guidance [50] or cell sheet technology [51] in developing the engineered large airway.

In all of these methodologies, cell choice is important to ensure that the models have the proper phenotype. Primary human bronchial epithelial and smooth muscle cells are commercially available, though costly. Induced pluripotent stem cells (iPSC) provide a more scalable solution for recellularizing airway scaffolds. iPSC may be cultured and differentiated into relevant cell types for the large airway model.

There is also a design choice as to which type of organ the decellularized tissue should be made. Decellularized tissue scaffolds were first made from the small intestinal submucosa or bladder from the pig, with many successes. The large airway could be amenable to being modeled with extracellular matrix derived from these or other organs instead of the airway. Badylak et al. examined the growth of human bronchial epithelial cells grown on decellularized scaffolds from porcine bladders vs. tracheas [52]. They also tested hydrogels derived from each extracellular matrix and concluded that the hydrogels from either tissue supported epithelial cell growth better than the intact decellularized scaffolds. Hydrogels have been made from bulk lung extracellular matrix from porcine [53, 54] or human [55]. Hydrogels provide a platform to include the extracellular matrix components in an environment amenable to 3D cell culture.

## Cellular, Synthetic, or Hybrid Biomaterial Approaches

Cellular-level models for the epithelium include cultures on Transwell plates with epithelium grown at an air-liquid interface for stratification and development of motile cilia. These culture models may be made more complex by introducing multiple cell types such as mesenchymal cells, including smooth muscle and fibroblasts, and immune cells, including macrophages, neutrophils, eosinophils, or T-cell subsets. These differentiated cell cultures may be grown on implantable materials such as collagen sponges [56] or de-epithelized matrix [57] for cellular transplant. Organoid approaches to recreating the lung utilize neonatal primary cells or iPSC to differentiate into multiple airway cell types in three dimensions. Multi-organoid fusion can be induced to form larger multicellular structures [58].

To achieve spatial location within the 3D cultures, additional biomaterial approaches may be utilized. Many of these materials use synthetic polymeric scaffolds alone or with naturally derived materials to retain bioactivity. Common synthetic scaffolds include poly(caprolactone) [59], poly-L-lactic acid, expanded poly(tetrafluoroethylene) commonly known as Gore-Tex, or poly(ethylene terephthalate) also called Dacron. Synthetic grafts vs. decellularized grafts have varied immune responses [60], making hybrid approaches that combine both synthetic and naturally derived polymers attractive [61].

### *6.5.2 Manufacturing Techniques for Large Airway Models*

The relatively simple, tube structure of the large airways makes advanced manufacturing techniques attractive for forming these physiological models. Microphysiological and airway-on-a-chip models are useful for examining small-scale cell-cell interactions of the airways [62], while three-dimensional (3D) printing techniques can be used for proper organization and larger-scale airway models. 3D printing can employ both synthetic and naturally derived polymers as described in the above section. Extracellular matrix reinforced bioinks provide excellent options for printing large airways [63]. Other models use 3D-printed polycaprolactone and collagen [64] or image-based 3D printing for proper anatomy [65]. 3D printing advances provide the spatial resolution to produce models of the large airway that can be anatomically personalized to the patient for laboratory study of engineered approaches or therapeutics.

## 6.6 Tools for Functional Assessment of Large Airway Models

In any of the above large airway models, physiological measurements must be made to show the functionality of the model. These functional assessments at the airway epithelium may include electrical impedance allografts to assess barrier function [66] or mucus clearance assays to assess ciliary function [67]. Smooth muscle function may be measured with fluorescence calcium imaging, load cells, or methacholine challenge in vivo. The overall tissue structure may be imaged with microCT for tracheal tissue engineering [49]. Additionally, these models may be tested mechanically with biomechanical testing methods, such as tensile testing, compression, torsion, burst pressure, microscale indentation, and/or atomic force microscopy. Mechanical strength is a key issue in engineered large airways. For example, the human trachea must withstand a maximal extension length of up to 134% and have elastic moduli between 12 and 25 MPa [68]. Bending and compression are also important mechanical properties of the trachea and large airways and vary with length of the airway segment [69]. Successful models of the large airways will include functional assessments such as the above.

## 6.7 Limitations and Future Considerations

There are a large number of exciting innovations in the arena of large airway models for lung pathophysiology and repair. The different approaches mentioned in this chapter each capture several characteristics of the large airway. However, there are limitations in that none of these models perfectly captures the complex nature of the disease states, such as cystic fibrosis, chronic obstructive pulmonary disease, or asthma. Furthermore, in the in vitro models, the main limitation is the lack of interaction with the immune system. While some models incorporate immune cells in culture, none of the models have the dynamic nature of the immune system found in vivo. Combinatorial modeling approaches that include computational, in vitro, and in vivo models of the large airway are the best way to overcome limitations of any single model for the large airway.

## References


1. Feher J. 6.1 – The Mechanics of Breathing. In: Feher J, editor. Quantitative Human Physiology (Second Edition) [Internet]. Boston: Academic Press; 2017 [cited 2022 Aug 4]. p. 623–32. Available from: https://www.sciencedirect.com/science/article/pii/B9780128008836000604
2. Eskandari M, Arvayo AL, Levenston ME. Mechanical properties of the airway tree: heterogeneous and anisotropic pseudoelastic and viscoelastic tissue responses. J Appl Physiol. 2018 Sep;125(3):878–88.
3. Kia'i N, Bajaj T. Histology, Respiratory Epithelium. In: StatPearls [Internet]. Treasure Island (FL): StatPearls Publishing; 2022 [cited 2022 Aug 4]. Available from: http://www.ncbi.nlm.nih.gov/books/NBK541061/
4. Shi HC, Deng WJ, Pei C, Lu D, Zhang XJ, Wang XH, et al. Biomechanical properties of adult-excised porcine trachea for tracheal xenotransplantation. Xenotransplantation. 2009 Jun;16(3):181–6.
5. Zhang B, Lu Y, Sun F, Wu Q, Shi H. Compression property of trachea: A key mechanical property for artificial trachea graft. Am J Otolaryngol. 2022 Jun;43(3):103413.
6. Eskandari M, Nordgren TM, O'Connell GD. Mechanics of Pulmonary Airways: Linking Structure to Function Through Constitutive Modeling, Biochemistry, and Histology. Acta Biomater. 2019 Oct 1;97:513–23.
7. Carbone JE, Marini JJ. Bronchodilatory Effect of Warm Air Inhalation During Quiet Breathing. West J Med. 1984 Mar;140(3):398–402.
8. Uzeloto JS, Ramos D, Silva BS de A, Lima MBP de, Silva RN, Camillo CA, et al. Mucociliary Clearance of Different Respiratory Conditions: A Clinical Study. Int Arch Otorhinolaryngol. 2021 Jan;25(1):e35–40.
9. Mertens TCJ, Karmouty-Quintana H, Taube C, Hiemstra PS. Use of airway epithelial cell culture to unravel the pathogenesis and study treatment in obstructive airway diseases. Pulm Pharmacol Ther. 2017 Aug;45:101–13.
10. Liu G, Betts C, Cunoosamy DM, Åberg PM, Hornberg JJ, Sivars KB, et al. Use of precision cut lung slices as a translational model for the study of lung biology. Respir Res. 2019 Jul 19;20(1):162.
11. Berrocal T, Madrid C, Novo S, Gutiérrez J, Arjonilla A, Gómez-León N. Congenital Anomalies of the Tracheobronchial Tree, Lung, and Mediastinum: Embryology, Radiology, and Pathology. RadioGraphics. 2004 Jan;24(1):e17–e17.

12. Rommens JM, Iannuzzi MC, Kerem B, Drumm ML, Melmer G, Dean M, et al. Identification of the cystic fibrosis gene: chromosome walking and jumping. Science. 1989 Sep 8;245(4922):1059–65.
13. Pini L, Pinelli V, Modina D, Bezzi M, Tiberio L, Tantucci C. Central airways remodeling in COPD patients. Int J Chron Obstruct Pulmon Dis. 2014 Sep 1;9:927–33.
14. Joseph C, Tatler AL. Pathobiology of Airway Remodeling in Asthma: The Emerging Role of Integrins. J Asthma Allergy. 2022;15:595–610.
15. Hough KP, Curtiss ML, Blain TJ, Liu RM, Trevor J, Deshane JS, et al. Airway Remodeling in Asthma. Front Med [Internet]. 2020 [cited 2022 Aug 5];7. Available from: https://www.frontiersin.org/articles/10.3389/fmed.2020.00191
16. McCarron A, Parsons D, Donnelley M. Animal and Cell Culture Models for Cystic Fibrosis: Which Model Is Right for Your Application? Am J Pathol. 2021 Feb 1;191(2):228–42.
17. Sun X, Olivier AK, Liang B, Yi Y, Sui H, Evans TIA, et al. Lung phenotype of juvenile and adult cystic fibrosis transmembrane conductance regulator-knockout ferrets. Am J Respir Cell Mol Biol. 2014 Mar;50(3):502–12.
18. Klymiuk N, Mundhenk L, Kraehe K, Wuensch A, Plog S, Emrich D, et al. Sequential targeting of CFTR by BAC vectors generates a novel pig model of cystic fibrosis. J Mol Med. 2012 May 1;90(5):597–608.
19. Ghorani V, Boskabady MH, Khazdair MR, Kianmeher M. Experimental animal models for COPD: a methodological review. Tob Induc Dis. 2017 May 2;15:25.
20. Adamo D, Galaverni G, Genna VG, Lococo F, Pellegrini G. The Growing Medical Need for Tracheal Replacement: Reconstructive Strategies Should Overcome Their Limits. Front Bioeng Biotechnol. 2022;10:846632.
21. Marconi S, De Lazzari C. In silico study of airway/lung mechanics in normal human breathing. Math Comput Simul. 2020 Nov;177:603–24.
22. Wang Y, Li Y, Chen W, Zhang C, Liang L, Huang R, et al. Deep Learning for Automatic Upper Airway Obstruction Detection by Analysis of Flow-Volume Curve. Respir Int Rev Thorac Dis. 2022 May 12;1–10.
23. Foy BH, Soares M, Bordas R, Richardson M, Bell A, Singapuri A, et al. Lung Computational Models and the Role of the Small Airways in Asthma. Am J Respir Crit Care Med. 2019 Oct 15;200(8):982–91.
24. Moslemi A, Kontogianni K, Brock J, Wood S, Herth F, Kirby M. Differentiating COPD and Asthma using Quantitative CT Imaging and Machine Learning. Eur Respir J. 2022 Feb 24;2103078.
25. DeBoer EM, Kimbell JS, Pickett K, Hatch JE, Akers K, Brinton J, et al. Lung inflammation and simulated airway resistance in infants with cystic fibrosis. Respir Physiol Neurobiol. 2021 Nov;293:103722.
26. Longest PW, Bass K, Dutta R, Rani V, Thomas ML, El-Achwah A, et al. Use of computational fluid dynamics deposition modeling in respiratory drug delivery. Expert Opin Drug Deliv. 2019 Jan;16(1):7–26.
27. Huang F, Zhu Q, Zhou X, Gou D, Yu J, Li R, et al. Role of CFD based in silico modelling in establishing an in vitro-in vivo correlation of aerosol deposition in the respiratory tract. Adv Drug Deliv Rev. 2021 Mar;170:369–85.
28. Talaat M, Si XA, Dong H, Xi J. Leveraging statistical shape modeling in computational respiratory dynamics: Nanomedicine delivery in remodeled airways. Comput Methods Programs Biomed. 2021 Jun;204:106079.
29. Phuong NL, Quang TV, Khoa ND, Kim JW, Ito K. CFD analysis of the flow structure in a monkey upper airway validated by PIV experiments. Respir Physiol Neurobiol. 2020 Jan;271:103304.
30. Dong SJ, Wang L, Chitano P, Vasilescu DM, Paré PD, Seow CY. Airway and parenchymal tissue resistance and elastance in ex vivo sheep lungs: effects of bronchochallenge and deep inspiration. Am J Physiol Lung Cell Mol Physiol. 2022 Jun 1;322(6):L882–9.
31. Patil LS, Varner VD. Toward Measuring the Mechanical Stresses Exerted by Branching Embryonic Airway Epithelial Explants in 3D Matrices of Matrigel. Ann Biomed Eng. 2022 Sep;50(9):1143–57.

32. Fernandes-Silva H, Alves MG, Araújo-Silva H, Silva AM, Correia-Pinto J, Oliveira PF, et al. Lung branching morphogenesis is accompanied by temporal metabolic changes towards a glycolytic preference. Cell Biosci. 2021 Jul 17;11(1):134.
33. Li G, Cohen JA, Martines C, Ram-Mohan S, Brain JD, Krishnan R, et al. Preserving Airway Smooth Muscle Contraction in Precision-Cut Lung Slices. Sci Rep. 2020 Apr 15;10(1):6480.
34. Van Dijk EM, Culha S, Menzen MH, Bidan CM, Gosens R. Elastase-Induced Parenchymal Disruption and Airway Hyper Responsiveness in Mouse Precision Cut Lung Slices: Toward an Ex vivo COPD Model. Front Physiol. 2016;7:657.
35. Lauenstein L, Switalla S, Prenzler F, Seehase S, Pfennig O, Förster C, et al. Assessment of immunotoxicity induced by chemicals in human precision-cut lung slices (PCLS). Toxicol Vitro Int J Publ Assoc BIBRA. 2014 Jun;28(4):588–99.
36. Lam M, Royce SG, Donovan C, Jelinic M, Parry LJ, Samuel CS, et al. Serelaxin Elicits Bronchodilation and Enhances β-Adrenoceptor-Mediated Airway Relaxation. Front Pharmacol. 2016;7:406.
37. Preuß EB, Schubert S, Werlein C, Stark H, Braubach P, Höfer A, et al. The Challenge of Long-Term Cultivation of Human Precision-Cut Lung Slices. Am J Pathol. 2022 Feb;192(2):239–53.
38. Bailey KE, Pino C, Lennon ML, Lyons A, Jacot JG, Lammers SR, et al. Embedding of Precision-Cut Lung Slices in Engineered Hydrogel Biomaterials Supports Extended Ex Vivo Culture. Am J Respir Cell Mol Biol. 2020 Jan;62(1):14–22.
39. Damiano G, Palumbo VD, Fazzotta S, Curione F, Lo Monte G, Brucato VMB, et al. Current Strategies for Tracheal Replacement: A Review. Life Basel Switz. 2021 Jun 25;11(7):618.
40. Baranovskii D, Demner J, Nürnberger S, Lyundup A, Redl H, Hilpert M, et al. Engineering of Tracheal Grafts Based on Recellularization of Laser-Engraved Human Airway Cartilage Substrates. Cartilage. 2022 Mar;13(1):19476035221075950.
41. Tsao CK, Hsiao HY, Cheng MH, Zhong WB. Tracheal reconstruction with the scaffolded cartilage sheets in an orthotopic animal model. Tissue Eng Part A. 2022 Feb 9;
42. Xu Y, Dai J, Zhu X, Cao R, Song N, Liu M, et al. Biomimetic Trachea Engineering via a Modular Ring Strategy Based on Bone-Marrow Stem Cells and Atelocollagen for Use in Extensive Tracheal Reconstruction. Adv Mater Deerfield Beach Fla. 2022 Feb;34(6):e2106755.
43. Chan C, Liu L, Dharmadhikari S, Shontz KM, Tan ZH, Bergman M, et al. A Multimodal Approach to Quantify Chondrocyte Viability for Airway Tissue Engineering. The Laryngoscope [Internet]. [cited 2022 Aug 5];n/a(n/a). Available from: https://onlinelibrary.wiley.com/doi/abs/10.1002/lary.30206
44. Ohno M, Fuchimoto Y, Higuchi M, Yamaoka T, Komura M, Umezawa A, et al. Long-term observation of airway reconstruction using decellularized tracheal allografts in micro-miniature pigs at growing stage. Regen Ther. 2020 Dec;15:64–9.
45. Martínez-Hernández NJ, Díaz-Cuevas A, Milián-Medina L, Sancho-Tello M, Roselló-Ferrando J, Morcillo-Aixelá A, et al. Decellularized tracheal prelamination implant: A proposed bilateral double organ technique. Artif Organs. 2021 Dec;45(12):1491–500.
46. Pennarossa G, Ghiringhelli M, Gandolfi F, Brevini TAL. Tracheal In Vitro Reconstruction Using a Decellularized Bio-Scaffold in Combination with a Rotating Bioreactor. Methods Mol Biol Clifton NJ. 2022;2436:157–65.
47. Young BM, Antczak LAM, Shankar K, Heise RL. A Two-Step Bioreactor for Decellularized Lung Epithelialization. Cells Tissues Organs. 2021;210(4):301–10.
48. Zhou Q, Ye X, Ran Q, Kitahara A, Matsumoto Y, Moriyama M, et al. Trachea Engineering Using a Centrifugation Method and Mouse-Induced Pluripotent Stem Cells. Tissue Eng Part C Methods. 2018 Sep;24(9):524–33.
49. Townsend JM, Weatherly RA, Johnson JK, Detamore MS. Standardization of Microcomputed Tomography for Tracheal Tissue Engineering Analysis. Tissue Eng Part C Methods. 2020 Nov;26(11):590–5.
50. Jin Y, Liu L, Yu P, Lin F, Shi X, Guo J, et al. Emergent Differential Organization of Airway Smooth Muscle Cells on Concave and Convex Tubular Surface. Front Mol Biosci. 2021;8:717771.

51. Dang LH, Hung SH, Tseng Y, Quang LX, Le NTN, Fang CL, et al. Partial Decellularized Scaffold Combined with Autologous Nasal Epithelial Cell Sheet for Tracheal Tissue Engineering. Int J Mol Sci. 2021 Sep 25;22(19):10322.
52. Ravindra A, D'Angelo W, Zhang L, Reing J, Johnson S, Myerburg M, et al. Human Bronchial Epithelial Cell Growth on Homologous Versus Heterologous Tissue Extracellular Matrix. J Surg Res. 2021 Jul;263:215–23.
53. Pouliot RA, Young BM, Link PA, Park HE, Kahn AR, Shankar K, et al. Porcine Lung-Derived Extracellular Matrix Hydrogel Properties Are Dependent on Pepsin Digestion Time. Tissue Eng Part C Methods. 2020 Jun;26(6):332–46.
54. Pouliot RA, Link PA, Mikhaiel NS, Schneck MB, Valentine MS, Kamga Gninzeko FJ, et al. Development and characterization of a naturally derived lung extracellular matrix hydrogel. J Biomed Mater Res A. 2016 Aug;104(8):1922–35.
55. de Hilster RHJ, Sharma PK, Jonker MR, White ES, Gercama EA, Roobeek M, et al. Human lung extracellular matrix hydrogels resemble the stiffness and viscoelasticity of native lung tissue. Am J Physiol Lung Cell Mol Physiol. 2020 Apr 1;318(4):L698–704.
56. Nakamura R, Katsuno T, Tsuji T, Oyagi S, Kishimoto Y, Suehiro A, et al. Airway ciliated cells regenerated on collagen sponge implants acquire planar polarities towards nearby edges of implanted areas. J Tissue Eng Regen Med. 2021 Aug;15(8):712–21.
57. Chen J, Mir SM, Pinezich MR, O'Neill JD, Guenthart BA, Bacchetta M, et al. Homogeneous Distribution of Exogenous Cells onto De-epithelialized Rat Trachea via Instillation of Cell-Loaded Hydrogel. ACS Biomater Sci Eng. 2022 Jan 10;8(1):82–8.
58. Liu Y, Dabrowska C, Mavousian A, Strauss B, Meng F, Mazzaglia C, et al. Bio-assembling Macro-Scale, Lumenized Airway Tubes of Defined Shape via Multi-Organoid Patterning and Fusion. Adv Sci Weinh Baden-Wurtt Ger. 2021 May;8(9):2003332.
59. Zhao Y, Tian C, Wu K, Zhou X, Feng K, Li Z, et al. Vancomycin-Loaded Polycaprolactone Electrospinning Nanofibers Modulate the Airway Interfaces to Restrain Tracheal Stenosis. Front Bioeng Biotechnol. 2021;9:760395.
60. Tan ZH, Dharmadhikari S, Liu L, Wolter G, Shontz KM, Reynolds SD, et al. Tracheal Macrophages During Regeneration and Repair of Long-Segment Airway Defects. The Laryngoscope. 2022 Apr;132(4):737–46.
61. Young BM, Shankar K, Allen BP, Pouliot RA, Schneck MB, Mikhaiel NS, et al. Electrospun Decellularized Lung Matrix Scaffold for Airway Smooth Muscle Culture. ACS Biomater Sci Eng. 2017 Dec 11;3(12):3480–92.
62. Lagowala DA, Kwon S, Sidhaye VK, Kim DH. Human microphysiological models of airway and alveolar epithelia. Am J Physiol Lung Cell Mol Physiol. 2021 Dec 1;321(6):L1072–88.
63. De Santis MM, Alsafadi HN, Tas S, Bölükbas DA, Prithiviraj S, Da Silva IAN, et al. Extracellular-Matrix-Reinforced Bioinks for 3D Bioprinting Human Tissue. Adv Mater Deerfield Beach Fla. 2021 Jan;33(3):e2005476.
64. Frejo L, Goldstein T, Swami P, Patel NA, Grande DA, Zeltsman D, et al. A two-stage in vivo approach for implanting a 3D printed tissue-engineered tracheal replacement graft: A proof of concept. Int J Pediatr Otorhinolaryngol. 2022 Apr;155:111066.
65. Oberoi G, Eberspächer-Schweda MC, Hatamikia S, Königshofer M, Baumgartner D, Kramer AM, et al. 3D Printed Biomimetic Rabbit Airway Simulation Model for Nasotracheal Intubation Training. Front Vet Sci. 2020;7:587524.
66. Peterson DM, Beal EW, Reader BF, Dumond C, Black SM, Whitson BA. Electrical Impedance as a Noninvasive Metric of Quality in Allografts Undergoing Normothermic Ex Vivo Lung Perfusion. ASAIO J Am Soc Artif Intern Organs 1992. 2022 Jul 1;68(7):964–71.
67. Antunes MB, Cohen NA. Mucociliary clearance--a critical upper airway host defense mechanism and methods of assessment. Curr Opin Allergy Clin Immunol. 2007 Feb;7(1):5–10.
68. Safshekan F, Tafazzoli-Shadpour M, Abdouss M, Shadmehr MB. Mechanical Characterization and Constitutive Modeling of Human Trachea: Age and Gender Dependency. Materials. 2016 Jun;9(6):456.
69. Huang Z, Wang L, Zhang CX, Cai ZH, Liu WH, Li WM, et al. Biomechanical strength dependence on mammalian airway length. J Thorac Dis. 2021 Feb;13(2):918–26.

# Chapter 7
# Engineering Large Airways

**Tehreem Khalid and Cian O'Leary**

## 7.1 Introduction

Currently, there remains an unmet clinical need in respiratory medicine due to low availability of viable replacement options for non-operable large tracheal defects. This is largely due to the issues that arise with complications from implanted constructs, such as inflammatory response, that can lead to stenosis of implanted constructs or the mismatch of mechanical properties. The native trachea has anisotropic mechanical properties, meaning the mechanical properties will differ when measured in different directions. However, the majority of engineering attempts focus on fabricating uniform solid tubular scaffolds, with the aim of mimicking the cartilaginous rings of the native tissue. These scaffolds often lack comprehensive mechanical analysis, which limits their outcome *in vivo*, as well as hindering comparisons of different scaffolds across literature. The aim of this chapter is to thus provide insight into the mechanical properties of native tracheal tissue, as well as to recommend key tests for consideration to influence design criteria for mechanically adequate trachea replacements.

T. Khalid · C. O'Leary (✉)
School of Pharmacy and Biomolecular Sciences, RCSI University of Medicine and Health Sciences, Dublin, Ireland

Tissue Engineering Research Group, RCSI, Dublin, Ireland

Advanced Materials & Bioengineering Research (AMBER) Centre, RCSI & Trinity College, Dublin, Ireland
e-mail: tehreemkhalid@rcsi.com; cianoleary@rcsi.com

 
C. M. Magin (ed.), *Engineering Translational Models of Lung Homeostasis and Disease*, Advances in Experimental Medicine and Biology 1413,
https://doi.org/10.1007/978-3-031-26625-6_7

## 7.2 Forces During Respiration and How They Can Influence Construct Design

The forces acting on the airways are critical in determining the appropriate parameters for the evaluation and comparison of different engineered trachea analogues, as well as native tissue. The mechanical forces on the trachea are numerous and complex, making precise airway modelling difficult. Understanding the different pressures that are exhibited during respiration within the whole respiratory system may provide insight on optimal tracheal construct design to better recapitulate the trachea by mitigating tracheal collapses and airway resistance. During normal respiration, a pressure gradient is formed, which results in air being pushed in and out of the lungs. During inspiration, the diaphragm contracts and the lungs expand; as the volume within the lungs increases, there is a concomitant decrease in pressure. Conversely, the diaphragm relaxes during expiration to reduce the volume of the lungs and increase pressure, resultantly driving air out of the lungs.

An equal pressure point is developed when maximal expiratory flow is reached and the pressure inside the airway is equal to the pressure outside (intrapleural pressure). For a long time, the equal pressure point was thought to directly correlate with the most probable location of airway collapse. However, it is now thought that airway collapse has a greater tendency to occur above the equal pressure point, if the cartilaginous support is weak [1, 2]. On the other hand, some argue that it is difficult to identify an exact position that is more prone to airway collapse as the transmural pressure (pressure difference across (inside to outside) a hollow structure) exhibited at any given site on the trachea varies greatly along its length [3, 4] (Fig. 7.1). Similar to the pleural pressure, the extratracheal pressure in the tracheal cavity is generally negative but can become positive during forced expiration [5]. To understand the extratracheal pressure, pleural pressure must also be understood. The pleural cavity is the space between the lungs and the pleural membrane, which is filled with fluid. Transpleural pressure during normal respiration ranges between 5 cm and 8 cm $H_2O$ [5]. The maximum pleural pressure that can be achieved is determined by an individual's muscle strength. The average maximum inspiratory and expiratory pressures are approximately 170 cm $H_2O$ and 114 cm $H_2O$ in males, respectively, and 76 cm $H_2O$ and 86 cm $H_2O$ in females, respectively [6]. The extratracheal pressure in the neck is approximately atmospheric; however, during respiration, the trachea slides between compartments during changes in intratracheal pressure, further complicating the system. Therefore, these anatomical shifts may cause the changes in transmural pressure at different regions along the trachea [4].

Under normal quiet breathing, fluctuations in the trachea's diameter are hardly measurable. However, during atypical conditions, collapse can occur during respiration, most commonly during expiration [4]. In the cases of abnormal airway resistance, collapse is driven by an increase in transmural pressure due to increasing respiratory efforts [4]. It has been reported that there are large increases in airway resistance in the trachea during expiration as opposed to inspiration, which explains the prevalence of expiratory collapse in children with tracheal defects [2, 4]. Bronchoscopic observations have also measured a decrease in the cross-sectional

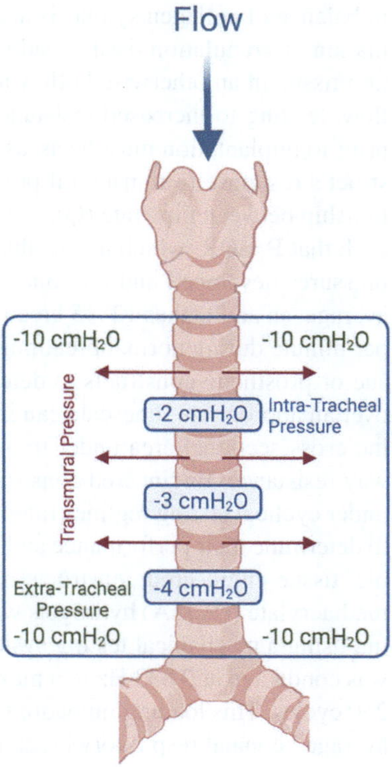


**Fig. 7.1** Schematic outlining respiratory exemplary pressure values exerted on the trachea during normal respiration. Intratracheal pressure is the pressure measured within the trachea during respiration. Transmural pressure is the pressure difference across the tracheal wall. Extratracheal pressure is within the neck and is approximately atmospheric. (Adapted from Wittenborg et al. [4])

area of the trachea during expiration and a partial trachea collapse during coughing, further indicating the difference in airway resistance between inspiration and expiration [7]. Collapse during inspiration can occur and is usually caused by the development of a negative intratracheal pressure than normal, whereas collapse during expiration is caused by the increase in respiratory effort and plural pressure. Therefore, to mitigate the possibility of collapse, ensuring your construct is able to withstand changes in pressures, similar to native tissue, is vital.

In addition to understanding these pressure gradients and forces within the respiratory system, knowledge of airway resistance in the trachea can provide insight into appropriate construct development. As such, increased airway resistance at any given position in the trachea can result in the development of uncharacteristically high transmural pressure; thus, constructs must be designed to alleviate increases in flow resistance. The formation of stenosis due to an inflammatory and fibrotic response in an otherwise mechanically adequate construct increases airway resistance and leads to airway collapse and construct failure [8]. An inflammatory response can occur following construct implantation if non-biocompatible materials are used or further injury to the tracheal tissue has occurred. This will trigger an inflammatory response leading to the production of numerous inflammatory factors and profibrotic cytokines, which are involved in fibroblast cell activation and cell proliferation, and deposition of extracellular matrix, ultimately leading to an imbalance of collagen synthesis and degradation [9]. This in turn results in the formation of granulation tissue, leading to tracheal stenosis. The presence of granulation tissue in an otherwise hollow and smooth tubular structure will disturb airway flow, leading to increased resistance. Therefore, the flow dynamics of a construct prior to implantation must be assessed to prevent airway collapse. Assessing a construct's response to transmural pressure is challenging in vitro; however, the relationship between flow rate (Q), airway resistance (R) and transmural pressure (P) is such that P = QR, which means that the respiratory rate affects both the transmural pressures developed and the time over which loads are normally applied [10]. On average, an adult takes 12–15 breathes per minute, while a newborn takes 38 breaths per minute during normal breathing [11–13]. The assessment of native tracheal tissue or prosthetic constructs to determine their ability to recoil from applied loads over an appropriate timescale can indicate their performance to resisting changes in the cross-sectional area under transmural pressure changes and thus reducing airway resistance. Engineered constructs can be loaded at physiological loading rates under cyclical testing regimes ranging from normal to stressed breathing conditions to determine their performance and their ability to recoil adequately [14]. For example, tissue-engineered constructs made from a series of patterned 2-hydroxyethyl methacrylate (HEMA) hydrogels were mechanically characterised by cyclical loading using a mechanical testing system with a 10 N load cell. Cyclical compression was conducted at 0.833 Hz to a maximum strain of 15% of the original diameter for 250 cycles. This loading/unloading rate was calculated in this instance to mimic the average neonatal respiratory breathing rate of 1.2 s/breath [14].

## 7.3 The Structure of the Trachea and Its Mechanical Properties

Albeit a seemingly simple tube-like structure, the trachea is in actuality a complex organ consisting of numerous components (Fig. 7.2) that not only fulfil multiple unique physiological functions but also possess distinct mechanical properties, which collectively provide the whole trachea with its complex anisotropic mechanical properties [10]. However, it is this complex feature of the trachea that provides a huge challenge in defining the mechanical requirements for a substitute graft. It has been difficult to establish the appropriate mechanical tests and define the applicable range of mechanical properties; as reported figures and testing techniques over the years have largely varied (Table 7.1), resulting in inconsistency [10, 15]. If the engineered construct's mechanical properties do not match the native tissue, granulation can occur at the anastomosis site, which can induce bleeding and also stenosis, leading to fatal obstruction [16]. Mechanical forces have been reported to be involved in fibrosis, through prolonged inflammation mediated by T-cells [8]. Therefore, understanding the mechanical features of each component of the trachea and how it can be assessed individually and as a whole is vital when developing a tracheal replacement construct.

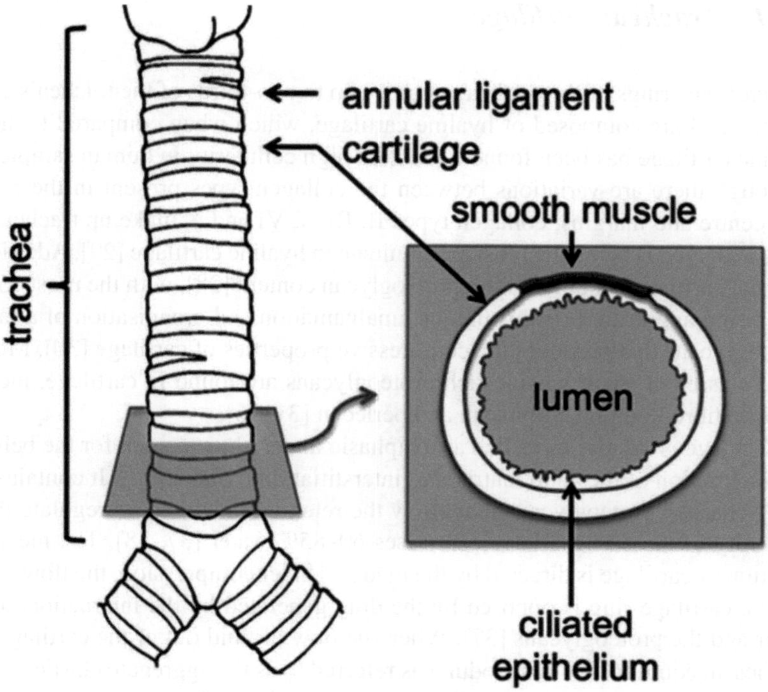


**Fig. 7.2** Cross-section of the trachea highlighting the structural relationship between each component. Reprinted with permission from Boazak et al., ACS Biomater Sci Eng. 2018;4(4):1272–84. Copyright 2018 American Chemical Society

**Table 7.1** Mechanical properties of tracheal cartilage, trachealis muscle and whole trachea

| Tissue type | Specimen type | Test type | Modulus | References |
|---|---|---|---|---|
| Tracheal cartilage | Human | Tensile | 16.92 MPa | [18] |
| | Human | Tensile | 1–15 MPa | [19] |
| | Porcine | Tensile | 5.62 MPa | [20] |
| | | Compression | 1.36 MPa | |
| | Ovine | Tensile | 10.6 MPa | [21] |
| Trachealis muscle | Human | Tensile | 5.7–364 KPa | [22] |
| | Human | Tensile | 0.84–34.15 KPa | [23] |
| Whole trachea | Human | Tensile | 12–20 MPa | [24] |
| | Human | Compression | 0.16–2 MPa | [25] |
| | Porcine | Compression | 1.10 MPa | [20] |
| | Ovine | Tensile | 1.5 MPa | [14] |
| | Rabbit | Compression | 1.72 MPa | [26] |
| | Rabbit | Three-point bending | 0.214 MPa | [27] |

### 7.3.1 Tracheal Cartilage

The cartilage rings of the trachea comprise up to two-thirds of the trachea's circumference and are composed of hyaline cartilage, which when compared to articular hyaline cartilage has been found to contain high cellularity in human samples [28]. Although there are variations between the collagen types present in the cartilage ring centre and margins, collagen types II, III, V, VI and X make up tracheal cartilage, with type II being the most predominate in hyaline cartilage [29]. Additionally, tracheal cartilage contains a high proteoglycan content [28], with the most common type being aggrecan. It is this unique amalgamation and organisation of a range of proteoglycans that produces the compressive properties of cartilage [30]. Members of all classes of small leucine-rich proteoglycans are found in cartilage, including decorin, fibromodulin, epiphycan and perlecan [31–36].

Cartilage is usually modelled as a biphasic material to account for the behaviour and interaction of the solid matrix and interstitial fluid phase [37]. It contains negatively charged proteoglycans that allow the retention of fluid and regulate the permeability of the tissue. It also comprises 60–85% water [37, 38]. The mechanical function of cartilage is directed by the matrix. Under compression, the flow of water out the cartilage ring is opposed by the drag generated by the interaction between water and the proteoglycans [37]. When the flow of fluid out of the cartilage tissue reaches an equilibrium, this modulus is referred to as the aggregate elastic modulus. As a result, the aggregate or equilibrium modulus primarily tests the tensile or compressive properties of the collagen network rather than the complex tissue response to mechanical loading. Mechanical evaluation of cartilage samples at a slow loading rate can attempt to capture the cartilage's elastic properties [21]. In vivo, the rings endure both sustained and transient loads, through contractions from the trachealis muscle and coughing, respectively; however, reported mechanical assessment of the cartilage rings has primarily looked to observe measurements applied from simple large uniaxial compressive or tensile loading [19]. The effects of transient loads on cartilage deformation are determined by the viscoelastic properties of the cartilage. The cartilage rings of the trachea are responsible for ensuring the trachea remains open during changes in inter-thoracic pressures that occur during respiration and thus should be mechanically robust; conversely, they must also be able to deform radially to tolerate fluctuations in the cross-sectional area during coughing and deglutition [11, 39].

Rings of adult human trachea have been reported to exhibit a tensile modulus between 1 and 20 MPa [19, 40, 41], while porcine trachea cartilage has been reported to have a compressive modulus of 1.3–17 MPa. This large range is due to the variations in measurement methods and interindividual variation in tracheas [20, 39]. The moduli of cartilage have also been reported to increase with age, as collagen within the cartilage will decrease with age, but the proteoglycan will remain unchanging [19, 42]. Overall, both the tensile and compressive properties of the trachea's cartilage rings are pertinent to its function, as bending of the rings generates tensile loads on the outer regions of the rings and compressive loads of the inner region, thus making it difficult to accurately define the biomechanics of the rings via either tensile or compressive testing.

### 7.3.2 *Trachealis Muscle*

The trachealis muscle is a smooth muscle that is responsible for changing the trachea diameter in response to stimuli, such as asthma or cough reflex. It is considered an active component of the trachea, as it responds to aforementioned external stimuli [22]. The muscle contains stretch receptors that respond to transverse strain, but not longitudinal [43]. During deep and forceful breathing, contraction of the muscle prevents tracheal compression and increases the maximum airflow through the trachea [3]. As the trachealis muscle is a highly cellularised tissue, the smooth muscle cells contribute to both its active and passive mechanical properties. The trachealis muscle has been found to generate a higher active tension than other smooth muscles and has highly non-linear mechanical properties. It has been widely reported that the trachealis muscle is stiffer longitudinally than circumferentially [10, 44]. Studies have measured a tensile modulus of between 0.8 and 34 kPa longitudinally and 0.34 and 35 kPa transversally in adult humans [24, 39]. The capacity of the trachealis muscle to deform contributes significantly to tracheal compliance and also permits extension of the trachea.

### 7.3.3 *Annular Ligament*

The annular ligament connects the cartilage rings of the trachea and allows for extensibility longitudinally and are predominately comprised of type I collagen [45]. It is challenging to measure the longitudinal mechanical properties of the annular ligaments due to their short inter-ring length. Thus, studies rely on modelling of the properties of the trachealis muscle as an approximation [46]. In a recent study, large samples of whole tracheal tissue were mechanically evaluated, and segments of cartilage were accounted for mathematically to calculate the values of the annular ligament. Notably, samples from older donors were approximated to be mechanically stiffer, although this was not statistically significant [18].

## 7.4 Mechanical Properties of the Whole Trachea and the Implications of Mechanical Property Mismatch

A critical factor for maintaining construct patency in tracheal replacement grafts is the fabrication of a tracheal construct with anisotropic properties approximate to native tissue. Given that normal tracheal function involves significant load and deformation in both longitudinal and radial planes, mechanical mismatch in either direction could result in excessive scarring. The compliance of the native trachea and its ability to extend longitudinally are important factors in native tissue, as they heavily influence respiration. The subsections that follow investigate the consequences of failing to account for the native radial and longitudinal properties.

### 7.4.1 Compliance

Although the components of the trachea all function to mitigate collapse, it does retain this capability under certain circumstances. For example, at particulate transmural pressures, it can deform to accelerate air during respiration and also clear mucus of the respiratory tract [47, 48]. As the most physically reactive component of the trachea, the trachealis muscle is most responsible for altering the cross-section of the trachea. Fluid-solid interface modelling has found that tracheal stents, which are mechanically rigid and not compliant, can avert the trachealis muscle from contracting, which in turn can alter the flow of air at the top of the stent, resulting in an accumulation of mucus [49]. Build-up of mucus can lead to obstruction of the airways and increased risk of infection [49, 50]. The transplantation of a tracheal construct that is incapable of change in its cross-sectional profile alongside native tissue would almost certainly result in unwanted airflow patterns. Additionally, not only would a non-compliant implant result in mucus build-up, but it may also result in mechanical mismatch particularly at the anastomosis sites, resulting in stenosis [8, 16].

Reduction in whole trachea diameters in adults can range from 11% to 61% during forced inhalation and expiration, and from 20% to 50% diameter change in children crying [4, 51]. In an aim to validate a construct's ability to radially deform, radial compressive testing is applicable. The force required to achieve relevant reductions in the trachea diameter as reported in native tracheal tissue, as well as the construct's capacity to recoil back to its original diameter successfully, should be considered.

Compliance testing is seen as a more physiologically relevant loading scenario and is defined as the change in airway volume over the change in transmural pressure [52]. To account for comparisons against different-sized tracheal samples, volume pressure curves commonly normalise the change in airway pressure against the original volume. Measurement of compliance requires airtight experimental systems, which have reported 1.8 cm $H_2 0^{1-}$ ($\Delta P = 0 \pm 10$ cm $H_2 0^{1-}$) in human infants aged 1–7 months to 24 cm $H_2 0^{1-}$ ($\Delta P = 0 \pm 10$ cm $H_2 0^{1-}$) in 69-year-old adults [53]. However, no data has been available for engineered constructs. Under testing, transmural pressure has been shown not to stabilise until up to 45 min after the trachea has been inflated or deflated [35]. Notably, the trachea has been found to be more compliant under early, minor changes in transmural pressure, wherein compliancy begins to plateau as transmural pressure increases. It is this response that allows for coughing to occur without complete collapse of the trachea. Interestingly, compliancy testing of thick-walled tubular scaffolds highlights a different response compared to native tissue. Initially, compliance is low, and then it dramatically increases after collapse [54]. This has provided interesting insights into design requirements of tubular constructs for trachea replacement as it has been reported that the pressure gradient required for the necessary compliance, such as partial collapsing during coughing, of a thick-walled tubular construct is affected by the bending stiffness of the construct wall [55]. The bending stiffness of a construct itself is defined by the wall thickness, indicating that the dimensions of a construct, such as wall thickness, are a very important factor in construct design that need to be considered in assessing the ability to resist collapsing. This also factors into selecting relevant animal tissues to mechanically evaluate, compare against and select for in vivo testing when optimising scaffold design. Ultimately, it is beneficial to select animal models that have similar-sized trachea to that of humans for successful translation of experimental results. For example, rabbits, commonly selected as animal models for trachea transplants, have lower tracheal compliance at low transmural pressure; thus, success in a rabbit model may not translate well into clinical settings [55].

Additional assessment of the radial geometry of constructs can be measured via cyclical radial compression. This allows for assessment of the construct's ability to recoil to its original shape and dimensions under constant loading and unloading over a range of time. An implanted construct, which cannot retain the ability to return to its original shape, would not only result in eventual collapse, but this would also have important implications for the mechanical properties of the neighbouring healthy cartilage rings. Cartilage is mechanosensitive and deteriorates under the absence of regular mechanical loading [19]. Thus, matching mechanical properties is vital; however, it is improbable to expect a scaffold to not degrade over time. The long-term success of a tracheal construct will rely on matching the degradation rate of the construct to the ability of cells to generate sufficient extracellular matrix to maintain the scaffold's initial mechanical properties [56].

### 7.4.2 Extension and Bending

Longitudinal extension and bending of the trachea are permitted by the annular ligaments, which connect the cartilage rings of the trachea together; this allows for movement of the neck, contracting of the diaphragm and inflation of the lungs. While mechanical evaluation of the individual components of the trachea can provide insight into construct design requirements, due to the anisotropic properties of the trachea because of the unique properties of each component, evaluation of the whole trachea can produce different values. For example, the tensile modulus of porcine tracheal cartilage bands and annular ligaments has been calculated to 1.10 MPa, which was significantly different than the tensile modulus of the cartilage rings alone [20]. Thus, it is beneficial to examine the individual and whole tissue properties of the trachea.

Although the assessment of scaffold bending properties and maximum longitudinal extensibility is required and oftentimes emphasised in the literature as the most important measurement, one overlooked feature of native tissue is the degree of normal tracheal extension. If possible, dog-bone-shaped sections of the construct would need to be cut out and secured on a mechanical tester with pneumatic grips. Grip can be increased by using high-grit sandpaper to prevent slippage. Testing in this manner would indicate the construct's extensibility properties and maximum load required for failure. A trachea can extend up to 20% in adults and 46% in neonates during normal respiration [57, 58]. Thus, fabricating a rigid tube will most likely not allow for normal breathing when implanted, as illustrated in a case study of vertically fused cartilage in four newborns with a rare congenital disease [58]. These patients have experienced recurring lower respiratory tract infections and ongoing respiratory problems as a likely consequence of the modified airflow dynamics. Accordingly, such cases with abnormal trachea development can provide insight into the implications of these defects on normal breathing, and this knowledge can be applied to the development of tracheal replacement designs. Consequently, the use of a rigid tubular construct would not be suitable and would cause substantial mechanical mismatch [57]. Additional factors to consider would be developing a prosthetic that can alter its dimension to increase its length to mitigate tension at the attachment sites and accommodate the movements of respiration [59]. Tension at the attachment sites can result in an inflammatory response and fibrotic stenosis, leading to increased airway resistance, which in turn would result in atypical mechanical loading conditions, causing graft collapse in an otherwise mechanically suitable implant [8, 16].

Furthermore, the use of a rigid tubular construct would not only generate inflammation and fibrosis due to mechanical mismatch but may also prevent stretching of the respiratory epithelium of the trachea. Extension of the trachea facilitates the regulation of the airway epithelium, and its secretory and immune functions as tracheal epithelium have been found to be mechanosensitive, responding to tensile and compressive forces [60–63]. Signalling molecules, eicosanoids, help regulate smooth muscle tone, permeability of vasculature, secretion of mucus and immune response and have been shown to be downregulated by the cyclical strain in a frequency-dependant manner [60, 64, 65]. Additionally, it has been reported that epithelial cells produce factors, which regulate the inflammatory environment of the trachea wall, and fibroblast recruitment and proliferation. Therefore, the lack or excess of mechanical loading on the respiratory epithelium may induce inflammation-mediated fibrosis and should be considered when designing tracheal constructs [66].

Mechanical assessment of the whole native trachea has generally been limited to quantitation of the compressive modulus under longitudinal compressive loads alone, with no radial assessment. Human tracheas have been found to exhibit a compressive modulus in the range of 0.16–2 MPa during compression testing [24, 25]. Examination of porcine trachea under compression testing has reported a modulus of 1.26 MPa [20]. Not only should the trachea be able to withstand longitudinal compressive forces, but it should also be capable of deforming radially, which can be assessed via cyclical lateral compression to observe its compliance. However, cyclical/fatigue testing has not been widely reported on, but studies on the flexibility of the native trachea show that it has a very low flexural modulus, in the range of 0.2–1 MPa [20, 67]. The flexural moduli can be assessed using a three-point bend testing rig on a mechanical testing system with the construct resting on two roller supports and being subjected to concentrated load at its centre. As the construct bends, it is subjected to numerous forces, such as tension, compression and shear, thus providing the reaction of the construct to realistic loading situations. Three-point bend testing can be done until construct failure or until 50% displacement is achieved. The flexural modulus of the trachea is very low, with no notable difference found between the force required to bend the trachea forwards verses left to right [67]. Due to the anisotropic mechanical properties of the whole trachea, a variety of mechanical tests would be suitable to truly understand the mechanical nature of the trachea. Studies in this field are highly variable in terms of testing methods and specimen type used, which provides significant challenges in understanding the mechanical properties of the native trachea in order to design potential replacement devices. However, the results from these studies can begin to form the foundation for the development of tracheal replacement constructs.

## 7.5 Key Considerations and Summary of Recommended Mechanical Tests

To better understand the mechanical response of tissue-engineered tracheas and reduce variability in in vitro testing, we recommend, if possible, robust mechanical assessment, including tensile testing, three-point bending, radial compression, compliance testing or cyclical loading (Fig. 7.3).

Tensile testing provides information about the extensibility of the trachea, which should be normalised to specimen cross-sectional area or, in the case of testing whole tracheas, the circumference multiplied by the wall thickness. In cases where tensile testing is not possible, the bending properties of the construct could be evaluated via three-point bend testing, which would examine the construct's bending stiffness. Compliance testing would be the most appropriate test type as it would allow for the direct evaluation of the ability of a construct to withstand varying physiological transmural pressures. During compliance testing, the sample length should be considered to avoid edge effects, and it should be noted that shorter constructs may exhibit artificially low compliance [49]. However, compliance testing may not be accessible to all; instead, radial compression is a straightforward alternative. While it may not be typical loading regime in vivo, it does offer quantitative

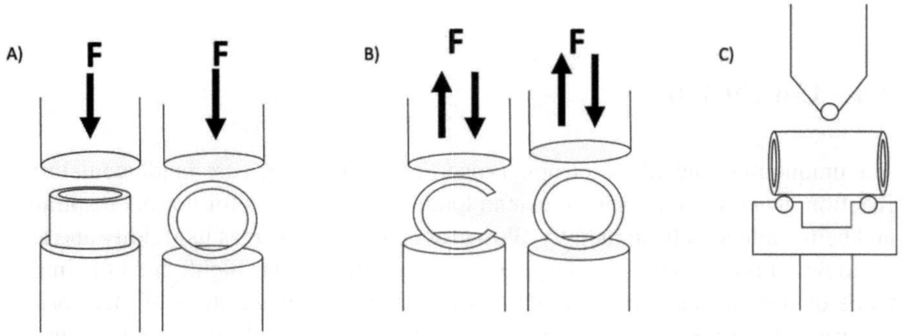


**Fig. 7.3** Schematic of some recommended mechanical testing regimes for tubular constructs. (**a**) Longitudinal and radial compression. (**b**) Radial cyclical loading. (**c**) Three-point bending

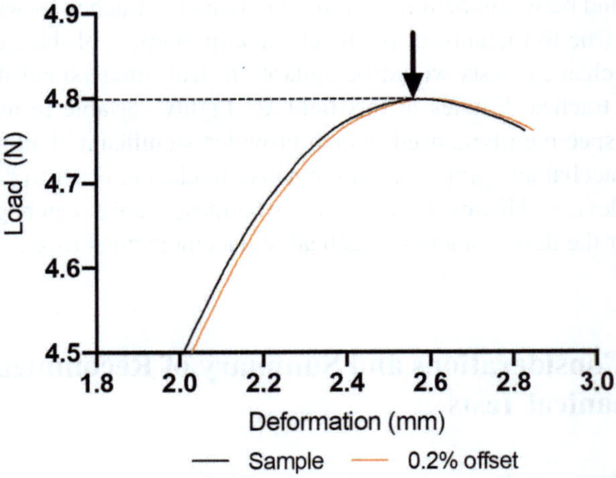


**Fig. 7.4** The offset method of yield point calculation. A typical load/deformation curve of radially compressed specimen (black) and the 0.2% offset curve (orange) used to measure the yield load (arrow), identified by the intersection point between the two curves

information of the tubular construct's resistance to collapse. The yield point of the construct can be calculated, which is the point at which the construct begins to deform plastically and would inform of the construct's resilience against native tissue; beyond this point, deformation to the construct will be permanent. This point can be calculated using the 0.2% offset method. The load-deformation curve from a sample is offset by 0.2% (Fig. 7.4; orange line), and the point of intersection between the original curve and offset curve is used as the yield point.

Again, displacement values from radial compressive testing should be normalised against the tube diameter. Additionally, cyclical testing is also a useful evaluation for assessing construct recoil and compliance. The testing regime should examine the construct's ability to return to its original dimensions without collapse or buckling at normal and elevated respiratory rates for more physiological relevant examination.

## 7.6 Conclusion

The unique mechanical anisotropic properties of the trachea are fundamental to its function, wherein being able to extend longitudinally allows for normal respiration and being able to deform radially allows for coughing, whereas its rigidity and ability to recoil both maintain airway patency. This chapter has highlighted the importance of the trachea's mechanics to its function and for the need of clear design criteria and metrics across all studies both of native tissue and engineered constructs.

## References


1. Mead J, Turner JM, Macklem PT, Little JB. Significance of the relationship between lung recoil and maximum expiratory flow. J Appl Physiol. 1967;22(1):95–108.
2. Mead J. Physical Properties of Human Lungs Measured During Spontaneous Respiration. J Appl Physiol. 1953;5:779–796.
3. Macklem PT, Mead J. Factors determining maximum expiratory flow in dogs. J Appl Physiol. 1968;25(2):159–69.
4. Wittenborg MH, Gyepes MT, Crocker D. Tracheal dynamics in infants with respiratory distress, stridor, and collapsing trachea. Radiology. 1967;88(4):653–62.
5. Rhoades RA, Bell DR. Medical Physiology Principles for Clinical Medicine 4th edition [Internet]. 5th ed. LWW; 2012. 819 p. Available from: http://books.google.com/books?id=Htc9AwAAQBAJ&pgis=1
6. Smyth RJ, Chapman KR, Rebuck AS. Maximal inspiratory and expiratory pressures in adolescents. Normal values. Chest [Internet]. 1984;86(4):568–72. Available from: https://doi.org/10.1378/chest.86.4.568
7. Di Rienzo S. Bronchial dynamism. Radiology. 1949;53(2):168–86.
8. Wong VW, Paterno J, Sorkin M, Glotzbach JP, Levi K, Januszyk M, et al. Mechanical force prolongs acute inflammation via T-cell-dependent pathways during scar formation. FASEB J. 2011;25(12):4498–510.
9. Huang Z, Peng WEI, Luoman GAN, Wentao LI, Zeng T, Caicheng QIN, et al. Protective effects of different anti-inflammatory drugs on tracheal stenosis following injury and potential mechanisms. Mol Med Rep. 2021;23(5):1–11.
10. Boazak EM, Auguste DT. Trachea Mechanics for Tissue Engineering Design. ACS Biomater Sci Eng. 2018;4(4):1272–84.
11. Arthur C Guyton JEH. Textbook of Medical Physiology,. 11th ed. Elsevier Saunders: Philadelphia, USA. 2006. 471–482 p.
12. McArdle WD, Katch FI, Katch VL. Essentials of Exercise Physiology. Lippincott Williams & Wilkins; 2006. 679 p.
13. Cook CD, SUTHERLAND JM, SEGAL S, CHERRY RB, MEAD J, MCILROY MB, et al. Studies of respiratory physiology in the newborn infant. III. Measurements of mechanics of respiration. J Clin Invest. 1957;36(3):440–8.
14. Mansfield EG, Jr VKG, Auguste DT, Greene VK, Auguste DT. Patterned, tubular scaffolds mimic longitudinal and radial mechanics of the neonatal trachea. Acta Biomater [Internet]. 2016;33:176–82. Available from: https://doi.org/10.1016/j.actbio.2016.01.034
15. Gilpin SE, Charest JM, Ren X, Tapias LF, Wu T, Evangelista-Leite D, et al. Regenerative potential of human airway stem cells in lung epithelial engineering. Biomaterials [Internet]. 2016;108:111–9. Available from: https://doi.org/10.1016/j.biomaterials.2016.08.055
16. Kawaguchi S, Nakamura T, Shimizu Y, Masuda T, Takigawa T, Liu Y, et al. Mechanical properties of artificial tracheas composed of a mesh cylinder and a spiral stent. Biomaterials. 2001;22(23):3085–90.
17. Kizirian A. The Respiratory System [Internet]. 2021 [cited 2022 Jan 5]. Available from: https://antranik.org/the-respiratory-system/
18. Safshekan F, Tafazzoli-Shadpour M, Abdouss M, Shadmehr B. M. Mechanical Characterization and Constitutive Modeling of Human Trachea: Age and Gender Dependency. Materials (Basel). 2016;9(6).
19. Rains JK, Bert JL, Roberts CR, Pare PD. Mechanical properties of human tracheal cartilage. J Appl Physiol. 1992;72(1):219–25.
20. Hoffman B, Martin M, Brown BN, Bonassar LJ, Cheetham J. Biomechanical and biochemical characterization of porcine tracheal cartilage. Laryngoscope. 2016;126(10):E325–31.
21. Kojima K, Vacanti CA. Tissue engineering in the trachea. Anat Rec. 2014;297(1):44–50.

22. Teng Z, Trabelsi O, Ochoa I, He J, Gillard JH, Doblare M. Anisotropic material behaviours of soft tissues in human trachea: An experimental study. J Biomech [Internet]. 2012;45(9):1717–23. Available from: https://doi.org/10.1016/j.jbiomech.2012.04.002
23. Trabelsi O, del Palomar AP, López-villalobos JL, Ginel A, Doblaré M. Experimental characterization and constitutive modeling of the mechanical behavior of the human trachea. Med Eng Phys. 2010;32(1):76–82.
24. Safshekan F. Investigation of the Mechanical Properties of the Human Tracheal Cartilage. 2017;16(2):107–14.
25. Wang J-Y, Mesquida P, Pallai P, Corrigan CJ, Lee TH. Dynamic Properties of Human Bronchial Airway Tissues. 2011;(November). Available from: http://arxiv.org/abs/1111.5645
26. Sun F, Jiang Y, Xu Y, Shi H, Zhang S, Liu X, et al. Genipin cross-linked decellularized tracheal tubular matrix for tracheal tissue engineering applications. Sci Rep. 2016 Apr 15;6.
27. Park HS, Lee JS, Jung H, Kim DY, Kim SW, Sultan MT, et al. An omentum-cultured 3D-printed artificial trachea: in vivo bioreactor. Artif Cells, Nanomedicine Biotechnol [Internet]. 2018;46(sup3):S1131–40. Available from: https://doi.org/10.1080/21691401.2018.1533844
28. Wachsmuth L, Söder S, Fan Z, Finger F, Aigner T. Immunolocalization of matrix proteins in different human cartilage subtypes. Histol Histopathol. 2006;21(4–6):477–85.
29. Kelly DJ, Crawford A, Dickinson SC, Sims TJ, Mundy J, Hollander AP, et al. Biochemical markers of the mechanical quality of engineered hyaline cartilage. J Mater Sci Mater Med. 2007;18(2):273–81.
30. Knudson CB, Knudson W. Cartilage proteoglycans. Semin Cell Dev Biol. 2001;12(2):69–78.
31. Hinderer S, Schesny M, Bayrak A, Ibold B, Hampel M, Walles T, et al. Engineering of fibrillar decorin matrices for a tissue-engineered trachea. Biomaterials [Internet]. 2012;33(21):5259–66. Available from: https://doi.org/10.1016/j.biomaterials.2012.03.075
32. Bianco P, Fisher LW, Young MF, Termine JD, Gehron Robey P. Expression and localization of the two small proteoglycans biglycan and decorin in developing human skeletal and non-skeletal tissues. J. J Histochem Cytochem. 1990;38(11):1549–63.
33. Plaas AHK, Neame PJ, Nivens CM, Reiss L. Identification of the keratan sulfate attachment sites on bovine fibromodulin. J Biol Chem. 1990;265(33):20634–40.
34. Melching LI, Roughley PJ. Modulation of keratan sulfate synthesis on lumican by the action of cytokines on human articular chondrocytes. Matrix Biol. 1999;18(4):381–90.
35. Johnson HJ, Rosenberg L, Choi HU, Garza S, Höök M, Neame PJ. Characterization of epiphycan, a small proteoglycan with a leucine-rich repeat core protein. J Biol Chem. 1997;272(30):18709–17.
36. SundarRaj N, Fite D, Ledbetter S, Chakravarti S, Hassell JR. Perlecan is a component of cartilage matrix and promotes chondrocyte attachment. J Cell Sci. 1995;108(7):2663–72.
37. Mow VC, Kuei SC, Lai WM, Armstrong CG. Biphasic creep and stress relaxation of articular cartilage in compression: Theory and experiments. J Biomech Eng. 1980;102(1):73–84.
38. Schulz RM, Bader A. Cartilage tissue engineering and bioreactor systems for the cultivation and stimulation of chondrocytes. Eur Biophys J. 2007;36(4–5):539–68.
39. Teng Z, Ochoa I, Li Z, Lin Y, Rodriguez JF, Bea JA, et al. Nonlinear mechanical property of tracheal cartilage: A theoretical and experimental study. J Biomech. 2008;41(9):1995–2002.
40. Lambert RK, Baile EM, Moreno R, Bert J, Pare PD. A method for estimating the Young's modulus of complete tracheal cartilage rings. J Appl Physiol. 1991;70(3):1152–1159.
41. Roberts CR, Rains JK, Paré PD, Walker DC, Wiggs B, Bert JL. Ultrastructure and tensile properties of human tracheal cartilage. J Biomech. 1997;31(1):81–6.
42. Roberts CR, Pare PD. Composition changes in human tracheal cartilage in growth and aging, including changes in proteoglycan structure. Am J Physiol – Lung Cell Mol Physiol. 1991;261(2 5-1).
43. Bartlett D, Jeffery P, Sant'ambrogio G, Wise JCM. Location of stretch receptors in the trachea and bronchi of the dog. J Geophys Res Ocean. 1975;258:409–20.
44. Sarma PA, Pidaparti RM, Meiss RA. Anisotropic properties of tracheal smooth muscle tissue. J Biomed Mater Res – Part A. 2003;65(1):1–8.

45. Amiel D, Frank C, Harwood F, Fronek J, Akeson W. Tendons and Ligaments: A Morphological and Biochemical Comparison A historical review of the evolution of the use of tendons as ligament substitutes reveals that these. J Orthop Res. 1984;1:251–65.
46. Trabelsi O, Pérez del Palomar A, Mena Tobar A, López-Villalobos JL, Ginel A, Doblaré M. FE simulation of human trachea swallowing movement before and after the implantation of an endoprothesis. Appl Math Model [Internet]. 2011;35(10):4902–12. Available from: https://doi.org/10.1016/j.apm.2011.03.041
47. Malvè M, Del Palomar AP, López-Villalobos JL, Ginel A, Doblaré M. FSI analysis of the coughing mechanism in a human trachea. Ann Biomed Eng. 2010;38(4):1556–65.
48. Sun F, Usón J, Ezquerra J, Crisóstomo V, Luis L, Maynar M. Endotracheal stenting therapy in dogs with tracheal collapse. Vet J. 2008;175(2):186–93.
49. Malvè M, Del Palomar AP, Chandra S, López-Villalobos JL, Finol EA, Ginel A, et al. FSI analysis of a human trachea before and after prosthesis implantation. J Biomech Eng. 2011;133(7).
50. Kim WD. Lung mucus: A clinician's view. Eur Respir J. 1997;10(8):1914–7.
51. Stern EJ, Graham CM, Richard Webb W, Gamsu G, Francisco S. Normal Trachea during Forced Expiration: Dynamic CT Measurements. Radiology. 1993;187:27–31.
52. Olsen CR, Stevens AE, McIlroy MB. Rigidity of tracheae and bronchi during muscular constriction. J Appl Physiol. 1967;23(1):27–34.
53. Croteau JR, COOK CD. Volume-pressure and length-tension measurements in human tracheal and bronchial segments. J Appl Physiol. 1961;16:170–2.
54. Kozlovsky P, Zaretsky U, Jaffa AJ, Elad D. General tube law for collapsible thin and thick-wall tubes. J Biomech [Internet]. 2014;47(10):2378–84. Available from: https://doi.org/10.1016/j.jbiomech.2014.04.033
55. Ten Hallers EJO, Rakhorst G, Marres HAM, Jansen JA, Van Kooten TG, Schutte HK, et al. Animal models for tracheal research. Biomaterials. 2004;25(9):1533–43.
56. Shelton JC, Bader DL, Lee DA. Mechanical conditioning influences the metabolic response of cell-seeded constructs. Cells Tissues Organs. 2003;175(3):140–50.
57. Macklin CC. The Musculature of the Bronchi and Lungs: A Retrospect. Can Med Assoc journal,. 1929;20(19):1035.
58. Inglis AF, Kokesh J, Siebert J, Richardson MA. Vertically Fused Tracheal Cartilage. Arch Otolaryngol – Head Neck Surg [Internet]. 1992;118:436–8. Available from: http://archotol.jamanetwork.com/
59. Grillo HC. Slide tracheoplasty for long-segment congenital tracheal stenosis. Ann Thorac Surg [Internet]. 1994;58(3):613–20. Available from: https://doi.org/10.1016/0003-4975(94)90714-5
60. Savla U, Sporn PHS, Waters CM. Cyclic stretch of airway epithelium inhibits prostanoid synthesis. Am J Physiol. 1997;273(5 PART 1):1013–9.
61. Ressler B, Lee RT, Randell SH, Drazen JM, Kamm RD. Molecular responses of rat tracheal epithelial cells to transmembrane pressure. Am J Physiol – Lung Cell Mol Physiol. 2000;278(6 22-6):1264–72.
62. Felix JA, Woodruff ML, Dirksen ER. Stretch Increases Inositol 1,4,5-trisphosphate Concentration in Airway Epithelial Cells. Am J Respir Cell Mol Biol. 1996;14(3):296–301.
63. Gao Y, Vanhoutte PM. Responsiveness of the guinea pig trachea to stretch: Role of the epithelium and cyclooxygenase products. J Appl Physiol. 1993;75(5):2112–6.
64. Adler KB, Fischer BM, Wright DT, Cohn LA, Becker S. Interactions between respiratory epithelial cells and cytokines: Relationships to lung inflammation. Ann N Y Acad Sci. 1994;725:128–45.
65. Savla U, Appel HJ, Sporn PHS, Waters CM. Prostaglandin E2 regulates wound closure in airway epithelium. Am J Physiol – Lung Cell Mol Physiol. 2001;280(3 24-3):421–31.
66. Tschumperlin DJ, Drazen JM. Mechanical stimuli to airway remodeling. Am J Respir Crit Care Med. 2001;164(10 Pt 2).
67. Shi HC, Deng WJ, Pei C, Lu D, Zhang XJ, Wang XH, et al. Biomechanical properties of adult-excised porcine trachea for tracheal xenotransplantation. Xenotransplantation. 2009;16(3):181–6.

# Part III
# Engineering and Modeling the Mesenchyme and Parenchyma

# Chapter 8
# Engineering and Modeling the Lung Mesenchyme

**Melinda E. Snitow, Fatima N. Chaudhry, and Jarod A. Zepp**

## 8.1 Introduction

The structure of the mammalian lung controls the flow of air through the airways and into the distal alveolar region where gas exchange occurs. Specialized cells in the lung mesenchyme produce the extracellular matrix (ECM) and growth factors required for lung structure. Historically, characterizing the mesenchymal cell subtypes was challenging due to their ambiguous morphology, overlapping expression of protein markers, and limited cell-surface molecules needed for isolation. The recent development of single-cell RNA sequencing (scRNA-seq) complemented with genetic mouse models demonstrated that the lung mesenchyme comprises transcriptionally and functionally heterogeneous cell-types. Bioengineering approaches that model tissue structure clarify the function and regulation of mesenchymal cell types. These experimental approaches demonstrate the unique abilities of fibroblasts in mechanosignaling, mechanical force generation, ECM production, and tissue regeneration. This chapter will review the cell biology of the lung mesenchyme and experimental approaches to study their function.

M. E. Snitow · F. N. Chaudhry
Division of Pulmonary and Sleep Medicine, Children's Hospital of Philadelphia, Philadelphia, PA, USA

J. A. Zepp (✉)
Division of Pulmonary and Sleep Medicine, Children's Hospital of Philadelphia, Philadelphia, PA, USA

Department of Pediatrics, University of Pennsylvania, Philadelphia, PA, USA
e-mail: zeppj@chop.edu



C. M. Magin (ed.), *Engineering Translational Models of Lung Homeostasis and Disease*, Advances in Experimental Medicine and Biology 1413,
https://doi.org/10.1007/978-3-031-26625-6_8

## 8.2 Advancing the Discovery of Fibroblast Heterogeneity

The lung mesenchyme includes fibroblasts and smooth muscle that are derived from the mesoderm during lung development. During lung development, the mesoderm provides cues, such as fibroblast growth factor 10 (FGF10) and ECM components (fibronectin and tenascin C), that direct cell differentiation, tissue patterning, and lung branching morphogenesis [1–9]. In the mature lung, varied responses in the mesenchyme are required for proper repair after injury. Fibroblasts and smooth muscle were recently identified as sources of growth factors required for epithelial regeneration. During injury repair, fibroblasts regulate the ECM by producing collagen, elastin, and matrix metalloproteinases (MMPs), promoting structural viability and inflammatory cell infiltration. While these attributes of fibroblasts are normally associated with tissue repair processes or wound healing, they may also contribute to a variety of diseases. Pulmonary fibrosis (PF) is a progressive disease characterized by pathogenic scar formation in the alveoli that impairs gas exchange [10–12]. There is no cure for PF, and current therapy slows disease progression and improves lung function but does not reverse the pathologic scarring of the lung [13]. Fibroblasts' contribution to fibrosis is the focus of many bioengineering assays [14].

Single-cell RNA sequencing (scRNA-seq) in complex tissues has demonstrated the vast heterogeneity of transcriptionally distinct cell types or cell states. These studies uncovered distinct fibroblast subsets that differentially express transcription factors, growth factors, and ECM components [15–21]. While gene expression profiles indicate that many distinct fibroblast subsets reside in lung tissue, further experiments are crucial to validate these subsets as functionally distinct populations. In the following sections, we will introduce a current catalog of lung fibroblast lineages and highlight bioengineering approaches to study and target distinct fibroblast subsets.

## 8.3 The Organization and Heterogeneity of Lung Fibroblasts

The lung is structurally complex, consisting of proximal airways and blood vessels (arteries and veins) that continuously branch and decrease in diameter until they terminate at the distal alveoli. Along this proximal-distal path, the lung's cellular composition also changes. A large body of research has thoroughly described the spatial heterogeneity of the lung epithelium and the functional roles of epithelial cell types and resident stem/progenitor cells (thoroughly reviewed elsewhere [8, 22, 23]). Based on recent observations, the lung mesenchyme is similarly arranged in a proximal-distal axis that parallels the airway and alveolar epithelium [15, 16, 24]. The conducting airways and large vessels are wrapped in airway smooth muscle (ASM) and vascular smooth muscle (VSM), respectively, and are readily distinguishable using common histological techniques. A heterogeneous milieu of fibroblasts reside between the smooth muscle, in the adventitial area surrounding proximal vessels, and extend out to the distal alveolar compartment (Fig. 8.1). The following section will cover the types of fibroblasts and their respective functions.

### 8.3.1 Platelet-Derived Growth Factor Receptor Alpha (PDGFRα)-Expressing Alveolar Fibroblasts 1 and 2

In 1970, O'Hare and Sheridan used electron microscopy to characterize interstitial fibroblasts in the alveoli of rat lungs [25]. Follow-up studies by Brody and Kaplan found that alveolar interstitial fibroblasts consist of two types of cells, fibroblasts with or without large lipid deposits [26, 27]. A lipid-laden and proteinaceous surfactant coats the surface of alveoli and is produced by lung alveolar resident epithelial type 2 (AT2) cells. Based on these studies and their association with the pulmonary surfactant-producing AT2 cells, these fibroblasts have been called lipofibroblasts (LF). More recently, subsequent studies using genetic knock-in mice have revealed that most alveolar fibroblasts are Pdgfra positive.

Transgenic and knock-in mouse strains have enabled researchers to identify, purify, and functionally characterize fibroblast subsets. In 2013, a seminal study showed Pdgfra+ alveolar fibroblasts reside near AT2 cells [28]. In addition to producing pulmonary surfactant, AT2 cells are progenitor cells that proliferate and repair damaged alveolar epithelium [28]. Flow cytometry-based

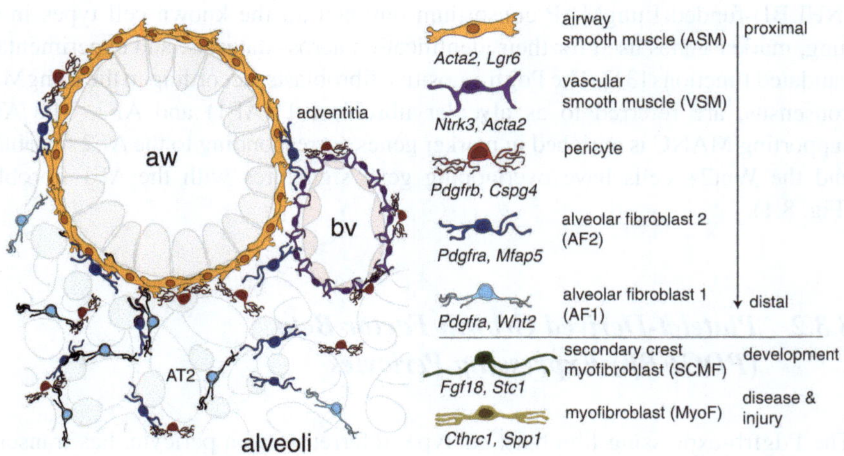


**Fig. 8.1** Lung mesenchymal cell composition is patterned in a proximal-to-distal manner. Large proximal airways and blood vessels are wrapped in distinct populations of airway or vascular smooth muscle. Fibroblast lineages, such as pericytes and alveolar fibroblast 2, reside in adventitial space around airways and vessels and also populate the distal alveoli along with alveolar fibroblast 1. Secondary crest myofibroblasts are restricted to transient stages of lung development, and myofibroblasts appear in disease and injury states. aw, airway; bv, blood vessel; AT2, alveolar epithelium type 2

fluorescence-activated cell sorting (FACS) was used to purify the AT2 cells and Pdgfra+ fibroblasts expressing a GFP reporter inserted into the endogenous Pdgfra gene. When cocultured in an ECM-rich gel (commercially known as Matrigel), the AT2 cells will form organoids (discussed in more detail in later section) but grow best in the presence of Pdgfra-positive fibroblasts [28]. These data show the utility of genetic mouse models to isolate and study lung fibroblasts. Furthermore, this study demonstrated that lung alveolar fibroblasts can serve as a niche cell by employing the ex vivo organoid model system.

Subsequent studies using scRNA-seq revealed that Pdgfra-positive fibroblasts consist of Wnt-responsive and Wnt2-expressing cell types. These fibroblasts were isolated for functional tests using either a Wnt-responsive reporter (*Axin2* promoter driven) or Wnt2 reporter [15]. In the alveolar organoid model, Wnt-responsive (Axin2+) Pdgfra+ fibroblasts supported the growth of the AT2 organoids and, therefore, were named mesenchymal alveolar niche cell (MANC). This MANC was uniquely enriched in growth factors and cytokines, such as Fgf7, IL-6, Fstl1, and Grem2, that were sufficient to enhance organoid growth [15]. Importantly, conditional knockouts of *Bmpr1a* in AT2 cells or of Wnt ligand secretion in Pdgfra-positive cells lead to impaired alveolar epithelial repair in vivo [29, 30]. Taken together, these studies indicate that alveolar Pdgfra-expressing fibroblasts consist of two distinct subsets with unique abilities and proximity to alveolar epithelial progenitors.

Though there are several names used to denote these fibroblasts and their unique abilities (alveolar niche cell, AT2 associated cells, adventitial/alveolar fibroblasts), a recent consensus published from the National Heart, Lung, and Blood Institute (NHLBI)-funded LungMAP consortium outlined all the known cell types in the lung, marker genes used for their identification across datasets, and experimentally validated functions [31]. The Pdgfra-positive fibroblasts, according to the LungMAP consensus, are referred to as alveolar fibroblast 1 (AF1) and AF2. The AT2-supporting MANC is enriched in marker genes corresponding to the AF2 fibroblast, and the Wnt2+ cells have overlapping gene signatures with the AF1 fibroblast (Fig. 8.1).

### 8.3.2 Platelet-Derived Growth Factor Beta (PDGFRβ)-Expressing Pericytes

The Pdgfrb-expressing fibroblast subtype, referred to as a pericyte, has transcriptional and spatial similarities to VSM. The general definition of the pericyte is a cell that is embedded within the vascular basement membrane [32, 33]. Much of what is understood about pericyte biology and function has been derived from developmental studies and injury models in different organs [34]. A canonical marker for pericytes is Pdgfrb, and its ligand, Pdgfb, is primarily expressed in endothelial cells [35]. Embryonic developments in *Pdgfrb* or *Pdgfb* knockout mice exhibit reduced mesenchymal proliferation and impaired VSM differentiation, resulting in large vessel dilation [36]. These studies suggest that VSM and pericytes function to control the tone and constriction of vessels and capillaries. However, the function of lung pericytes remains poorly characterized.

A major focus of the research conducted on lung pericytes has been on their capacity to differentiate into myofibroblasts in mouse models of lung fibrosis. It was shown that fibroblasts expressing *Foxd1* (a marker for pericytes as well as other fibroblasts) can differentiate into damage-induced myofibroblasts in vivo [37]. Recently, it was shown that Pdgfrb-expressing pericytes generate a significant amount of myofibroblasts [38]. Similar findings were generated previously using lineage tracing of Axin2-positive Wnt-responsive cells [15]. However, conflicting results were obtained using a restrictive pericyte marker, *Cspg4* (NG2) [39]. This study showed that very few of these cells differentiate into myofibroblasts. Although these data suggest a possible role for pericytes in generating myofibroblasts in injury models, future studies and more specific lineage-tracing tools will be needed to clarify contradictory results.

### *8.3.3 Airway and Vascular Smooth Muscle (ASM and VSM)*

The large airways and vessels of the lung are wrapped in contractile smooth muscle. Consistent with their contractile properties, the ASM and VSM can constrain the luminal size of the airways and vessels and thereby regulate air and blood flow in and out of the lungs. Recent work demonstrated that ASM is also an important source of signals for the proper repair of the airway epithelium [16, 40, 41]. The airway epithelium harbors progenitors, basal cells and club cells, that proliferate in response to injury and regenerate lost basal, club, and ciliated cell types. Mouse reporter lines demonstrated that Wnt signaling is critical for epithelial repair. After injury, Wnt signaling in ASM promotes the production of growth factors such as Fgf10, which signals to the adjacent epithelial progenitors [40]. This was further demonstrated by targeted ablation of Lgr6+ ASM, which impaired epithelial repair after chemical injury [16].

## 8.4 Other Fibroblast Subtypes

### *8.4.1 Developmental Secondary Crest Myofibroblasts (SCMF)*

The late embryonic stages of lung development prior to birth and proceeding into early postnatal life are critical for the transition to breathing air. During these stages, the alveoli form and mature, specifying and differentiating alveolar cell types, including fibroblasts. Alveolar development includes formation of alveolar septal ridges and juxtaposition of capillary endothelium with thin AT1 epithelium. The formation of septal ridges is driven by poorly understood processes in the endothelium and mesenchyme. In the mesenchyme, alveolar secondary crest myofibroblasts (SCMF) arise in response to epithelial-produced Shh and Pdgf ligands [17, 42–44]. The SCMFs have the unique capacity among other alveolar fibroblasts to exert force, demonstrating their active role in the sculpting of alveolar septa. Traction force microscopy of purified SCMFs measured greater traction force exertion than by non-SCMF Pdgfra+ fibroblasts [17]. Moreover, deletion of the myosin light chain kinase in Pdgfra-expressing fibroblasts, including SCMFs, reduced alveolar septa formation, thus demonstrating their contractile function in vivo [45]. Unlike other fibroblast subsets, the SCMF is a transient lineage with peak abundance in the first 7–10 days of postnatal life, followed by significant decline via apoptosis [17, 46].

### *8.4.2 Fibrotic Disease-Associated Myofibroblasts (MyoF)*

Fibroblasts have well-documented activities in the wound-healing response. Upon tissue damage, resident fibroblasts proliferate and produce matrix-degrading enzymes, including MMP and ADAMTS family members. ECM degradation facilitates the influx of inflammatory cells. Fibroblasts are also an important source of ECM, including collagens. Excessive ECM production and apoptosis of the activated fibroblasts results in the formation of a scar. In the lung, excessive ECM production and stiffening of the alveolar region results in deleterious fibrosis, which impairs lung function. The fibroblasts associated with fibrotic lung disease are termed myofibroblasts. Historically, the myofibroblasts are identified by their expression of alpha-smooth muscle actin (aSMA) using standard histological techniques. The myofibroblasts are thought to also be the main producers of excessive collagen. However, new mouse models and single-cell RNA sequencing have indicated that multiple fibroblast populations may contribute to fibrosis and the scar-like lesions that can persist in the distal portion of the lung.

## 8.5 Bioengineering Approaches to Characterize Complex Fibroblast Behaviors

Modeling fibroblast behaviors ex vivo can isolate variables that are difficult to manipulate in vivo. Ex vivo modeling allows real-time access to developmental processes and disease progression and experimentation on human tissue that cannot be performed in vivo. Multiple ex vivo models of lung development and disease enable characterization of complex fibroblast behaviors (Fig. 8.2).

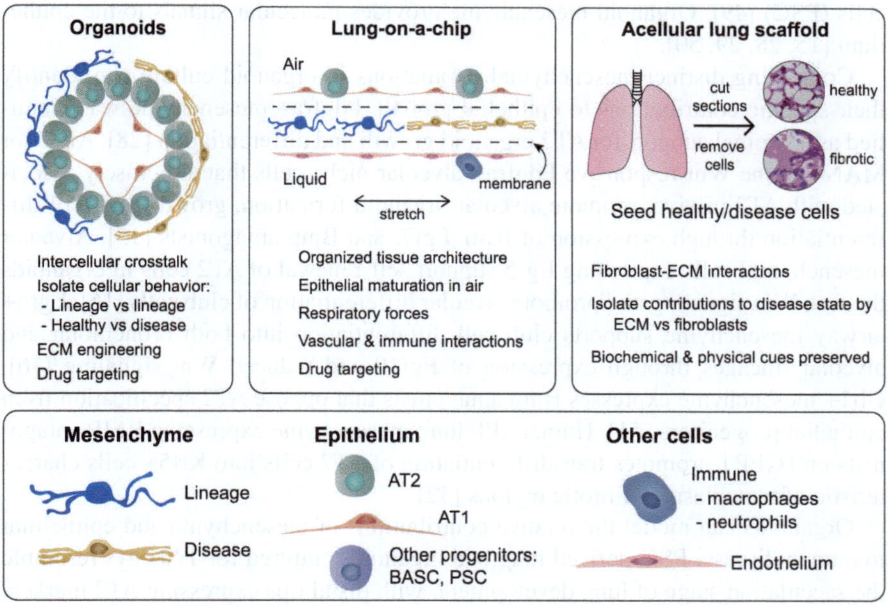


**Fig. 8.2** Ex vivo modeling of the lung mesenchyme. Organoids self-assemble from AT2 cells or other progenitors into AT2 and AT1 epithelial cells with mesenchymal support. Support of alveolar epithelial growth and differentiation in organoids can be tested across mesenchymal lineages or disease states, ECM components, and pharmacological intervention. Lung-on-a-chip seeds epithelial and endothelial cells on either side of a stretchable porous membrane with an air-liquid interface to model physiological environmental forces. Fibroblasts from different lineages or disease states can be added, as can immune cells, for experiments on barrier function, cell metabolism, immune responses, and drug screening. Acellular lung scaffolds are the biophysical architecture remaining after removing cells from lung tissue. The scaffolds can come from healthy or fibrotic lungs and can be reseeded with cells from specific lineages or disease states to test cell-ECM interactions

## 8.5.1 *Organoids to Model Mesenchymal-Epithelial Interactions*

Lung organoids are self-assembling 3D structures of the lung epithelium, usually with support from mesenchymal cells and embedded in a matrix, such as Matrigel. Organoids model aspects of lung biology, such as intercellular interactions, effects of biochemical and biophysical cues, and etiology of lung diseases. Therapeutic targets can be identified and tested in organoids, which allow ex vivo modeling of human alveoli unachievable in 2D culture.

Organoids model two key steps of lung epithelial repair, progenitor proliferation and differentiation. Alveolar organoids are characterized by AT2 proliferation and differentiation into AT1 cells. AT2 cells can be isolated from adult lungs or can be induced from bronchioalveolar stem cells (BASCs) and HopX+ AT1 cells [47]. AT2 cells can also be differentiated from fetal tissue progenitors [48] or pluripotent stem cells (PSC) [49]. Organoid mesenchyme provides molecular signals to the epithelium [15, 28, 29, 50].

Comparing distinct mesenchymal populations in organoid culture can identify their specific contributions to epithelial growth. Pdgfra+ mesenchyme was identified as a stromal support for AT2 organoid growth and differentiation [28]. AF2s (or MANCs), the Wnt-responsive Pdgfra+ alveolar niche cells that are closely associated with AT2 in vivo, promote alveolar organoid formation, growth, and AT1 differentiation through expression of IL-6, Fgf7, and Bmp antagonists [15]. Alveolar mesenchymal cells expressing Lgr5 support self-renewal of AT2 cells in organoids through Wnt signaling and promote alveolar differentiation of club cells [16]. Lgr6+ airway mesenchyme supports club cell differentiation into both bronchiolar and alveolar lineages through expression of Fgf10 and reduced Wnt signaling [16]. Gli1+ mesenchyme expresses Bmp antagonists that oppose AT2 specification from epithelial progenitors [51]. Human IPF lung mesenchyme expressing BMP antagonists or TGFB1 promotes transdifferentiation of AT2 cells into Krt5+ cells characteristic of metaplasia in fibrotic regions [52].

Organoids can model the relative contributions of mesenchyme and epithelium to human disease. PSC-derived lung bud organoids cultured for 170 days resemble the sacculation stage of lung development, with distal tips expressing AT2 markers and associating with mesenchyme. Modeling Hermansky-Pudlak syndrome (HPS) by deleting the *HPS1* gene demonstrated that profibrotic changes can begin during lung development. Organoid modeling revealed that mutant epithelium can transform wild-type mesenchyme through expression of IL-11, a novel putative target [53, 54].

Lung fibroblasts sense and exert mechanical forces that influence lung development, repair, and fibrosis. Cross talk between fibroblast contractility and lung epithelium can be modeled with organoids. Fibroblast contractility and the mechanoresponsive transcription factor YAP are required for distal epithelial tubes to grow out of bronchiolar organoids [55]. The lung epithelium also influences fibroblast behavior, decreasing contractility and attenuating profibrotic proliferation and collagen deposition [56]. Forces can also be exerted on lung organoids through the 3D matrix. Respiratory forces that influence alveolar development are modeled by cyclic stretch, increasing expression of the myofibroblast marker aSMA in lung fibroblasts isolated during alveologenesis [57].

Engineered 3D matrices can isolate biochemical and biophysical cues, pattern tissue shape, and allow growth. Synthetic hydrogels modified with ECM components decouple physical properties, such as stiffness and viscoelasticity, from biochemical cues, such as growth factors and adhesion molecule binding sites. Microfabrication into specific geometries or patterns can guide tissue organization. Organoid growth over time can be accommodated with biodegradable elements, such as reversible cross-links or MMP recognition sites, or with a pore size that allows tissue invasion and morphogenesis [58, 59].

### 8.5.2 Lung-on-a-Chip to Model Human Lung Architecture and Environmental Forces

Lung-on-a-chip technology aims to model human lung biology ex vivo using an engineered microfluidic device with human lung epithelium and endothelium on either side of porous membrane. The epithelium is exposed to air, and liquid culture medium continuously flows through the endothelial "vessel," aiming to mimic physiological tissue architecture and environmental conditions. Microfluidics allow addition of pharmacological agents for drug screening, pathogens and immune cells for disease modeling, and metabolic assessment of spent media. Chips are optically clear for live imaging and made of a flexible polymer to allow physiological stretch provided by lateral vacuum channels [60]. Lung-on-a-chip can mimic the mechanical forces of breathing, leading to findings that chip models of edema, immune responses to infection, and nebulized nanoparticle infiltration into the vasculature are sensitive to cyclic strain [60–62]. A limitation of these studies is the use of cell lines that may not recapitulate in vivo physiology, but subsequent studies have used primary cells [62–64]. Recent efforts to improve alveolar chips incorporate 3D architecture resembling the geometry of the alveolus to better model 3D physiological stretch and nonlinear airflow [63, 65].

Most lung-on-a-chip models do not incorporate mesenchyme, and alveolar chip models still lack incorporation of alveolar fibroblasts and their influence in alveolar biology. Fibroblasts and smooth muscle have been added in airway-on-a-chip models as an intermediate layer between epithelial and endothelial layers [64, 66]. Another model incorporated fibroblasts in the endothelial channel, with IPF or normal lung fibroblasts seeded in additional channels to compare their effects on the airway epithelium [67]. Mesenchymal populations can be compared for their effects on the epithelium, endothelium, and immune response [68].

### 8.5.3 Acellular Tissue Scaffolds to Model Fibroblast and ECM Interactions

Acellular lung tissue scaffolds are created by removing cells from a whole organ, resected tissue, or precision-cut sections up to 500um thick. Tissue scaffolds preserve the ECM and tissue architecture that provide biochemical and biophysical cues to cells. They provide a 3D culture system for modeling interactions between ECM and fibroblast behaviors that avoid artifacts of 2D culture, such as altered cell morphology, mechanotransduction, adhesion, and migration. The microenvironment of acellular lung scaffolds determines the shape, migration, and proliferation of immortalized mouse lung fibroblasts, differing across lung regions and distinct from 2D culture [69].

The pathogenic contributions of tissue matrix and cell-intrinsic behaviors can be modeled by combining cells and scaffolds from patient and control lungs [70–72].

Fibrotic acellular lungs maintain disease-specific characteristics and promote myofibroblast differentiation when reseeded with healthy lung fibroblasts [71]. Alternatively, healthy fibroblasts and scaffolds can be treated with fibrosis-associated factors to determine their contribution to fibrosis and identify targets for therapeutic intervention [73].

ECM can be extracted from acellular scaffolds to avoid cellular contamination. ECM from healthy and fibrotic lungs can be extracted and analyzed by mass spectrometry to identify disease-associated components [71]. Extracted ECM can coat culture dishes for in vitro study of its biochemical cues or be mixed with hydrogels of varying stiffness to determine the relative contributions of biochemical and biophysical cues on cell behavior, such as myofibroblast transformation characteristic of fibrosis [74].

Each approach to modeling complex fibroblast behavior ex vivo has unique features. Organoids model epithelial proliferation and differentiation, epithelial-mesenchymal interactions, comparisons of fibroblast lineages or disease state, contributions of additional cell types, mechanotransduction, and matrix-fibroblast interactions. Lung-on-a-chip matures the epithelium in an air-liquid interface, allowing investigation of barrier function and gas exchange. Lung-on-a-chip also models respiratory forces, vascular interactions with airways, and can model infections and edema. However, chip models lag in incorporation of the mesenchyme. Acellular lung scaffolds model fibroblast interactions with matrix composition and architecture. The relative contributions of cells and scaffolds to disease phenotypes can be modeled using healthy fibroblasts in disease-specific scaffolds and vice versa. These model systems continue to develop, borrowing features from each other to increase complexity and control.

## 8.6 Targeting Fibroblasts with Nanoparticles as Strategy for Intervention

Nanoparticles are engineered particles with a diameter of less than 100nm, whose size, shape, and surface features allow physical interactions with biomolecules. Their high surface area-to-mass ratio exponentially increases as particle diameter decreases, increasing tissue exposure to therapeutic agents. Nanoparticles can be loaded with hydrophobic and hydrophilic compounds, peptides, and nucleic acids and can be directed to specific cell types by incorporating ligands or antibodies. Inhaled nanoparticles are prepared as liquid solutions, suspensions, or dry powder, which can be aerosolized and delivered to the lungs by nebulizers or inhalers. Inhaled nanoparticle delivery to the lung concentrates the therapy locally, allowing lower overall dosage than systemic delivery [75–77].

Alveolar fibroblasts are not directly exposed to inhaled or intravascular delivery: they must be reached through an epithelial or endothelial barrier (Fig. 8.3). Potential mechanisms for reaching alveolar fibroblasts with inhaled nanoparticles include

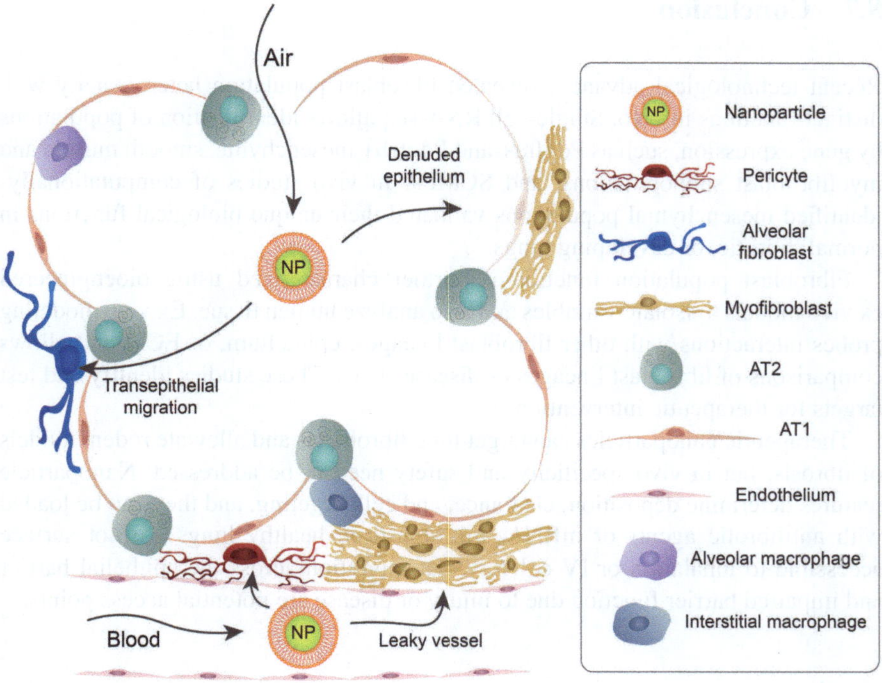


**Fig. 8.3** Nanoparticle delivery to lung fibroblasts. Nanoparticles must reach alveolar fibroblasts or myofibroblasts by migration across the epithelial or endothelial barrier. Fibroblasts may be more directly accessible to nanoparticle delivery in an injury or disease state that compromises these barriers. NP, nanoparticle; AT1, alveolar epithelium type 1; AT2, alveolar epithelium type 2

transepithelial migration or deposition onto injured tissue denuded of the epithelium [78]. Intravascular delivery of nanoparticles is possible to lung tumors with permeable vasculature [79], and leaky vasculature in injured tissue may enable delivery to fibroblasts [78, 80].

Pulmonary fibroblasts can internalize nanoparticles in vitro, but evidence for in vivo targeting is indirect. Isolated IPF lung fibroblasts internalized glycol chitosan nanoparticles by macropinocytosis, a form of endocytosis. Uptake was increased by culturing the fibroblasts on collagen, suggesting relative specificity for fibrotic lesions high in collagen content [81]. Fibroblasts expressing the surface receptor CD44 internalized liposome nanoparticles incorporating a CD44 ligand, while CD44-negative fibroblasts did not [82]. Several studies used nanoparticles carrying antifibrotic agents to attenuate fibrosis in vivo in animal models. These studies demonstrated efficacy against myofibroblast transformation in isolated lung fibroblasts in vitro and reduced fibrosis in vivo. However, they did not identify target cells in vivo and lack direct evidence for uptake by fibroblasts [83–85]. Identifying where nanoparticles localize in vivo is also important for toxicology, because inhaled nanoparticles can cause profibrotic inflammation and fibrosis in the lung and can also translocate and accumulate in other organs [76, 86, 87].

## 8.7 Conclusion

Recent technological advances revealed fibroblast population heterogeneity with distinct functions in vivo. Single-cell RNA-seq allows identification of populations by gene expression, such as Pdgfra+ and Pdgfrb+ mesenchyme, smooth muscle and myofibroblast subpopulations, and SCMFs. In vivo studies of computationally-identified mesenchymal populations validated their unique biological functions in normal, fibrotic, or developing lungs.

Fibroblast population function is further characterized using bioengineered ex vivo models to isolate variables and also analyze human tissue. Ex vivo modeling probes interactions with other fibroblast lineages, epithelium, or ECM and allows comparisons of fibroblast lineages or disease states. These studies identify and test targets for therapeutic intervention.

Therapeutic nanoparticles can target lung fibroblasts and alleviate rodent models of fibrosis, but in vivo specificity and safety need to be addressed. Nanoparticle features determine deposition, clearance, and cell targeting, and they can be loaded with antifibrotic agents or mRNA. Fibroblasts in healthy lungs are not surface accessible to inhalation or IV delivery, but migration across the epithelial barrier and impaired barrier function due to injury or disease are potential access points.

## References


1. Bellusci, S., Grindley, J., Emoto, H., Itoh, N. & Hogan, B.L. Fibroblast growth factor 10 (FGF10) and branching morphogenesis in the embryonic mouse lung. *Development* **124**, 4867–4878 (1997).
2. Daley, W.P. & Yamada, K.M. ECM-modulated cellular dynamics as a driving force for tissue morphogenesis. *Curr Opin Genet Dev* **23**, 408–414 (2013).
3. De Langhe, S.P., *et al.* Dickkopf-1 (DKK1) reveals that fibronectin is a major target of Wnt signaling in branching morphogenesis of the mouse embryonic lung. *Dev Biol* **277**, 316–331 (2005).
4. Gebb, S.A., Fox, K., Vaughn, J., McKean, D. & Jones, P.L. Fetal oxygen tension promotes tenascin-C-dependent lung branching morphogenesis. *Dev Dyn* **234**, 1–10 (2005).
5. McCulley, D., Wienhold, M. & Sun, X. The pulmonary mesenchyme directs lung development. *Curr Opin Genet Dev* **32**, 98–105 (2015).
6. Roth-Kleiner, M., Hirsch, E. & Schittny, J.C. Fetal lungs of tenascin-C-deficient mice grow well, but branch poorly in organ culture. *Am J Respir Cell Mol Biol* **30**, 360–366 (2004).
7. Volckaert, T., *et al.* Localized Fgf10 expression is not required for lung branching morphogenesis but prevents differentiation of epithelial progenitors. *Development* **140**, 3731–3742 (2013).
8. Whitsett, J.A., Kalin, T.V., Xu, Y. & Kalinichenko, V.V. Building and Regenerating the Lung Cell by Cell. *Physiol Rev* **99**, 513–554 (2019).
9. Zepp, J.A. & Morrisey, E.E. Cellular crosstalk in the development and regeneration of the respiratory system. *Nat Rev Mol Cell Biol* **20**, 551–566 (2019).
10. Nho, R.S., Ballinger, M.N., Rojas, M.M., Ghadiali, S.N. & Horowitz, J.C. Biomechanical Force and Cellular Stiffness in Lung Fibrosis. *Am J Pathol* **192**, 750–761 (2022).
11. Wolters, P.J., *et al.* Time for a change: is idiopathic pulmonary fibrosis still idiopathic and only fibrotic? *Lancet Respir Med* **6**, 154–160 (2018).

12. Noble, P.W., Barkauskas, C.E. & Jiang, D. Pulmonary fibrosis: patterns and perpetrators. *J Clin Invest* **122**, 2756–2762 (2012).
13. Somogyi, V., *et al.* The therapy of idiopathic pulmonary fibrosis: what is next? *Eur Respir Rev* **28**(2019).
14. Davidson, M.D., Burdick, J.A. & Wells, R.G. Engineered Biomaterial Platforms to Study Fibrosis. *Adv Healthc Mater* **9**, e1901682 (2020).
15. Zepp, J.A., *et al.* Distinct Mesenchymal Lineages and Niches Promote Epithelial Self-Renewal and Myofibrogenesis in the Lung. *Cell* **170**, 1134–1148 e1110 (2017).
16. Lee, J.H., *et al.* Anatomically and Functionally Distinct Lung Mesenchymal Populations Marked by Lgr5 and Lgr6. *Cell* **170**, 1149–1163 e1112 (2017).
17. Zepp, J.A., *et al.* Genomic, epigenomic, and biophysical cues controlling the emergence of the lung alveolus. *Science* **371**(2021).
18. Liu, X., *et al.* Categorization of lung mesenchymal cells in development and fibrosis. *iScience* **24**, 102551 (2021).
19. Tsukui, T., *et al.* Collagen-producing lung cell atlas identifies multiple subsets with distinct localization and relevance to fibrosis. *Nat Commun* **11**, 1920 (2020).
20. Adams, T.S., *et al.* Single-cell RNA-seq reveals ectopic and aberrant lung-resident cell populations in idiopathic pulmonary fibrosis. *Sci Adv* **6**, eaba1983 (2020).
21. Habermann, A.C., *et al.* Single-cell RNA sequencing reveals profibrotic roles of distinct epithelial and mesenchymal lineages in pulmonary fibrosis. *Sci Adv* **6**, eaba1972 (2020).
22. Basil, M.C., *et al.* The Cellular and Physiological Basis for Lung Repair and Regeneration: Past, Present, and Future. *Cell Stem Cell* **26**, 482–502 (2020).
23. Tata, P.R. & Rajagopal, J. Plasticity in the lung: making and breaking cell identity. *Development* **144**, 755–766 (2017).
24. Wang, C., *et al.* Expansion of hedgehog disrupts mesenchymal identity and induces emphysema phenotype. *J Clin Invest* **128**, 4343–4358 (2018).
25. O'Hare, K.H. & Sheridan, M.N. Electron microscopic observations on the morphogenesis of the albino rat lung, with special reference to pulmonary epithelial cells. *Am J Anat* **127**, 181–205 (1970).
26. Brody, J.S. & Kaplan, N.B. Proliferation of alveolar interstitial cells during postnatal lung growth. Evidence for two distinct populations of pulmonary fibroblasts. *Am Rev Respir Dis* **127**, 763–770 (1983).
27. Maksvytis, H.J., *et al.* In vitro characteristics of the lipid-filled interstitial cell associated with postnatal lung growth: evidence for fibroblast heterogeneity. *J Cell Physiol* **118**, 113–123 (1984).
28. Barkauskas, C.E., *et al.* Type 2 alveolar cells are stem cells in adult lung. *J Clin Invest* **123**, 3025–3036 (2013).
29. Chung, M.I., Bujnis, M., Barkauskas, C.E., Kobayashi, Y. & Hogan, B.L.M. Niche-mediated BMP/SMAD signaling regulates lung alveolar stem cell proliferation and differentiation. *Development* **145**(2018).
30. Nabhan, A.N., Brownfield, D.G., Harbury, P.B., Krasnow, M.A. & Desai, T.J. Single-cell Wnt signaling niches maintain stemness of alveolar type 2 cells. *Science* **359**, 1118–1123 (2018).
31. Sun, X., *et al.* A census of the lung: CellCards from LungMAP. *Dev Cell* **57**, 112–145 e112 (2022).
32. Sims, D.E. The pericyte--a review. *Tissue Cell* **18**, 153–174 (1986).
33. Weibel, E.R. On pericytes, particularly their existence on lung capillaries. *Microvasc Res* **8**, 218–235 (1974).
34. Armulik, A., Genove, G. & Betsholtz, C. Pericytes: developmental, physiological, and pathological perspectives, problems, and promises. *Dev Cell* **21**, 193–215 (2011).
35. Perros, F., *et al.* Platelet-derived growth factor expression and function in idiopathic pulmonary arterial hypertension. *Am J Respir Crit Care Med* **178**, 81–88 (2008).

36. Hellstrom, M., Kalen, M., Lindahl, P., Abramsson, A. & Betsholtz, C. Role of PDGF-B and PDGFR-beta in recruitment of vascular smooth muscle cells and pericytes during embryonic blood vessel formation in the mouse. *Development* **126**, 3047–3055 (1999).
37. Hung, C., *et al.* Role of lung pericytes and resident fibroblasts in the pathogenesis of pulmonary fibrosis. *Am J Respir Crit Care Med* **188**, 820–830 (2013).
38. Chandran, R.R., *et al.* Distinct roles of KLF4 in mesenchymal cell subtypes during lung fibrogenesis. *Nat Commun* **12**, 7179 (2021).
39. Rock, J.R., *et al.* Multiple stromal populations contribute to pulmonary fibrosis without evidence for epithelial to mesenchymal transition. *Proc Natl Acad Sci U S A* **108**, E1475–1483 (2011).
40. Volckaert, T., *et al.* Parabronchial smooth muscle constitutes an airway epithelial stem cell niche in the mouse lung after injury. *J Clin Invest* **121**, 4409–4419 (2011).
41. Moiseenko, A., *et al.* Identification of a Repair-Supportive Mesenchymal Cell Population during Airway Epithelial Regeneration. *Cell Rep* **33**, 108549 (2020).
42. Gouveia, L., *et al.* Lung developmental arrest caused by PDGF-A deletion: consequences for the adult mouse lung. *Am J Physiol Lung Cell Mol Physiol* **318**, L831–L843 (2020).
43. Kugler, M.C., *et al.* Sonic Hedgehog Signaling Regulates Myofibroblast Function during Alveolar Septum Formation in Murine Postnatal Lung. *Am J Respir Cell Mol Biol* **57**, 280–293 (2017).
44. Li, C., *et al.* Secondary crest myofibroblast PDGFRalpha controls the elastogenesis pathway via a secondary tier of signaling networks during alveologenesis. *Development* **146**(2019).
45. Li, R., Li, X., Hagood, J., Zhu, M.S. & Sun, X. Myofibroblast contraction is essential for generating and regenerating the gas-exchange surface. *J Clin Invest* **130**, 2859–2871 (2020).
46. Hagan, A.S., *et al.* Generation and validation of novel conditional flox and inducible Cre alleles targeting fibroblast growth factor 18 (Fgf18). *Dev Dyn* **248**, 882–893 (2019).
47. Jain, R., *et al.* Plasticity of Hopx(+) type I alveolar cells to regenerate type II cells in the lung. *Nat Commun* **6**, 6727 (2015).
48. Nikolic, M.Z., *et al.* Human embryonic lung epithelial tips are multipotent progenitors that can be expanded in vitro as long-term self-renewing organoids. *Elife* **6**(2017).
49. Gotoh, S., *et al.* Generation of alveolar epithelial spheroids via isolated progenitor cells from human pluripotent stem cells. *Stem Cell Reports* **3**, 394–403 (2014).
50. Liberti, D.C. & Morrisey, E.E. Organoid models: assessing lung cell fate decisions and disease responses. *Trends Mol Med* **27**, 1159–1174 (2021).
51. Cassandras, M., *et al.* Gli1(+) mesenchymal stromal cells form a pathological niche to promote airway progenitor metaplasia in the fibrotic lung. *Nat Cell Biol* **22**, 1295–1306 (2020).
52. Kathiriya, J.J., *et al.* Human alveolar type 2 epithelium transdifferentiates into metaplastic KRT5(+) basal cells. *Nat Cell Biol* **24**, 10–23 (2022).
53. Chen, Y.W., *et al.* A three-dimensional model of human lung development and disease from pluripotent stem cells. *Nat Cell Biol* **19**, 542–549 (2017).
54. Strikoudis, A., *et al.* Modeling of Fibrotic Lung Disease Using 3D Organoids Derived from Human Pluripotent Stem Cells. *Cell Rep* **27**, 3709–3723 e3705 (2019).
55. Tan, Q., Choi, K.M., Sicard, D. & Tschumperlin, D.J. Human airway organoid engineering as a step toward lung regeneration and disease modeling. *Biomaterials* **113**, 118–132 (2017).
56. Tan, Q., *et al.* Nascent Lung Organoids Reveal Epithelium- and Bone Morphogenetic Protein-mediated Suppression of Fibroblast Activation. *Am J Respir Cell Mol Biol* **61**, 607–619 (2019).
57. Joshi, R., Batie, M.R., Fan, Q. & Varisco, B.M. Mouse lung organoid responses to reduced, increased, and cyclic stretch. *Am J Physiol Lung Cell Mol Physiol* **322**, L162–L173 (2022).
58. Dye, B.R., *et al.* Human lung organoids develop into adult airway-like structures directed by physico-chemical biomaterial properties. *Biomaterials* **234**, 119757 (2020).
59. Hofer, M. & Lutolf, M.P. Engineering organoids. *Nat Rev Mater* **6**, 402–420 (2021).
60. Huh, D., *et al.* Reconstituting organ-level lung functions on a chip. *Science* **328**, 1662–1668 (2010).
61. Huh, D., *et al.* A human disease model of drug toxicity-induced pulmonary edema in a lung-on-a-chip microdevice. *Sci Transl Med* **4**, 159ra147 (2012).

62. Stucki, A.O., *et al.* A lung-on-a-chip array with an integrated bio-inspired respiration mechanism. *Lab Chip* **15**, 1302–1310 (2015).
63. Zamprogno, P., *et al.* Second-generation lung-on-a-chip with an array of stretchable alveoli made with a biological membrane. *Commun Biol* **4**, 168 (2021).
64. Sellgren, K.L., Butala, E.J., Gilmour, B.P., Randell, S.H. & Grego, S. A biomimetic multicellular model of the airways using primary human cells. *Lab Chip* **14**, 3349–3358 (2014).
65. Huang, D., *et al.* Reversed-engineered human alveolar lung-on-a-chip model. *Proc Natl Acad Sci U S A* **118**(2021).
66. Nesmith, A.P., Agarwal, A., McCain, M.L. & Parker, K.K. Human airway musculature on a chip: an in vitro model of allergic asthmatic bronchoconstriction and bronchodilation. *Lab Chip* **14**, 3925–3936 (2014).
67. Mejias, J.C., Nelson, M.R., Liseth, O. & Roy, K. A 96-well format microvascularized human lung-on-a-chip platform for microphysiological modeling of fibrotic diseases. *Lab Chip* **20**, 3601–3611 (2020).
68. Lagowala, D.A., Kwon, S., Sidhaye, V.K. & Kim, D.H. Human microphysiological models of airway and alveolar epithelia. *Am J Physiol Lung Cell Mol Physiol* **321**, L1072–L1088 (2021).
69. Burgstaller, G., *et al.* Distinct niches within the extracellular matrix dictate fibroblast function in (cell free) 3D lung tissue cultures. *Am J Physiol Lung Cell Mol Physiol* **314**, L708–L723 (2018).
70. Gilpin, S.E. & Wagner, D.E. Acellular human lung scaffolds to model lung disease and tissue regeneration. *Eur Respir Rev* **27**(2018).
71. Booth, A.J., *et al.* Acellular normal and fibrotic human lung matrices as a culture system for in vitro investigation. *Am J Respir Crit Care Med* **186**, 866–876 (2012).
72. Wagner, D.E., *et al.* Comparative decellularization and recellularization of normal versus emphysematous human lungs. *Biomaterials* **35**, 3281–3297 (2014).
73. Zhou, Y., *et al.* Chitinase 3-like 1 suppresses injury and promotes fibroproliferative responses in Mammalian lung fibrosis. *Sci Transl Med* **6**, 240ra276 (2014).
74. Sava, P., *et al.* Human pericytes adopt myofibroblast properties in the microenvironment of the IPF lung. *JCI Insight* **2**(2017).
75. Skibba, M., Drelich, A., Poellmann, M., Hong, S. & Brasier, A.R. Nanoapproaches to Modifying Epigenetics of Epithelial Mesenchymal Transition for Treatment of Pulmonary Fibrosis. *Front Pharmacol* **11**, 607689 (2020).
76. Praphawatvet, T., Peters, J.I. & Williams, R.O., 3rd. Inhaled nanoparticles-An updated review. *Int J Pharm* **587**, 119671 (2020).
77. Velino, C., *et al.* Nanomedicine Approaches for the Pulmonary Treatment of Cystic Fibrosis. *Front Bioeng Biotechnol* **7**, 406 (2019).
78. Deng, Z., Kalin, G.T., Shi, D. & Kalinichenko, V.V. Nanoparticle Delivery Systems with Cell-Specific Targeting for Pulmonary Diseases. *Am J Respir Cell Mol Biol* **64**, 292–307 (2021).
79. Kang, H., *et al.* Size-Dependent EPR Effect of Polymeric Nanoparticles on Tumor Targeting. *Adv Healthc Mater* **9**, e1901223 (2020).
80. Probst, C.K., Montesi, S.B., Medoff, B.D., Shea, B.S. & Knipe, R.S. Vascular permeability in the fibrotic lung. *Eur Respir J* **56**(2020).
81. Yhee, J.Y., *et al.* The effects of collagen-rich extracellular matrix on the intracellular delivery of glycol chitosan nanoparticles in human lung fibroblasts. *Int J Nanomedicine* **12**, 6089–6105 (2017).
82. Pandolfi, L., *et al.* Liposomes Loaded with Everolimus and Coated with Hyaluronic Acid: A Promising Approach for Lung Fibrosis. *Int J Mol Sci* **22**(2021).
83. Zhang, G., *et al.* Pulmonary delivery of therapeutic proteins based on zwitterionic chitosan-based nanocarriers for treatment on bleomycin-induced pulmonary fibrosis. *Int J Biol Macromol* **133**, 58–66 (2019).
84. Kim, J., *et al.* Lung-targeted delivery of TGF-beta antisense oligonucleotides to treat pulmonary fibrosis. *J Control Release* **322**, 108–121 (2020).

85. Keum, H., *et al.* Biomimetic lipid Nanocomplexes incorporating STAT3-inhibiting peptides effectively infiltrate the lung barrier and ameliorate pulmonary fibrosis. *J Control Release* **332**, 160–170 (2021).
86. Kreyling, W.G., *et al.* Translocation of ultrafine insoluble iridium particles from lung epithelium to extrapulmonary organs is size dependent but very low. *J Toxicol Environ Health A* **65**, 1513–1530 (2002).
87. Oberdorster, G. Safety assessment for nanotechnology and nanomedicine: concepts of nanotoxicology. *J Intern Med* **267**, 89–105 (2010).

# Chapter 9
# Engineering Dynamic 3D Models of Lung

**Rachel Blomberg, Rukshika S. Hewawasam, Predrag Šerbedžija, Kamiel Saleh, Thomas Caracena, and Chelsea M. Magin**

## 9.1 Introduction

The lung parenchyma comprises the anatomical structures necessary for gas exchange, most notably alveoli, which are bundled together into acinar structures, and surrounded by pulmonary capillaries. The cells that comprise these functional tissue units reside upon or within an interwoven extracellular matrix (ECM) [95]. Alveoli are created by two cell types, alveolar epithelial type 1 (AT1) cells, which comprise most of the surface area for gas exchange, and alveolar epithelial type 2 (AT2) cells, which secrete protective surfactant and are capable of transdifferentiating into AT1 cells in case of injury. Resident immune and inflammatory cells, particularly macrophages, patrol the air-exposed apical surface of alveoli to eliminate any inhaled pathogens or particulates. The major cell types of the lung and their relative orientations were originally determined by cellular morphology and staining in cross-sections, and modern research continues to add to our understanding and defining of cell subsets through sophisticated methods such as single-cell sequencing, which has begun to unveil even more specialized cell subtypes within

R. Blomberg · R. S. Hewawasam · P. Šerbedžija · K. Saleh · T. Caracena
Department of Bioengineering, University of Colorado, Denver | Anschutz Medical Campus, Aurora, CO, USA

C. M. Magin (✉)
Department of Bioengineering, University of Colorado, Denver | Anschutz Medical Campus, Aurora, CO, USA

Department of Pediatrics, University of Colorado, Anschutz Medical Campus, Aurora, CO, USA

Division of Pulmonary Sciences and Critical Care Medicine, Department of Medicine, University of Colorado, Anschutz Medical Campus, Aurora, CO, USA
e-mail: chelsea.magin@cuanschutz.edu

 
C. M. Magin (ed.), *Engineering Translational Models of Lung Homeostasis and Disease*, Advances in Experimental Medicine and Biology 1413,
https://doi.org/10.1007/978-3-031-26625-6_9

each of the broader historical categories, such as AT2 cells more likely dedicated to surfactant production relative to those with more regenerative and stem-cell like signaling [144]. At the minimal gas-exchange surfaces, alveolar epithelial cells share a basement membrane with pulmonary endothelial cells, creating a thin barrier layer for gas diffusion [152]. Yet across much of the parenchyma, this barrier also includes a thicker interstitial layer containing ECM—largely comprised of fibrous collagens, elastin, and glycosaminoglycans—along with the mesenchymal cells that synthesize and maintain this network [139, 152]. The interstitial ECM is as critical as the cellular components, as it provides not only biochemical cues for cell growth and behaviors, but also physical support and biomechanical cues (Fig. 9.1).

The parenchymal ECM is central to maintaining lung homeostasis and changes significantly during pathogenesis of several common lung diseases. One possible initiating event for idiopathic pulmonary fibrosis (IPF) is injury to the alveolar epithelium, which results in the release of factors such as transforming growth factor beta (TGFβ) that activate mesenchymal fibroblasts in a wound-healing response. If this response does not resolve naturally, activated fibroblasts will continue to synthesize dense ECM, which in turn promotes fibroblast activation in a positive feedback loop ultimately resulting in a prolonged fibroinflammatory response and lung stiffening [156]. Conversely, pulmonary emphysema—often a symptom of chronic obstructive pulmonary disease (COPD)—results from destruction of elastin fibers

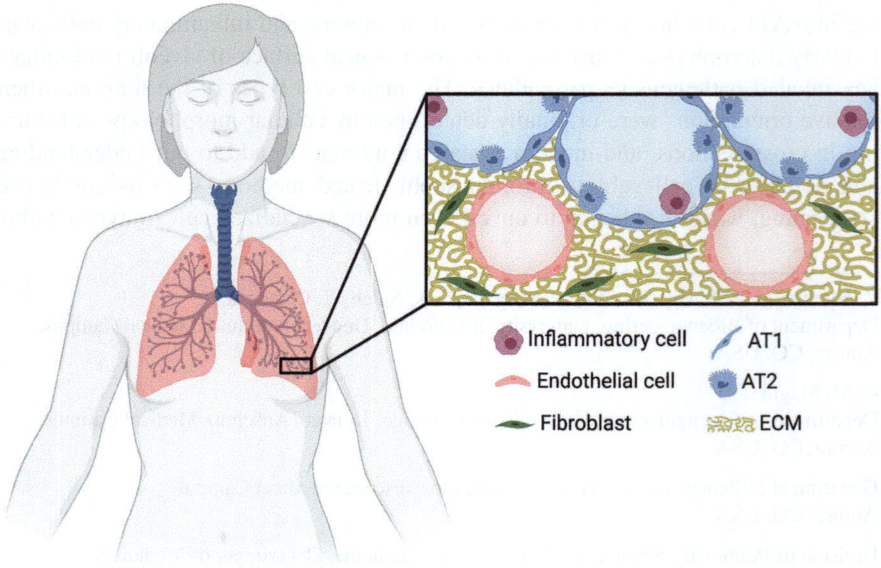


**Fig. 9.1** Cellular and extracellular components of the lung parenchyma. As the primary site for gas exchange in the lung, the parenchyma contains a variety of cell types to form and maintain alveolar and capillary structures. In addition, the parenchymal ECM provides both structural support and chemical cues to inform cellular behavior

in the parenchymal ECM, most classically by overexpression of proteases by alveolar macrophages. The destruction of supportive ECM leads to breakdown of alveolar walls, widening of airspaces, and impaired gas exchange [62]. These contrasting disease states highlight the critical role of carefully maintained parenchymal ECM to overall lung health. Since ECM plays such a key role in maintenance of lung health or development of disease, it is critical to incorporate ECM elements into engineered models of lung, since cells growing without the natural cues provided by ECM will behave very differently than they do in vivo [110].

Lung architecture is also critical to cell behavior. Much of what we know about the specific geometries and structures of the parenchyma comes from stereology, which is a technique for calculating three dimensional (3D) values from two dimensional (2D) cross-sections [106]. For example, Ochs et al. used a stereological approach to estimate the number and size of alveoli in human lung. By counting both alveolar openings and connection points between alveolar walls in adjacent tissue sections, they were able to build a 3D model of parenchymal tissue and thus determine that the average size of a single alveolus is $4.2 \times 10^6$ μm$^3$, and that the average lung contains approximately $480 \times 10^6$ alveoli. They also noted that larger lungs—typically male—contained more, rather than larger, alveoli [107]. Tissue geometry plays a critical role in lung development as well as cellular differentiation and behavior. During development, the local geometry of lung buds can affect the intracellular tension and access of growth factors to individual cells, which may act together to determine local proliferation and ultimately branch patterns within the lung [39, 101]. Geometry can also help determine cell fate; lung progenitor cells cultured in narrow tubes tend to develop more mature distal epithelial cell marker than those cultured in large tubes or on flat surfaces [135]. Similarly, AT2 cells encapsulated in a hydrogel patterned to mimic alveolar clusters will transdifferentiate into AT1-like cells with unique spatial patterning of differentiation based on the number of alveoli in the cluster [70]. The curvature of a surface cells grown upon can strongly affect intracellular tension and polarity, resulting in differences in proliferation and migration, patterns which can be different depending on cell type [8, 18]. For example, epithelial cells grown inside tubes tend to form monolayer sheets perpendicular to the surface curvature, reflective of their physiological role in barrier formation, while fibroblasts align parallel to the curvature and exhibit motility along straight tracks [18]. These examples also highlight the relationship between tissue geometry and biomechanics.

Various biomechanical parameters are essential to lung parenchyma function, and it is thus becoming increasingly recognized that any research models of lung tissue must account for these biomechanics. Respiration creates recurring cyclic stretch that influences the entire organ. In alveoli, normal tidal breathing causes linear distention up to 5%, while forced inflation can increase linear distention up to 40% [122]. On a cellular level, these constant cyclic deformations can alter the conformation of integrins, ion channels, or cytoskeletal components, and thus change intracellular signaling [118]. The alveolar barrier experiences shear forces created by airflow along one side and fluid flow along the other at an air-liquid interface at the cellular level, which helps externally regulate nutrient availability or exposure to pollutants, and internally regulate cell polarization [160]. The ECM provides a highly regulated stiffness to support three-dimensional structures in a parenchyma, and elasticity to manage the constant stretch associated with breathing [139, 140]. As mentioned above, increased ECM deposition during IPF results in a positive feedback loop of fibroblast activation; a critical part of this activation loop is enhanced mechanosignaling in fibroblasts induced by increased tissue stiffness [156]. There are various methods for measuring biomechanical parameters in the lung. Whole-lung analysis can be performed via spirometry and ventilation, yielding measurements such as total lung volume or elasticity [111]. Yet these whole-lung measurements are not always reflective of the local conditions any given cell within the lung senses [111, 140]. More regional measurements can be made on isolated tissue strips or slices, such as atomic force microscopy or rheology to study elastic modulus, or calculation of a stress-strain curve by application of tensile force [140].

While we have learned much about lung health and disease from classical model systems, there are also weaknesses in these methodologies, making it critical to engineer more sophisticated models of lung tissue to study interactions and pathways that classical models are not well suited to explore. Animal models have the benefit of containing all the various cellular, extracellular, and biomechanical components that might be relevant to any given phenotype, and yet because of that it can be extremely difficult to disentangle key mechanisms from the overall environment. Reductionist in vitro models can provide this clarity, yet traditional cell culture—in monolayers on hard plastic and immersed in static media—fails to recapitulate critical environmental parameters. While some modifications, such as culture in transwell inserts to create an air-liquid interface, can begin to mimic the parenchymal environment, generally much more sophisticated modeling is required to address complex, multicellular, and/or biomechanically driven phenotypes [5]. This chapter highlights a diverse array of engineered model systems used in parenchymal research to mimic the ECM, build lung geometry, and incorporate dynamic mechanical forces, along with opportunities to improve our ability to recapitulate complex in vivo aspects of pulmonary tissues in vitro (Fig. 9.2).

## 9.2 Building the Extracellular Microenvironment

Ensuring that cells grown in vitro faithfully recapitulate the behavior of cells in vivo requires a microenvironment that replicates the ECM. ECM is comprised of a complex network of proteins, proteoglycans, and glycosaminoglycans that provides both structural support to the tissue as well as being a source of extracellular signaling moieties and mechanosignaling cues [139]. Key structural components of the lung ECM include the fibrillar collagens 1 and 3, elastin, fibronectin, and basement membrane components such as laminin and non-fibrillar collagens. Many of these molecules contain binding sites for cell-surface receptors, especially integrins, which also allows them to directly influence cell signaling. For example, AT2 cells

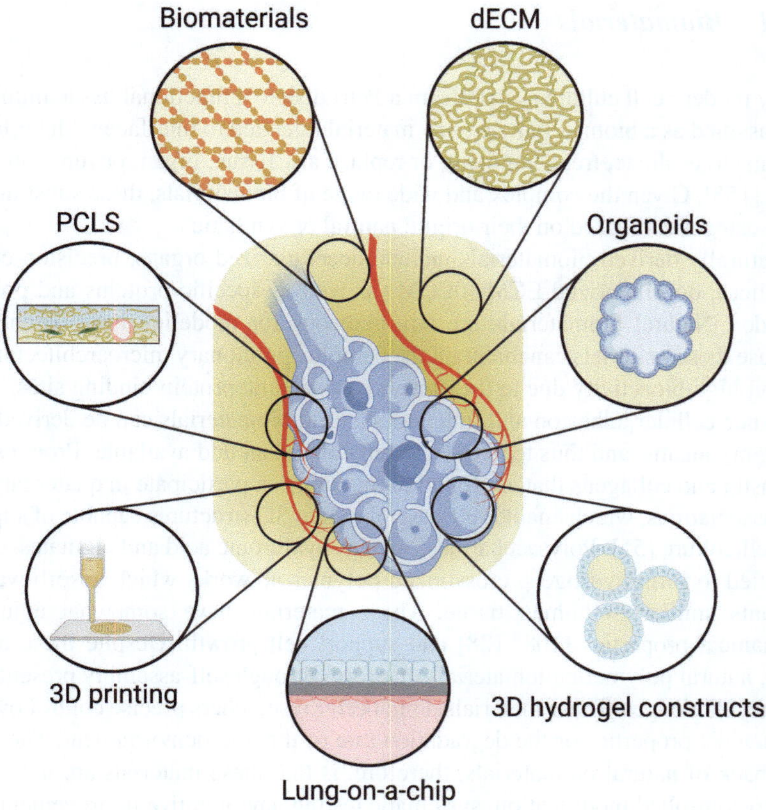


**Fig. 9.2** Bioengineering strategies for modeling the lung parenchyma. Multiple complementary methods exist for modeling varying aspects of distal lung tissue, including biomaterials which recapitulate the ECM, various 3D cell culture systems that mimic alveolar geometry, whole tissue explants, and multi-layer lung-on-a-chip devices which recreate the gas exchange barrier

cultured on a predominantly collagen/fibronectin matrix tend to transdifferentiate into AT1 cells, while culture with laminin-5 preserves AT2 cell phenotype [60]. Associated with these structural ECM components are a variety of growth factors and signaling molecules, such as TGFβ and thrombospondin 1, which can bind both to structural ECM and to cells, thus regulating their activity [96]. In terms of biomechanics, increases in the accumulation of ECM components, or changes to overall composition can increase tissue stiffness. Healthy human lung has an elastic modulus in the range of 1–5 kPa, while diseases such as fibrosis can increase the modulus to 10 kPa and above [31]. Such changes can alter the conformation of signaling molecules and thus increase cellular mechanosignaling, resulting in increased proliferation and/or cellular activation [139]. In this way, the ECM is a critical provider of both structure and signaling in tissues, and therefore modeling the pulmonary ECM in cell culture systems is a critical first step to modeling lung parenchyma, and can be accomplished by incorporating a diverse array of biomaterials into in vitro systems.

### 9.2.1 Biomaterials

Every modern cell culture system from a Petri dish to a functional tissue mimic can be classified as a biomaterial, i.e., any material intended to interface with biological systems to evaluate, treat, augment, or replace any tissue, organ, or function of the body [155]. Given the complex and wide range of biomaterials, these substrates are often categorized based on their origin: natural or synthetic.

Naturally derived biomaterials include decellularized organs, precision-cut tissue slices, decellularized ECM (dECM) as well as specific proteins and polysaccharides. Natural biomaterials are advantageous for modeling lung parenchyma because these materials can maintain the complex pulmonary microarchitecture and exhibit high bioactivity due to the presence of specific protein binding sites, which influence cellular adhesion and function. Natural biomaterials can be derived from living organisms, and thus tend to be readily abundant and available. Proteins such as elastin and collagens that are found in the lung can participate in quaternary protein associations, which enable self-assembly into 3D structures capable of supporting cell culture [55]. Polysaccharides, such as hyaluronic acid and alginates, can be modified to form hydrogels, crosslinked polymer networks which absorb water in amounts similar to human tissue. These materials have somewhat adjustable mechanical properties [108, 128] and support cell growth. Despite these advantages, natural polymeric biomaterials that form through self-assembly present engineering challenges. These materials do not offer researchers precise control over the mechanical properties or the degradation rate of the microenvironment. The major drawback of natural biomaterials, therefore, is that these materials are not conducive to controlled modification, systematic testing, and iterative improvement [97].

Synthetic biomaterials, which are prepared through organic chemical reactions, can be engineered to have specific mechanical properties, chemical composition, and rate of degradation. These material properties can be independently tuned, systematically tested, and adjusted to optimize the cell culture microenvironment. The major weakness of synthetic biomaterials is the lack of endogenous cellular adhesion properties and thus relatively low bioactivity. Short peptide sequences that mimic cell-binding regions of ECM proteins are often added to these materials to overcome this limitation. Synthetic biomaterials can be engineered to degrade through hydrolysis, i.e., exposure to an aqueous environment such as cell culture medium, by selecting specific polymer backbone sequences and side groups. These materials can also be made degradable by enzymes that encapsulated cells secrete via crosslinking with peptide sequences that are degradable by specific matrix metalloproteinases (MMPs) [57, 119]. Synthetic polymers commonly used in pulmonary research include polyacrylamide, poly(dimethylsiloxane) (PDMS), and polyethylene glycol (PEG).

Combining natural materials with synthetic polymers to fabricate hybrid-hydrogels is a promising way to facilitate control over biomaterial mechanical properties and maintain bioactivity. Engineering approaches for modeling the pulmonary ECM in vitro encompass a variety of biomaterial-based approaches from decellularizing whole organs for use in culture to synthesizing polymers that are biologically inert, but which can be decorated with cell-adhesive ligands.

### 9.2.2 Lung Decellularization and Recellularization

As a fully natural biomaterial, tissue-derived ECM can be used for studying cell-ECM interactions in parenchymal research through the process of decellularization. Physical, chemical, enzymatic, and apoptosis-driven approaches can be used to effectively decellularize tissues. Where lung tissue is concerned, most decellularization protocols employ a combination of sequential transcardial and intratracheal perfusions with detergents (e.g., triton, sodium dodecyl sulfate, sodium deoxycholate), osmotic shock agents (e.g., sodium chloride), and enzymatic digestion of residual DNA [146, 147]. The outcome is a comprehensive cell lysis and clearance of cellular material, leaving behind an intact dECM scaffold consisting largely of collagen, elastin, and laminin, which closely recapitulates the molecular composition, mechanical properties, and architecture of in vivo lung tissue. This scaffold can then be either digested further for incorporation into hydrogel-based cell culture substrates or repopulated with cells via recellularization to powerfully model both cell-ECM and cell-cell interactions, while maintaining the complex 3D lung architecture. One disadvantage of the decellularization process is the potential for physical disruption of ECM architecture and compromised basement membrane integrity. The degree of disruption depends on detergents used as well as the route and manner of perfusion. Application of a milder detergent such as potassium laureate [105] coupled with controlled constant-rate rather than variable perfusion pressure [41] can more fully preserve the dECM scaffold architecture. Over the past decade, detailed protocols have been published for decellularizing lung tissue from multiple species, including mice, pigs, and humans, as well as disease states such as fibrosis [17, 146, 153]. Various quality control measures have been described to assess the effectiveness and completeness of lung decellularization while keeping the ECM architecture intact, including histology, fluorescence microscopy, and quantification of residual detergents and DNA using sensitive colorimetric assays [24, 99, 161]. In 2011, Crapo et al. described a successfully decellularized scaffold as one that exhibits <50 ng DNA per mg dry weight by colorimetric DNA assay, no evidence of DNA fragments >200 bp by gel electrophoresis, and lack of nuclear staining by histology [24]. Importantly, dECM is both physically stable and biologically responsive, capable of promoting growth and proliferation of cells used to repopulate the scaffold [26]. Scaffolds can be amenable to cryopreservation [40] or immediate use, either for incorporation into in vitro hydrogel matrix models, or for recellularization.

Once decellularized, a dECM scaffold can be recellularized with new resident cells of choice. This can be performed to study the interaction between engrafted cells and the native ECM that has retained its biophysical properties, but also has therapeutic potential, either in repurposing tissue deemed unsuitable for transplant, or possibly in creating xenografts with human cells. Considerable progress has been made to date where recellularized lung scaffolds support survival and proliferation of mesenchymal stem cells, and even subsequent differentiation into mature lung epithelial cells. Lung scaffolds from rats make up the bulk of recellularization efforts published to date and examples include intratracheal delivery of mouse alveolar epithelial cells [41], human or rat endothelial and adipose-derived stem cells [47, 126], amniotic fluid mesenchymal stem cells [148], and human lung cancer cells [92]. In these studies, dECM scaffolds were reported to support survival, growth, and/or differentiation of repopulating cell populations. For example, a decellularized murine lung scaffold was capable of supporting growth of fetal AT2 cells for 7 days [117]. Similarly, porcine dECM scaffolds supported growth and proliferation of several human cell types including fibroblasts, epithelial, and mesenchymal stem cells over an entire 1-month period [114]. Recellularization has also been performed in human decellularized lung scaffolds, with similar outcomes using primary human fibroblasts [15] or mesenchymal stem cells [43]. Stem cells are at the forefront of recellularization efforts because of their ability to differentiate into mature lung populations and repair tissue. Pioneering studies have demonstrated successful engrafting, proliferation, and differentiation of stem cells, whether the scaffold originated from healthy or diseased lung [11, 73]. Importantly, fibroblasts and pericytes used to repopulate scaffolds that originated from IPF lungs expressed high levels of fibrotic markers such as alpha-smooth muscle actin [125]. Likewise, human fibroblasts used to repopulate a scaffold that originated from fibrotic human lung tissue resulted in TGFβ-independent myofibroblast differentiation that was not observed in a healthy matrix, underscoring the importance of the interplay between ECM and resident cells in etiology of pulmonary fibrosis [13]. To provide more control over the biomechanical properties of dECM, synthetic crosslinkers can be used to tune the elastic modulus of the scaffold [130]. While regain-of-function efforts following recellularization have so far failed to live up to clinical significance, initial experiments, albeit far-reaching, have laid the foundations for future work.

Decellularized tissues used as grafts to improve wound healing after injury are already in use in clinics today [89]. Decellularized lung scaffolds harbor enormous potential in regenerative medicine by virtue of recellularization and transplantation [150]. With the aim to generate a scaffold suitable for recellularization and future clinical lung transplantation, a pioneering study decellularized porcine lung using a combination of dextrose, sodium lauryl ether sulfate, and triton X-100. The scaffold was reported to have largely intact ECM composition, capable of supporting exogenous cells [80]. Taking this a step further, Kitano et al. seeded cadaveric porcine decellularized lung scaffolds with human airway epithelial and umbilical vein endothelial cells, then orthotopically transplanted the recellularized scaffold into 3 porcine recipients. Engraftment of cells was observed, and amazingly transplant perfusion and ventilation were successful, with recorded gas exchange immediately following conclusion of transplantation [68]. Although short-lived, this pioneering experiment demonstrated the crucial functionality of the recellularized scaffold and its potential in organ replacement. While various cell types can adhere to scaffolds, differentiate, and proliferate, full repopulation and especially tissue functionality remains very limited [150]. Lung tissue consists of more cell types (e.g., alveolar macrophages, basal cells, smooth muscle cells, etc.) than those which have been reintroduced in recellularization efforts to date, and optimal arrangements and combinations for lung function are poorly understood, while the basement membrane integrity remains compromised. Various approaches towards generating chimeric lung constructs may overcome some of the current barriers in creating re-cellularized lung transplants. Porcine dECM scaffolds are anatomically, mechanically, and biochemically similar enough to human lung to support human cell engraftment and thus have potential for hybrid xenotranplantation [104]. More localized decellularization techniques might also be able to create hybrid organs for transplantation. Dorello et al. developed a method of airway delivery of decellularization reagents that resulted in de-epithelialization of rat lungs, while leaving both basement membrane and vasculature intact. These scaffolds could then be recellularized with human lung epithelial cells [34]. In a follow up study, Guenthart et al. performed the same local decellularization on human lungs and recellularized with patient-derived epithelial and mesenchymal cells to create chimeric organs [43]. Such localized decellularization approaches can thus preserve structures, such as vasculature, which are difficult to regrow during recellularization, while allowing for the incorporation of healthy human epithelium. Overall, great progress has been made in recent years and lung replacement via de- and recellularization may be within grasp. In addition to these exciting translational results in using dECM as a scaffold for recellularization, dECM can also be incorporated into 3D hydrogel-based cell culture platforms.

### 9.2.3 dECM Hydrogels

Hydrogels can be engineered to present a specific 3D cell-instructive microenvironment that mimics the physical properties of native tissues and facilitates the diffusion of nutrients and cell-secreted factors [14]. These biomaterials can consist of a broad range of natural and/or synthetic polymers, including dECM. Biomimetic hydrogels that incorporate dECM to recapitulate lung tissue matrix are uniquely advantageous compared to intact organs because they offer a more controlled environment for studying cell-cell and cell-ECM interactions in 3D. In preparation for integration into a hydrogel, decellularized scaffolds are commonly lyophilized, milled, and acid- or pepsin-digested to increase solubility [116, 123]. This solution can then self-assemble into 3D hydrogels with elastic moduli that are dECM concentration- and composition-dependent, which support cell adhesion and viability [115]. Hydrogels produced with porcine, rat, mouse, and human dECM can support growth and proliferation of a variety of resident lung and lung-related cells, including fibroblasts [26, 112], mesenchymal stem cells [115], epithelial cell line A549, microvascular endothelial cells, umbilical vein epithelial cells [81], and small-cell lung cancer cells [64]. Pouliot et al. reported a protocol for fabricating hydrogels derived from porcine lung dECM that supported mesenchymal stem cell culture in vitro, but these biomaterials were limited by mechanical properties that did not recapitulate values found in healthy or diseased lung tissue ($E \sim 30$ to 120 Pa) [115]. More recently, hydrogels have been successfully produced for the first time with dECM from healthy, COPD, and IPF human lung, which were found to reflect the stiffness values of native tissue [31]. This ability to mimic disease-modeling potential of dECM hydrogels represents another venue in pursuit of therapeutic agents for treatment of yet incurable lung diseases. Yet a still higher level of control over the biophysical microenvironment can be achieved using synthetic or hybrid materials.

### *9.2.4 Synthetic Hydrogels*

Like natural biomaterials, synthetic biomaterials can be used to create cell-culture compatible hydrogels. Some of the earliest cell biology studies using hydrogels involved culturing cells on top of 2D polyacrylamide hydrogels, functionalized for cell attachment by coating with various ECM molecules [65]. For example, Liu et al. fabricated polyacrylamide hydrogels, functionalized with collagen 1 to encourage cell attachment, with stiffness gradients formed by photopolymerization to illustrate the effects of variable stiffness on lung fibroblasts. This study demonstrated that the low median stiffness of normal murine lung tissue ($E = 0.5$ kPa) strongly reduced fibrotic fibroblast characteristics such as proliferation and migration and elucidated a novel mechanism whereby matrix stiffening enhances fibroblast activation through inhibition of endogenous inflammatory pathways [82]. Similarly, Marinković et al. illustrated that polyacrylamide hydrogels of human physiological lung tissue stiffness ($E = 1$ kPa) reduced the proliferative and contractile response of IPF fibroblasts [86]. To decouple dECM composition from mechanical properties, Sava et al. coated polyacrylamide-based hydrogels of modulus values ranging from 1.8 kPa to 23.7 kPa with healthy and IPF human-lung dECM [125]. The results of these studies demonstrated that changes in alpha-smooth muscle actin (αSMA) expression and organization were mechanosensitive regardless of composition [125]. While these experiments were certainly a breakthrough in modeling fibrotic disease in vitro, the same strategy cannot be employed to encapsulate cells within 3D matrices to mimic pulmonary tissue architecture and epithelial-fibroblast interactions, critical mediators of fibroblast activation [23, 50].

Advances in synthetic hydrogel design have fueled a growing interest in using PEG-based hydrogels to create 3D cell-instructive microenvironments that model pulmonary tissues. A pivotal study conducted by the Anseth research group demonstrated that interstitial fibroblasts isolated from heart valve leaflets undergo significant changes in gene expression when cultured on 2D PEG-based hydrogels as compared to culture within 3D PEG hydrogels. Some of the gene networks affected included those involved in cytoskeletal organization and contractility, focal adhesions, TGFβ signaling, and matrix remodeling [83], all of which are critical in pulmonary fibrosis [121] and must be recapitulated in high-fidelity in vitro models of

IPF. PEG hydrogels are biologically inert but can be engineered to exhibit many of the same biophysical qualities as the pulmonary ECM, such as high-water content, tissue-relevant modulus values, and diffusion of oxygen and other nutrients. PEG-based polymers are readily modified using several cytocompatible reactions to introduce biochemical moieties to allow for cellular adhesion, signaling, and controlled biodegradation. One attribute of particular interest to control within in vitro models of lung homeostasis and disease is biomaterial modulus. Dual-stage polymerization systems are an innovative way to recreate heterogeneous dynamic matrix stiffening in vitro [44]. Light-induced processes such as photopolymerization and photodegradation, temperature, electromagnetic fields, and ultrasound are user-controlled stimuli that elicit these dynamic responses [67]. Functional groups can be incorporated into the polymer backbone or crosslinker to trigger changes in the hydrogel microenvironment. Thiol and acrylate groups are commonly used in photo-initiated reactions. For example, Guvendiren et al. implemented one of the first protocols for in situ hydrogel stiffening in the presence of cells. A methacrylated hyaluronic acid polymer backbone was first reacted off-stoichiometry via Michael addition with a short dithiothreitol (DTT) crosslinker to create a soft hydrogel ($E \sim 3$ kPa) for cellular encapsulation. Next, excess methacrylate moieties were sequentially crosslinked via light-initiated radical polymerization, increasing the stiffness modulus ($E \sim 30$ kPa) [44].

Thiol-ene and thiol-yne click reactions, such as the one described above, are one common way to provide dynamic substrate stiffening with cellular encapsulation for 3D culture. Click reactions are generally rapid and require mild reaction conditions, which is ideal for hydrogel formation in the presence of biological materials [14]. Thiol-ene reactions proceed following a step-growth mechanism where hydrogen abstraction from the thiol by an initiator can form a thiyl radical. Thiyl radicals then undergo addition to the alkene species in a stepwise manner. Moreover, the rate of the polymerization and polymerization mechanism depends on the choice of the alkene molecule. Thiol-ene reactions are generally executed with electron-rich and/or strained alkenes such as vinyl ethers, acrylates, and norbornenes. By taking advantage of the orthogonality of this chemical reactivity to alkene homopolymerization, hydrogels can be engineered to present dynamic scaffolds for recapitulating key features of lung anatomy and studying changes in microenvironmental mechanical properties that occur during lung disease progression. Although dynamic synthetic hydrogels are a relatively new tool in pulmonary medicine, these biomaterials have been demonstrated to be effective for studying the initiation of fibrosis in other tissues. PDMS, PEG, and polyacrylamide hydrogels are commonly used to probe fibrosis to understand the mechanobiology of the disease [29]. Sundarakrishnan et al., showed mechanical tunability of in vitro culture systems is an important criterion for modeling fibrotic lung disease using mechanically tunable silicone substrates, polyacrylamide, and collagen type I hydrogels to study the effect of matrix stiffness on myofibroblast–ECM signaling [141].

To also study dynamic substrate softening, hydrogels can be made from photoresponsive polymers containing *O*-nitrobenzyl ethers or coumarins, which will photodegrade at specific wavelengths [67]. This attribute enables researchers to grow cells in initially supraphysiologically stiff microenvironments and investigate cellular and molecular responses that occur when returning the microenvironment to physiological levels. A series of studies conducted by the Anseth research group demonstrated that interstitial fibroblasts isolated from heart valve leaflets undergo significant changes in gene expression when cultured on 2D surfaces that were softened [69, 151]. Wang et al, utilized the dynamic behavior of photodegradable PEG to understand which signals regulate de-activation of myofibroblasts during normal tissue repair [151]. Furthermore, a photodegradable monomer/crosslinker was used for synthesizing PEG-based dynamic substates to probe the effects of real-time microenvironment elasticity modulation on dynamic cellular processes [69]. The dynamic mechanical control of these kinds of polymers can be combined with the biochemistry of natural biomaterials to create sophisticated hybrid materials.

### *9.2.5 Hybrid-Hydrogels*

Combining natural materials with synthetic polymers to fabricate a hybrid-hydrogel capable of 3D cell encapsulation is a promising way to leverage the strengths of both natural and synthetic biomaterials (Fig. 9.3). High biological activity, tunable mechanical properties, and biostability for tissue engineering applications can be achieved with these biomaterials. Of particular importance to lung research are hybrid-hydrogels containing dECM and featuring mechanical properties that can be adjusted by photoinitiation following hybridization to hyaluronic acid [64], controlled incorporation of naturally occurring crosslinker genipin [81], or crosslinking to a synthetic polymer such as PEG α-methacrylate (PEGαMA) [112]. For instance, Kosmala et al. demonstrated decreased hydrophobicity and increased cell compatibility of poly(ε-caprolactone) (PCL) by aminolysis followed by immobilization of gelatin [72]. More recently, a biphasic copolymeric membrane has been developed with gelatin and PCL which mimics the microenvironment of alveolar epithelial cells with respect to mechanical, biophysical, and bioactive properties [35]. Decellularized human IPF ECM was combined with a polyacrylamide hydrogel to study the mechanoselective process in IPF by decoupling cellular responses and matrix composition. These hydrogels could be mechanically tuned to low ($E = 1.8 \pm 0.5$ kPa), medium ($E = 4.4 \pm 0.5$ kPa), or high ($E = 23.7 \pm 2.3$ kPa) stiffness, which recapitulated healthy, transitioning, and fibrotic human lung, respectively [125]. In another study, a clickable porcine dECM crosslinker was incorporated into a dynamically responsive PEGαMA hybrid-hydrogel to recreate ECM remodeling in vitro. When exposed to UV light, the initially soft hybrid-hydrogel ($E = 3.6$ kPa) was stiffened to resemble fibrotic lung tissue ($E = 13.4$ kPa) while allowing the matrix to retain the repertoire of cell adhesion proteins and proteoglycans found in vivo. This study demonstrated precise spatiotemporal control over fibroblast activation in response to changes in microenvironmental mechanical properties, as measured by expression patterns of collagen 1 and αSMA [112]. Using the same chemistry, dECM from either healthy or bleomycin-injured mouse

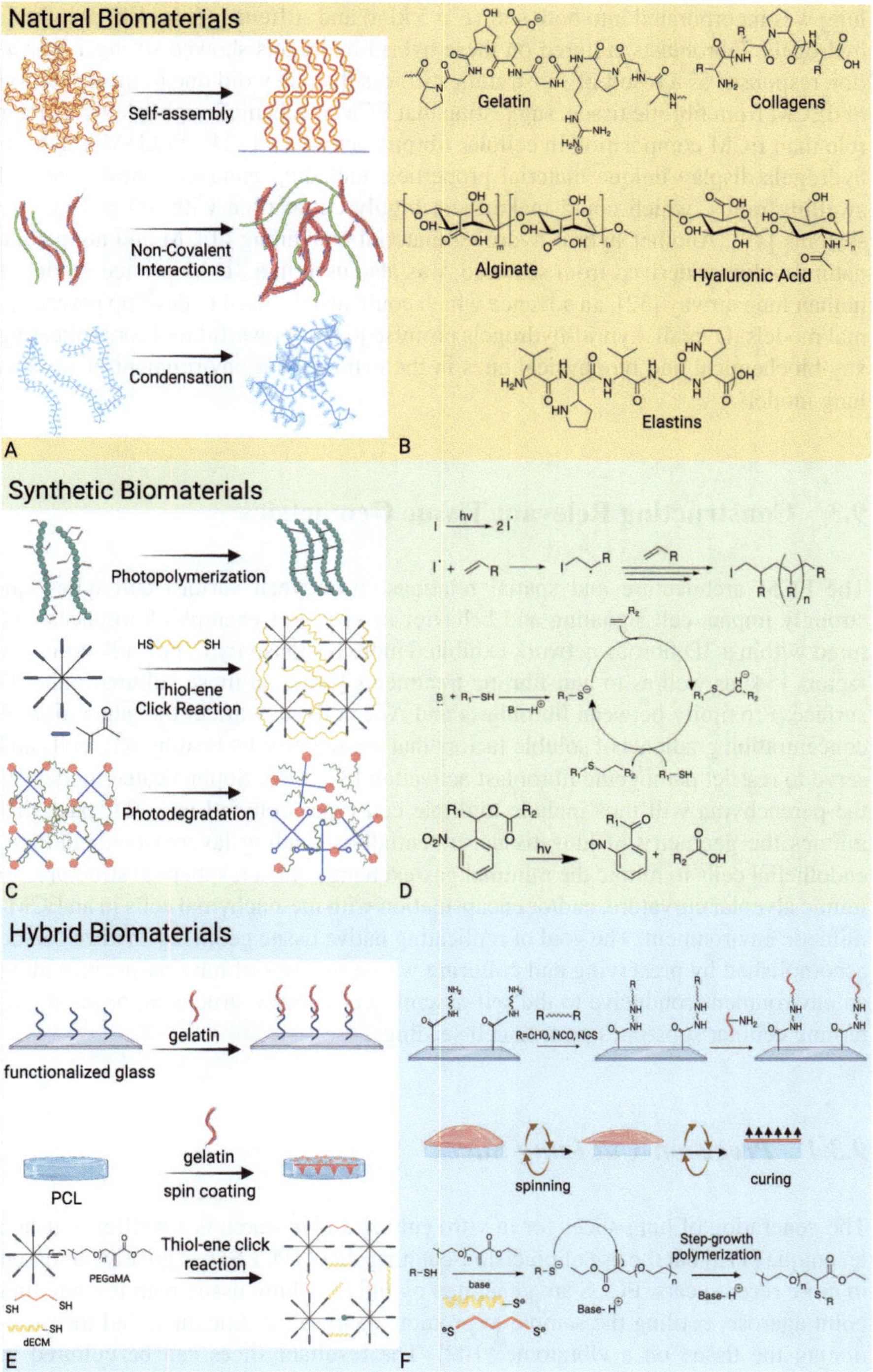


**Fig. 9.3** Crosslinking chemistries for hydrogel fabrication. Natural, synthetic, and hybrid-hydrogels can be developed to rely on diverse chemical reactions that give different levels of control over hydrogel mechanics

lung was incorporated into both soft ($E = 5$ kPa) and stiffened ($E = 11$ kPa) hybrid-hydrogels. Fibroblasts cultured on these hybrid-hydrogels showed stronger activation responses as a result of substratum stiffness than they did due to incorporation of dECM from fibrotic tissue, suggesting that ECM mechanics might player a larger role than ECM composition in cellular fibrotic activation [124]. PEGαMA hybrid-hydrogels display unique material properties, including enhanced mesh size and swelling ratios, which could make them highly compatible with 3D cell culture systems [49]. Another hybrid-hydrogel material containing dECM and alginate, a natural polymer derived from seaweed, was also used in a 3D bioprinted model of human lung airway [32], an advance which could also be used to develop parenchymal models. Overall, hybrid-hydrogels promise to be a powerful tool for replicating key biochemical and biophysical cues in the extracellular environment of in vitro lung models.

## 9.3 Constructing Relevant Tissue Geometries

The ECM architecture and spatial relationship between various cell types can strongly impact cell signaling and behavior in vivo. For example, fibroblasts cultured within a 3D fibrous network exhibited more responsivity to pro-inflammatory factors [54] as well as to anti-fibrotic treatments [88] than those cultured on a 2D surface. Proximity between fibroblasts and AT2 cells may affect the generation of concentration gradients of soluble factors that are secreted by healthy AT2 cells and serve to restrict pathogenic fibroblast activation [37, 143]. Sophisticated models of the parenchyma will thus include multiple cell types cultured in a 3D space that mimics the geometry of lung tissue, potentially including layered epithelial and endothelial cells to mimic the minimal gas-exchange barrier, spherical structures to mimic alveolar curvature, and/or encapsulation with mesenchymal cells in an ECM-mimetic environment. The goal of replicating native tissue geometry in vitro can be accomplished by preserving and culturing whole sections of lung tissue, providing an environment conducive to the self-assembly of cellular structures, or micropatterning cellular substrates prior to cell seeding.

### 9.3.1 *Precision-Cut Lung Slices*

The generation of lung slices for in vitro culture and research is a well-established technique [113] but the use of precision-cut lung slices (PCLS) has greatly increased in more recent years. PCLS are generated by inflating lung tissue with low-melting point agarose, cooling the sample to promote agarose solidification, and then sectioning the tissue on a vibratome [109]. The resultant slices can be cultured in standard tissue culture conditions, under which the agarose washes away and leaves behind a 3D section of lung tissue with all the representative cellular and extracellular components of lung tissue present in their native architecture. Since multiple PCLS can be generated from one animal or human tissue sample, use of PCLS can maximize the utility of tissue samples and reduce biological variability while still allowing for the study of complex multicellular interactions that more reductionist cell culture approaches do not [4].

The major limitation of PCLS is viability in long-term culture. PCLS retain physiological similarities for only 3–7 days [20, 120]. After this time, overall cell viability begins to decrease and more readily cultured cell types—such as mesenchymal cells—begin to outgrow less resilient types—such as AT2 cells [6]. Any use of PCLS, therefore, requires researchers to demonstrate that the slices retain overall viability and maintain baseline levels of any relevant cellular phenotypes over the course of the experiment. Multiple methods exist for determining overall viability, including assaying culture supernatant for metabolites [20, 120], live-dead staining [28], or staining for tissue architecture [6]. Methods do exist for prolonging the ex vivo viability of PCLS. For example, simply increasing slice thickness from 250 μm to 500 μm can increase the viability of PCLS for several days [120]. Embedding PCLS in supportive, biocompatible hydrogels can preserve overall viability and mesenchymal to AT2 cell ratios for a least three weeks in culture [6].

The major advantage of PCLS is the ability to study biological processes that depend on multiple cell types, maintenance of lung architecture, and/or involve multiple regions of the lung. In studies relevant to lung parenchyma, PCLS from fibrotic lungs have been used to screen potential anti-fibrotic drugs or probe relevant cell-signaling pathways. PCLS can also be used as a fully ex vivo model of lung fibrosis, as treatment of PCLS in culture with various fibrotic instigators, such as TGFβ, causes the development of fibrotic phenotypes [56, 66, 76]. Using a fibrotic PCLS model, Lehmann et al. demonstrated that the anti-fibrotic drug nintedanib, in addition to having direct effects on fibrotic fibroblasts, also reduced pro-fibrotic signaling in lung epithelium [76]. In addition to biochemical treatments, exposure to nanostructures can also induce fibrotic changes in PCLS, a model which was used to compare gene expression changes induced by nanostructures to those caused by bleomycin treatment, uncovering potential targets such as Arg1 which may be relevant across a broad range of fibrotic etiologies [120]. Similarly, PCLS have been used as a model for COPD by treatment with elastase, which acts directly on parenchyma through degradation of elastin but resulted in narrowing of airways, demonstrating the interconnectivity between parenchyma and upper airway in the progression of COPD [33]. Recently, tissue-engineered models of lung cancer premalignancy were created using hydrogel-embedded human PCLS [12].These data highlight the potential utility of PCLS, yet even more focused pathway studies can be performed using slightly more reductionist cell culture models.

## 9.3.2 Organoids

The term organoid has been applied to a broad range of 3D cell culture models, from spheroid monocultures that promote the survival and maintenance of sensitive cell types [129] to isolated and cultured primary organ fragments [127]. Organoids of intermediate complexity can involve co-culture of multiple mature cell types, or differentiation of stem cells into a multicellular structure. All these models share an emphasis on growing cells in a physiologically relevant geometry and in a manner that promotes cell-cell contact. There are also multiple methods available to form organoids. Spheroids of robust cell types can be formed by placing cells in a non-adhesive environment, such as hanging drop culture, upon which they will self-aggregate [98]. More complex geometries, such as luminal or branching structures, require embedding cells in more biologically active hydrogel, classically Matrigel [129, 142]. The major advantages of using organoids are that they provide a much more physiologically relevant environment for cell behavior than classical monolayer culture and allow for the study of phenotypes mediated by cell-cell contact. Organoid-based differentiation of stem cells into mature lung cell types also allows for the study of cells that are otherwise difficult to isolate or maintain, such as AT2 cells [61].

Organoids have been used to demonstrate some interesting parenchymal phenotypes. Organoids formed from a combination of epithelial cells, endothelial cells, and fibroblasts and subsequently treated with TGFβ showed reduced budding structures and increased αSMA expression, consistent with observations of lung fibrosis. Interestingly, some αSMA expression was observed in epithelial cells, while classic epithelial markers showed reduced expression, supporting the idea that epithelial cells undergo unique phenotypic changes during fibrosis that might be missed in more fibroblast-focused studies [142]. Similarly, genetic deletion of the HSP1 gene in human embryonic stem cells resulted in the formation of organoids with increased mesenchymal cell markers and endogenous ECM production, providing a potential multi-cell type model for lung fibrosis [21]. Organoid co-culture of primary epithelial cells and lung fibroblasts revealed that pre-treating fibroblasts with TGFβ impaired the ability of cells to form organoids; in this study the authors used organoid formation as a surrogate for healthy epithelial repair [102]. Similarly, in epithelial cell/fibroblast co-culture, WNT-5A and 5B, factors increased in the lungs of COPD patients, inhibited organoid formation, which was again interpreted as impairing healthy epithelial repair [157]. Since in other contexts organoid formation is used as a surrogate for tumorigenesis, these results do highlight the importance of carefully interpreting any results from what is still a very reductionist in vitro system.

Despite these diverse uses and advantages, organoids do have some drawbacks, particularly in terms of parenchymal research. Organoid formation largely depends on cells' ability to self-assemble, and this process can be stochastic. Several studies report that not all the organoids seeded develop the mature structures most relevant to their research questions [75, 102, 129]. Timing is also a factor, since the development of more mature structures, e.g., branching, can take months. Lung organoids derived from pluripotent stem cells require an initial differentiation step into endoderm, a process which can take days to weeks depending on the complexity of the model, followed by maturation and further differentiation into mature lung cell types [21, 36, 61], which often takes several months. Use of lung-resident stem cells can remove the need for this first differentiation, but then may limit the mature cell types present in the organoid; for example, EpCAM$^{high}$CD24$^{low}$Sca-1$^{+}$ epithelial progenitor cells can give rise to a diverse array of lung epithelial cell types, but not mesenchymal or endothelial cells [149]. In addition to development of structures of interest, it can take a very long time for differentiated stem cells to begin to display mature cell type markers, and even after some markers develop it can be questionable whether the organoid truly replicates mature lung, or is more reflective of developing lung [21]. While this makes stem-cell derived organoids an even more powerful tool for developmental studies, it may limit their utility for investigating mature parenchyma. Another major limitation across the field is the reliance on Matrigel in organoid formation protocols. Matrigel is a basement membrane extract from murine cancer cells that, while readily commercially available, may not be fully reflective of healthy lung ECM, has extremely high batch-to-batch variability, and raises concerns about cross-species interactions when used with human cells [2].

To create robust organoid models of the parenchyma, researchers need to consider what cell types they are starting with, and what cell types those might give rise to along the course of the experiment. Different culture substrates and supplements might be necessary to induce or maintain particular phenotypes, and there is not always consistency across the field in knowing what the best of these supplements are. Still, with careful experimental design and interpretation, organoids can provide a relatively simple but physiologically relevant in vitro lung model, particularly as the field develops more robustly reproducible protocols.

### 9.3.3 Engineered 3D Hydrogel Constructs

One aspect of lung biology that has been challenging to reproduce in vitro is the complex geometry of the alveolar region, an important feature given the close relationship between form and function in the lung. Recently, cellular organoid systems have been combined with hydrogel biomaterials to accurately replicate physiological lung microarchitecture in vitro. Hydrogels offer a robust solution to this challenge because hydrogel solutions can be polymerized into predetermined shapes through a variety of processes including molding, emulsification, and 3D printing [1, 7, 53]. Recent developments have exemplified the immense potential these techniques possess for accurately modeling physiologic systems [42]. The unique properties of different polymer backbones make some hydrogel solutions more optimal for certain applications than others. For example, thermosensitive polymers are well suited to molding and extrusion techniques, while photosensitive polymers are more applicable to emulsion or photolithography [7].

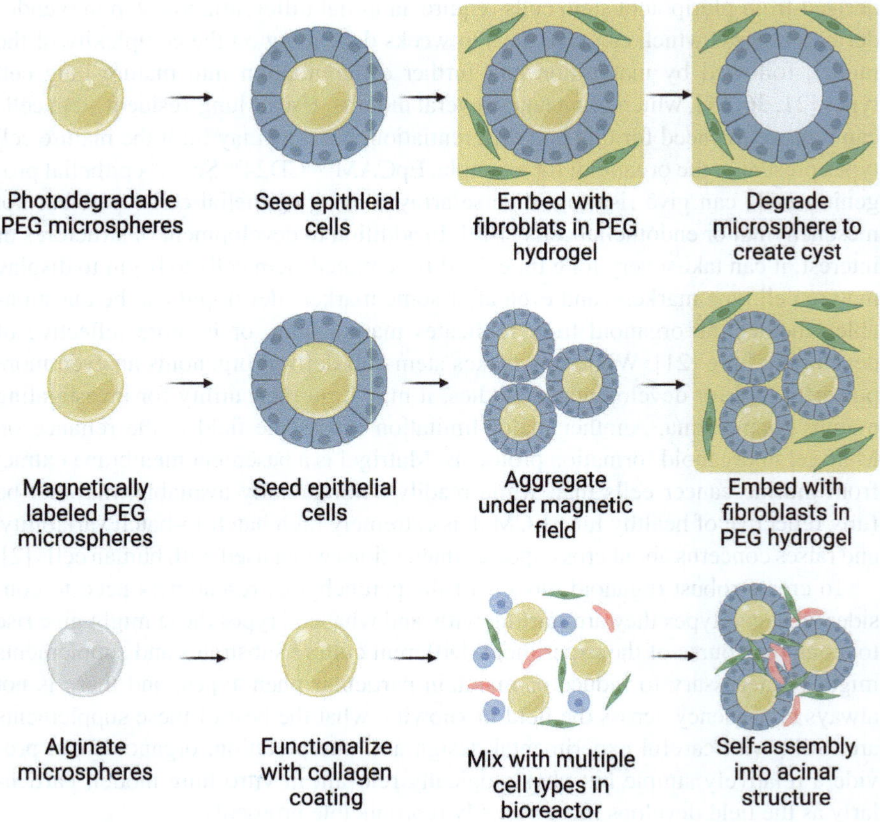


**Fig. 9.4** Models for mimicking alveolar geometry. Three strategies for using hydrogel microspheres as a template for 3D cell culture. Lewis et al. used single photodegradable spheres to create alveolar cyst structures in an encapsulating hydrogel. Caracena et al. used magnetic aggregation to generate acinar structures in an encapsulating hydrogel. Sucre et al. used a bioreactor to promote self-assembly of free-floating acinar structures

These techniques have been employed with different hydrogel polymers to create complex, multi-faceted cell culture scaffolds (Fig. 9.4). These scaffolds can be tailored to specifically mimic the distal lung by controlling the geometric and mechanical properties of the hydrogel to mimic those of native lung tissue. An emulsification polymerization technique has been used to create photodegradable PEG-based microspheres roughly 120 μm in diameter [78]. Lung epithelial cells were seeded in a monolayer around these microspheres in one example and then encased within a nondegradable hydrogel. Removal of the photodegradable microspheres by exposure to UV light resulted in the formation of alveoli-like cyst structures [78]. This template was used as the basis for epithelial cell and fibroblast co-culture and demonstrated that cancerous epithelial cells enhanced the migration and ECM remodeling activity of otherwise healthy fibroblasts, similar to in vivo observations [77]. In another study, 160-μm-diameter alginate microbeads were created by an electrostatic droplet generator, functionalized with collagen, and combined with fibroblasts in a rotating bioreactor. Fibroblast binding acted as a bridge between microspheres and created an aggregate acinar structure [137]. This model was also used with both pluripotent stem cells or multiple mature cell types to generate micropatterned organoids containing epithelial, endothelial, and mesenchymal cells [154]. This study found the 3D geometry of this model to be highly informative in visualizing the in-vitro morphological changes facilitated by fibroblast-mediated ECM remodeling and contraction. To model larger-order acinar structures, Caracena et al. used emulsification polymerization to generate 170 μm PEG-based microspheres with incorporated magnetic nanoparticles. These microspheres could thus be aggregated with primary murine AT2 cells and the resulting acinar structure embedded in a fibroblast-studded encapsulated hydrogel. This model demonstrated the differentiation of AT2 cells to AT1, with a subpopulation arrested in a transitional state. The fibroblasts in this model displayed enhanced activation phenotype when cultured with the primary epithelial cells in a stiff microenvironment, suggesting synergistic effects of cell-cell and cell-matrix interactions in the development of fibrotic cellular phenotypes [19]. Models such as these demonstrate that simultaneously controlling the geometry, mechanics, and chemical composition of the cell culture environment enables the creation of more accurate in vitro tissue models than have previously been possible.

### 9.3.4 3D Bioprinting

Bioprinting is a promising tissue fabrication technology that has developed alongside advances in traditional additive manufacturing processes [22] and has potential to engineer even more sophisticated geometries relevant to the lung parenchyma. It operates by depositing cell-laden printing material (often referred to as bioink) layer by layer, in an effort to recapitulate complex tissue architecture in 3D [84]. Bioprinting has been utilized in the engineering of cardiac tissue to create vasculature models with complex bifurcations and curvature, and to develop biomimetic disease models [85, 91]. Additionally, bioprinting has shown promise in printing transplantable materials, such as bone, skin, and cartilage [84].

Bioprinting systems fall into four broad categories, each with unique strengths and limitations: inkjet, laser-assisted, extrusion bioprinting, and stereolithography. Inkjet bioprinting was the first developed bioprinting technology and makes use of a bioink cartridge and computer-controlled printer head [145]. The bioink cartridge contains a solution of cells and hydrogel pre-polymer, which is fed directly into the printer head. As the printer head goes into position, it squeezes out particles of a controlled size via the assistance of either a thermal or piezoelectric actuator. These systems are relatively cheap to implement and generally have good cell viability [25]. However, inkjet printers are not able to print viscous and cell-dense bioinks. Laser-assisted printing offers even greater cell viability than the traditional inkjet system and works well with viscous materials. In laser-assisted printing, a laser is applied to a metal film, with the bioink solution suspended below. Below this setup is a receiving substrate. As the laser pulses against the metal film, spherical particles of bioink are generated and captured along the substrate, enabling cell-level printing resolution. Despite the advantages of this system, it can be expensive and challenging to implement. Extrusion printing is a derivative of inkjet printing that has been modified to print high viscosity bioinks. Rather than deforming the printer head to dispense bioink droplets, extrusion printers apply force (via air pump or plunger) to push out cylindrical lines. These printers offer the advantage of printing a larger range of materials, but expose cells to significantly greater mechanical stress. As a result, cell viability is decreased [94]. Lastly, stereolithography techniques have been adapted to support high-resolution bioprinting functionality. In this printing method, light is applied to a thin layer of photo-curable polymer liquid in a specific pattern defined by an array of micromirrors until it has formed a solid structure [79]. The build platform then moves down, and another layer is exposed to light to additively build a structure.

In lung research, the Miller lab demonstrated the feasibility of modeling alveolar topography with a surrounding vascular structure using stereolithography [42]. In this landmark tissue engineering accomplishment, scientists were able to not only print complex 3D vascularized alveolar topology but demonstrated the ability to inflate the printed alveolus with oxygen and perfuse oxygen into adjacent blood vessel mimics. Decellularized and solubilized extracellular matrix derived from native tissue has been used as a bioink with tissue-specific composition, which allows for building more complex and realistic tissue architecture. However, using only ECM as the bioink causes slow gelation kinetics which limits the precision of constructs. To obtain consistency and rapid gelation, Wagner and colleagues combined human lung dECM and alginate to form a versatile bioink. Using this material, they generated a 3D bioprinted human airway incorporating regionally specified primary human lung cells, which can differentiate towards mature human airway epithelial cell types [32]. In an effort to manufacture more complex 3D anatomical architectures, the Feinberg lab pioneered the freeform reversible embedding of suspended hydrogels (FRESH) printing method for printing soft tissue structures [51, 131]. The printed material is embedded within a thermoreversible hydrogel support bath; creating structures at a print resolution of ~200 μm with elastic materials such as alginate, collagen, and fibrin. In fact, the utility of this printing approach expanded beyond naturally derived materials, when Hinton et al. demonstrated FRESH printing of the organosilicone elastomer, polydimethylsiloxane (PDMS) (Hinton 2017). In the field of cardiac bioprinting, the FRESH method has been used to print anatomically relevant heart structure in which cardiomyocytes can undergo synchronized contraction [74]. For studying lung biology, 3D printed tubes have been generated to mimic the pulmonary artery, and adventitial fibroblasts printed within these models display stiffness-dependent proliferation and activation [30]. Despite these exciting advances, there is still room for improvement in current bioprinting approaches, including the continued development of bioinks which mimic both the mechanical and chemical properties of tissue, improving spatial resolution for complex ECM structures, achieving greater fidelity of cell deposition, and generating vascularization and innervation in organ models. Nonetheless, with advances in bioink, mechanical reinforcement and multi-material bioprinting, it may well become more feasible to print higher complexity lung structures that better replicate native tissue geometries and cellular interactions.

## 9.4 Incorporating Dynamic Mechanical Forces

The lung is a tissue constantly in motion. Various cell types in the lung experience airflow, blood flow, interstitial fluid pressure, and of course the constant motion of breathing. These physical cues can strongly affect cellular behavior. For example, physiological cyclic stretch of lung organoids in vitro resulted in mesenchymal gene expression changes important for postnatal lung development [63]. In contrast, non-physiological cycle stretch, increases alveolar epithelial cell permeability resulting in ventilator-induced lung injury [28]. Therefore, models that incorporate dynamic mechanical forces recapitulate critical aspects of lung physiology. Such dynamic systems can include application of cyclic stretch to mimic breathing and microfluidics to provide flow.

### *9.4.1 Biomechanical Modeling*

Biomechanics is a field of study that uses mechanical principles to understand the form and function of living structures. Traditional biomechanics research has usually involved applying a mechanical perturbation to a tissue or organ and observing any effects. Observations of macroscopic lung tissue structures under various loading regimes have led to models of linear elastic lung behavior [138] that describe quantitative relationships between the tissue constituent molecules and architecture and overall function. In linear elastic models, stress or force applied per unit area is assumed to be proportional to strain or the measured deformation of the tissue. In this regime, the tissue sample will return to the original shape when it is unloaded, and the stress-strain behavior is independent of loading rate. Although these traditional biomechanical models apply only under static conditions, results from these analyses create baselines for normal versus diseased tissue behavior. Linear elastic models have been used in many cases to test the uniaxial stress and strain behavior of lung parenchymal tissue. For example, Al Jamal et al. used a linear elastic model to explain the contributions of glycosaminoglycans to lung tissue viscoelasticity. Subpleural parenchymal strips were attached on one end to a force transducer and on the other to a lever arm that applied oscillating strains. Applying a linear elastic model to this setup resulted in quasi-static stress-strain curves that suggested that a reduction in tissue elasticity followed heparan sulfate degradation, i.e., the loss of glycosaminoglycans [3].

Despite the utility of linear elastic models, lung tissue mechanics are nonlinear. Measured stress values are not directly proportional to applied strains in lung tissue. Lungs are subject to continuous cyclic stretch during respiration and display dynamic mechanical properties including, hysteresis due to alveolar recruitment and derecruitment. Studies have demonstrated that healthy lung tissue mechanics vary in time- and frequency-domains. Moreover, in the case of diseased lungs, the mechanical behavior becomes increasingly aberrant, and it is critical to accurately model the disrupted alveolar mechanical environment. To accurately model alveolar micromechanics, the Smith lab uses a cyclic ventilator setup [71]. In this methodology, tissue elastance is repetitively measured and partial, quasistatic pressure-volume loops are recorded during oscillatory ventilation. Such studies have revealed that cyclic ventilation can cause ventilator-induced lung injury, which results in both mechanical changes to the elastance of the lung [134], as well as breakdown of the cellular blood-gas barrier [133]. PCLS have also been used in multiple studies of mechanical stretch by attaching PCLS to deformable synthetic membranes and then stretching the membrane [27, 28, 93]. These systems can be used to study biological processes in response to physiological stretch (i.e., breathing) or disease states such as ventilator-induced lung injury. Such studies have demonstrated that cyclic stretch comparable to breathing can induce cell signaling pathways including increases in intracellular calcium [27], nuclear translocation of p65 in AT1 cells [28], and increased secretion of IL-1β [93]. Additionally, PCLS embedded in tunable-stiffness hydrogels can be used to study the effects of varying modulus on integrated tissue responses [6].

In another approach, computational methodologies are often employed for lung simulation and prediction of lung behavior during disease, by utilizing respiratory mechanics data [45, 87, 134]. A commonly used computational method is the compartmental model, due to its computational efficiency and clinical applications for prediction of diseases such as ARDS [100]. Compartmental models place lung properties and anatomies into interconnected compartments and relate these properties through simple mathematical formulas [10]. For example, Bates et al. used a simple compartmental model to simulate the time-dependent nature of alveolar recruitment and derecruitment, wherein volume and pressure are related through an innovative application of the Salazer-Knowles relationship [9].

Understanding lung properties through biomechanical modeling is a complex and evolving discipline. Linear elastic models paved the way for understanding basic mechanical behaviors of lung parenchymal tissue. Since then, developments in lung ventilation have coincided with computational modeling methodologies to create clinically applicable simulations. Future models will aim to combine compartmental modeling strategies with clinical imaging data for high-fidelity, patient specific outcomes. These experiments demonstrate how understanding lung mechanics in both health and disease can contribute to studies of integrated tissue responses to disease.

### 9.4.2 Lung-on-a-Chip

The study of cellular responses to various biomechanical forces can also be studied in complex in vitro systems. Organ-on-a-chip technology involves microfabricated cell culture platforms with precise control over biomechanical parameters such as fluid flow or tensile stretch. Simple microfluidic devices may contain multiple chambers for cell culture connected by channels that allow for directional flow of cell-conditioned media between chambers. This arrangement enables experiments involving paracrine effects of soluble factors within a closed system where fresh media can be supplied without disturbing the co-culture [46]. More complex lung-on-a-chip platforms can create a functional alveolar barrier by building an air-liquid interface with epithelial cells on the apical side of a deformable membrane and endothelial cells on the basal side. By creating a vacuum adjacent to the culture chamber, the deformable membrane undergoes stretch, creating a breath-like movement of the cellular bilayer [58, 136]. The major advantage of these sophisticated culture systems is the ability to integrate multiple cell types in physiologically relevant 3D geometries, and under the controlled application of physiologically relevant biomechanical forces, including fluid and air flow and cyclic stretch.

The major weaknesses of lung-on-a-chip technology are the specialized fabrication required and the resulting presence of non-physiological materials inside the cell culture system. To date, most lung-on-a-chip models have been custom made, and while it is possible and indeed common to fit multiple systems in parallel on one device, truly high-throughput models are rare [90]. Another issue is the use of non-physiological materials. Many lung-on-a-chip systems involve a PDMS membrane at the center, coated with ECM components into order to facilitate adhesion of differing cell types in a monolayer to either side. While coated PDMS can be used to control nutrient diffusion, cellular adhesion, and membrane deformation, it is a non-physiological material that can alter the availability of nutrients and drugs within a culture system [59]. As use of organ-on-a-chip technology advances, however, such devices are becoming both more commercially available—through companies such as Emulate and AlveoliX—and more physiologically relevant, particularly via the inclusion of ECM-based membranes [159].

For studying the parenchyma, multiple existing lung-on-a-chip devices have been developed to model dual epithelial/endothelial layers, and some have also incorporated ECM and/or mesenchymal cells. In one such system, the membrane between cell culture chambers was created by electrospinning poly(lactic-co-glycolic acid), creating a fibrous network that mimicked the geometry of native ECM and was used to study the invasive capacity of tumor cells [158]. In another study, suspended hydrogels incorporating ECM extracts were used in place of a membrane between cell layers. This study then replicated upper airway biology by placing smooth muscle cells opposite epithelial cells, but similar hydrogel technologies could be used to incorporate parenchymal cell types. In the upper airway model, varying the composition of the suspended hydrogel in terms of what ECM molecules were incorporated revealed increased cellular adhesion into collagen gels rather than matrigel [59]. Zamprogno et al. created an ECM-based membrane system that was fully stretchable by adding a mixture of collagen and elastin to a thin gold mesh. The pores in the mesh were small enough to hold the solution in place via surface tension while it formed hydrogels, but large enough for cells seeded on either side to fully interact with the biological components, and these membranes stretched without breaking under vacuum [159]. These studies all demonstrated a capacity to replace flat, synthetic membranes with more physiological ECM-based layers into dual-chamber lung-on-a-chip devices. To incorporate mesenchymal cells into their lung-on-a-chip, Mejías et al. mixed lung fibroblasts with endothelial cells in a lower hydrogel layer, upon which epithelial cells were cultured at an air-liquid interface. By using microfluidics to create a fibroblast-conditioned media gradient, endothelial cells underwent directional and three-dimensional vascularization within the hydrogel. This system was then used to recapitulate fibrotic lung by incorporating TGFβ treatment and fibroblasts from human fibrotic lung, which demonstrated enhanced fibroblast activation and putative epithelial cell differentiation in the fibrotic conditions [90]. For modeling other lung diseases, lung-on-a-chip devices have been used to demonstrate that cyclic stretch enhances therapeutic resistance in cancer cells [48] and impaired re-epithelialization in a wound healing assay [38]. Designing the most relevant lung-on-a-chip model requires precise knowledge of the most relevant biochemical and biomechanical parameters for any given question but can provide a very sophisticated model of local lung tissue.

## 9.5 Conclusion

Technologies for engineering models of lung parenchyma have vastly expanded in recent years. Biomaterials provide a way to precisely mimic the biochemical and biomechanical extracellular microenvironment sensed by cells in vivo and are thus of great value in modeling cell/ECM interactions. Various modalities for assembling cellular structures allow for the growth of cells in 3D geometries specifically modeled on parenchymal anatomy, making them ideal for studying interactions between multiple cell types. Incorporation of cyclic stretch, fluid flow, and other biophysical forces creates models that recapitulate the dynamic motion of lung tissue and enable study of mechanosensing in lung parenchyma. All these technologies exhibit individual strengths and weaknesses but can be used in complementary studies to improve our ability to model lung homeostasis and disease in vitro. For example, culturing stem cells within microstructured hydrogels, a combination of organoid and biomaterials technologies, yields a model with multiple cell types present in a physiologically relevant geometry [154]. Investigating PCLS in models of cyclic stretch preserves both the native architecture and mechanical stimulus of intact lung [27, 28, 93]. Such combinatorial approaches can even be used to look beyond the lung and study multi-organ interactions. In a 2017 study, Skardal et al. created a linked microfluidic system modeling lung, liver, and cardiac tissue. The liver and cardiac modules consisted of 3D bioprinted organoids within native ECM hydrogels, while the lung module consisted of epithelial and endothelial layers on a porous membrane, as in a lung-on-a-chip device. This multi-organ system was employed to detect signs of cardiac toxicity when the lung module was exposed to bleomycin [132]. Even more ambitiously, in 2020 Novak et al. designed a linked microfluidic system with eight discrete organ modules in a whole-body-on-a-chip model [103]. This study was carried out by 57 authors across nine institutions from five different countries in an impressive example of collaboration and ingenuity.

The future of the field is promising and may lead in multiple directions. First, we envision the refinement of existing models based on the most advanced and up-to-date characterization of lung tissue will occur rapidly over time. The more that is learned about lung architecture, mechanics, composition, and function from in vivo studies, the better-informed in vitro modeling can become. There is also space here for in silico computational models of lung, to be both built off of existing knowledge and to further guide engineered models for cell culture, disease modeling, and tissue engineering [16]. Individual labs must be guided by focused research questions and use the models that are equipped to replicate the parameters most relevant to the research in question, even in a reductionist system. Finally, researchers of different specialties should continue to seek interdisciplinary collaborations to combine expertise, learn new perspectives, and together build ever more advanced translational models of lung disease and repair that will advance our understanding of and ability to treat human disease.

## References


1. Aimar, A., Palermo, A., & Innocenti, B. (2019). The Role of 3D Printing in Medical Applications: A State of the Art. *J Healthc Eng*, 2019, 5340616. https://doi.org/10.1155/2019/5340616.
2. Aisenbrey, E. A., & Murphy, W. L. (2020). Synthetic alternatives to Matrigel. *Nature reviews. Materials*, 5(7). https://doi.org/10.1038/s41578-020-0199-8.
3. Al Jamal, R., Roughley, P., & Ludwig, M. (2001). Effect of glycosaminoglycan degradation on lung tissue viscoelasticity. *American journal of physiology. Lung cellular and molecular physiology*, 280(2). https://doi.org/10.1152/ajplung.2001.280.2.L306.
4. Alsafadi, H. N., Uhl, F. E., Pineda, R. H., Bailey, K. E., Rojas, M., Wagner, D. E., et al. (2020). Applications and Approaches for Three-Dimensional Precision-Cut Lung Slices. Disease Modeling and Drug Discovery. *American journal of respiratory cell and molecular biology*, 62(6). https://doi.org/10.1165/rcmb.2019-0276TR.
5. Bailey, K. E., Floren, M. L., D'Ovidio, T. J., Lammers, S. R., Stenmark, K. R., & Magin, C. M. (2019). Tissue-informed engineering strategies for modeling human pulmonary diseases. *American journal of physiology. Lung cellular and molecular physiology*, 316(2). https://doi.org/10.1152/ajplung.00353.2018.
6. Bailey, K. E., Pino, C., Lennon, M. L., Lyons, A., Jacot, J. G., Lammers, S. R., et al. (2020). Embedding of Precision-Cut Lung Slices in Engineered Hydrogel Biomaterials Supports Extended Ex Vivo Culture. *American journal of respiratory cell and molecular biology*, 62(1). https://doi.org/10.1165/rcmb.2019-0232MA.
7. Bajaj, P., Schweller, R. M., Khademhosseini, A., West, J. L., & Bashir, R. (2014). 3D biofabrication strategies for tissue engineering and regenerative medicine. *Annu Rev Biomed Eng*, 16, 247–76. https://doi.org/10.1146/annurev-bioeng-071813-105155.

8. Baptista, D., Teixeira, L., Blitterswijk, C. van Giselbrecht, S., & Truckenmüller, R. (2019). Overlooked? Underestimated? Effects of Substrate Curvature on Cell Behavior. *Trends in biotechnology*, 37(8). https://doi.org/10.1016/j.tibtech.2019.01.006.
9. Bates, J. H. T., & Irvin, C. G. (2002). Time dependence of recruitment and derecruitment in the lung: a theoretical model. *Journal of Applied Physiology*, 93, 705–713. https://doi.org/10.1152/japplphysiol.01274.2001.
10. Ben-Tal, A. (2006). Simplified models for gas exchange in the human lungs. *Journal of Theoretical Biology*, 238, 474–495. https://doi.org/10.1016/j.jtbi.2005.06.005.
11. Bilodeau, C., Shojaie, S., Goltsis, O., Wang, J., Luo, D., Ackerley, C., et al. (2021). TP63 basal cells are indispensable during endoderm differentiation into proximal airway cells on acellular lung scaffolds. *NPJ Regenerative medicine*, 6(1). https://doi.org/10.1038/s41536-021-00124-4.
12. Blomberg, R., Sompel, K., Hauer, C., Peña, B., Driscoll, J., Hume, P. S., Merrick, D. T., Tennis, M. A., & Magin, C. M. (2023). Tissue-engineered models of lung cancer premalignancy. bioRxiv. https://doi.org/10.1101/2023.03.15.532835
13. Booth, A. J., Hadley, R., Cornett, A. M., Dreffs, A. A., Matthes, S. A., Tsui, J. L., et al. (2012). Acellular normal and fibrotic human lung matrices as a culture system for in vitro investigation. *American journal of respiratory and critical care medicine*, 186(9). https://doi.org/10.1164/rccm.201204-0754OC.
14. Brown, T. E., & Anseth, K. S. (2017). Spatiotemporal hydrogel biomaterials for regenerative medicine. *Chemical Society reviews*, 46(21). https://doi.org/10.1039/c7cs00445a.
15. Burgstaller, G., Gerckens, M., Eickelberg, O., & Königshoff, M. (2021). Decellularized Human Lung Scaffolds as Complex Three-Dimensional Tissue Culture Models to Study Functional Behavior of Fibroblasts. *Methods in molecular biology (Clifton, N.J.)*, 2299. https://doi.org/10.1007/978-1-0716-1382-5_30.
16. Burrowes, K. S., Swan, A. J., Warren, N. J., & Tawhai, M. H. (2008). Towards a virtual lung: multi-scale, multi-physics modelling of the pulmonary system. *Philosophical transactions. Series A, Mathematical, physical, and engineering sciences*, 366(1879). https://doi.org/10.1098/rsta.2008.0073.
17. Bölükbas, D. A., Santis, M. M. D., Alsafadi, H. N., Doryab, A., & Wagner, D. E. (2019). The Preparation of Decellularized Mouse Lung Matrix Scaffolds for Analysis of Lung Regenerative Cell Potential. *Methods in molecular biology (Clifton, N.J.)*, 1940. https://doi.org/10.1007/978-1-4939-9086-3_20.
18. Callens, S. J. P., Uyttendaele, R. J. C., Fratila-Apachitei, L. E., & Zadpoor, A. A. (2020). Substrate curvature as a cue to guide spatiotemporal cell and tissue organization. *Biomaterials*, 232. https://doi.org/10.1016/j.biomaterials.2019.119739.
19. Caracena, T., Blomberg, R., Hewawasam, R. S., Fry, Z. E., Riches, D. W. H., & Magin, C. M. (2022). Alveolar epithelial cells and microenvironmental stiffness synergistically drive fibroblast activation in three-dimensional hydrogel lung models. *Biomaterials Science*. https://doi.org/10.1039/d2bm00827k.
20. Cedilak, M., Banjanac, M., Belamarić, D., Radičević, A. P., Faraho, I., Ilić, K., et al. (2019). Precision-cut lung slices from bleomycin treated animals as a model for testing potential therapies for idiopathic pulmonary fibrosis. *Pulmonary pharmacology & therapeutics*, 55. https://doi.org/10.1016/j.pupt.2019.02.005.
21. Chen, Y.-W., Huang, S. X., Carvalho, A. L. R. T. de Ho, S.-H., Islam, M. N., Volpi, S., et al. (2017). A three-dimensional model of human lung development and disease from pluripotent stem cells (Text). https://doi.org/10.1038/ncb3510.
22. Chimene, D., Kaunas, R., & Gaharwar, A. K. (2020). Hydrogel Bioink Reinforcement for Additive Manufacturing: A Focused Review of Emerging Strategies. *Advanced materials (Deerfield Beach, Fla.)*, 32(1). https://doi.org/10.1002/adma.201902026.
23. Correll, K. A., Edeen, K. E., Zemans, R. L., Redente, E. F., Serban, K. A., Curran-Everett, D., et al. (2019). Transitional human alveolar type II epithelial cells suppress extracellular matrix and growth factor gene expression in lung fibroblasts. *American Journal of Physiology-*

*Lung Cellular and Molecular Physiology*, 317(2), L283–L294. https://doi.org/10.1152/ajplung.00337.2018.
24. Crapo, P. M., Gilbert, T. W., & Badylak, S. F. (2011). An overview of tissue and whole organ decellularization processes. *Biomaterials*, 32(12). https://doi.org/10.1016/j.biomaterials.2011.01.057.
25. Cui, X., Boland, T., D'Lima, D. D., & Lotz, M. K. (2012). Thermal inkjet printing in tissue engineering and regenerative medicine. *Recent patents on drug delivery & formulation*, 6(2). https://doi.org/10.2174/187221112800672949.
26. Dabaghi, M., Saraei, N., Carpio, M. B., Nanduri, V., Ungureanu, J., Babi, M., et al. (2021). A Robust Protocol for Decellularized Human Lung Bioink Generation Amenable to 2D and 3D Lung Cell Culture. *Cells*, 10(6). https://doi.org/10.3390/cells10061538.
27. Dassow, C., Wiechert, L., Martin, C., Schumann, S., Müller-Newen, G., Pack, O., et al. (2010). Biaxial distension of precision-cut lung slices. *Journal of applied physiology (Bethesda, Md.: 1985)*, 108(3). https://doi.org/10.1152/japplphysiol.00229.2009.
28. Davidovich, N., Huang, J., & Margulies, S. S. (2013). Reproducible uniform equibiaxial stretch of precision-cut lung slices. *American journal of physiology. Lung cellular and molecular physiology*, 304(4). https://doi.org/10.1152/ajplung.00224.2012.
29. Davidson, M. D., Burdick, J. A., & Wells, R. G. (2020). Engineered Biomaterial Platforms to Study Fibrosis. *Advanced healthcare materials*, 9(8). https://doi.org/10.1002/adhm.201901682.
30. Davis-Hall, D., Thomas, E., Peña, B., & Magin, C. M. (2023). 3D-bioprinted, phototunable hydrogel models for studying adventitial fibroblast activation in pulmonary arterial hypertension. *Biofabrication*, 15(1), 015017. https://doi.org/10.1088/1758-5090/aca8cf.
31. deHilster, R. H. J., Sharma, P. K., Jonker, M. R., White, E. S., Gercama, E. A., Roobeek, M., et al. (2020). Human lung extracellular matrix hydrogels resemble the stiffness and viscoelasticity of native lung tissue. *American journal of physiology. Lung cellular and molecular physiology*, 318(4). https://doi.org/10.1152/ajplung.00451.2019.
32. DeSantis, M. M., Alsafadi, H. N., Tas, S., Bölükbas, D. A., Prithiviraj, S., Silva, I. A. N. D., et al. (2020). Extracellular-Matrix-Reinforced Bioinks for 3D Bioprinting Human Tissue. *Advanced materials (Deerfield Beach, Fla.)*. https://doi.org/10.1002/adma.202005476.
33. Dijk, E. M. V., Culha, S., Menzen, M. H., Bidan, C. M., & Gosens, R. (2017). Elastase-Induced Parenchymal Disruption and Airway Hyper Responsiveness in Mouse Precision Cut Lung Slices: Toward an Ex vivo COPD Model. *Frontiers in physiology*, 7. https://doi.org/10.3389/fphys.2016.00657.
34. Dorrello, N. V., Guenthart, B. A., O'Neill, J. D., Kim, J., Cunningham, K., Chen, Y.-W., et al. (2017). Functional vascularized lung grafts for lung bioengineering. *Science advances*, 3(8). https://doi.org/10.1126/sciadv.1700521.
35. Doryab, A. (2021). A Biomimetic, Copolymeric Membrane for Cell-Stretch Experiments with Pulmonary Epithelial Cells at the Air-Liquid Interface – Doryab – 2021 – Advanced Functional Materials – Wiley Online Library. https://doi.org/10.1002/adfm.202004707.
36. Dye, B. R., Hill, D. R., Ferguson, M. A. H., Tsai, Y.-H., Nagy, M. S., Dyal, R., et al. (2015). In vitro generation of human pluripotent stem cell derived lung organoids. *eLife*, 4. https://doi.org/10.7554/eLife.05098.
37. Epa, A. P., Thatcher, T. H., Pollock, S. J., Wahl, L. A., Lyda, E., Kottmann, R. M., et al. (2015). Normal Human Lung Epithelial Cells Inhibit Transforming Growth Factor-β Induced Myofibroblast Differentiation via Prostaglandin E2. *PloS one*, 10(8). https://doi.org/10.1371/journal.pone.0135266.
38. Felder, M., Trueb, B., Stucki, A. O., Borcard, S., Stucki, J. D., Schnyder, B., et al. (2019). Impaired Wound Healing of Alveolar Lung Epithelial Cells in a Breathing Lung-On-A-Chip. *Frontiers in bioengineering and biotechnology*, 7. https://doi.org/10.3389/fbioe.2019.00003.
39. George, U. Z., & Lubkin, S. R. (2018). Tissue geometry may govern lung branching mode selection. *Journal of theoretical biology*, 442. https://doi.org/10.1016/j.jtbi.2017.12.031.

40. Gharenaz, N. M., Movahedin, M., & Mazaheri, Z. (2021). Comparison of two methods for prolong storage of decellularized mouse whole testis for tissue engineering application: An experimental study. *International journal of reproductive biomedicine*, 19(4). https://doi.org/10.18502/ijrm.v19i4.9058.
41. Girard, E. D., Jensen, T. J., Vadasz, S. D., Blanchette, A. E., Zhang, F., Moncada, C., et al. (2013). Automated procedure for biomimetic de-cellularized lung scaffold supporting alveolar epithelial transdifferentiation. *Biomaterials*, 34(38). https://doi.org/10.1016/j.biomaterials.2013.09.055.
42. Grigoryan, B., Paulsen, S. J., Corbett, D. C., Sazer, D. W., Fortin, C. L., Zaita, A. J., et al. (2019). Multivascular networks and functional intravascular topologies within biocompatible hydrogels. *Science*, 364(6439), 458–464. https://doi.org/10.1126/science.aav9750.
43. Guenthart, B. A., O'Neill, J. D., Kim, J., Fung, K., Vunjak-Novakovic, G., & Bacchetta, M. (2019). Cell replacement in human lung bioengineering. *The Journal of heart and lung transplantation: the official publication of the International Society for Heart Transplantation*, 38(2). https://doi.org/10.1016/j.healun.2018.11.007.
44. Guvendiren, M., & Burdick, J. A. (2012). Stiffening hydrogels to probe short- and long-term cellular responses to dynamic mechanics (Article). *Nature Communications*, 3, 792. https://doi.org/10.1038/ncomms1792. https://www.nature.com/articles/ncomms1792#supplementary-information.
45. Hamlington, K. L., Smith, B. J., Allen, G. B., & Bates, J. H. T. (2016). Predicting ventilator-induced lung injury using a lung injury cost function. *Journal of Applied Physiology*, 121(1), 106–114. https://doi.org/10.1152/japplphysiol.00096.2016 (Accessed 2023/03/20).
46. Hao, Y., Zhang, L., He, J., Guo, Z., Ying, L., Xu, Z., et al. (2013). Functional investigation of NCI-H460-inducible myofibroblasts on the chemoresistance to VP-16 with a microfluidic 3D co-culture device. *PloS one*, 8(4). https://doi.org/10.1371/journal.pone.0061754.
47. Hashimoto, Y., Tsuchiya, T., Doi, R., Matsumoto, K., Higami, Y., Kobayashi, E., et al. (2019). Alteration of the extracellular matrix and alpha-gal antigens in the rat lung scaffold reseeded using human vascular and adipogenic stromal cells. *Journal of tissue engineering and regenerative medicine*, 13(11). https://doi.org/10.1002/term.2923.
48. Hassell, B. A., Goyal, G., Lee, E., Sontheimer-Phelps, A., Levy, O., Chen, C. S., et al. (2017). Human Organ Chip Models Recapitulate Orthotopic Lung Cancer Growth, Therapeutic Responses, and Tumor Dormancy In Vitro. *Cell reports*, 21(2). https://doi.org/10.1016/j.celrep.2017.09.043.
49. Hewawasam, R. S., Blomberg, R., Šerbedžija, P., & Magin, C. M. (2023). Chemical modification of human decellularized extracellular matrix for incorporation into phototunable hybrid-hydrogel models of tissue fibrosis. *ACS Applied Materials & Interfaces*. https://doi.org/10.1021/acsami.2c18330.
50. Hill, C., Jones, M. G., Davies, D. E., & Wang, Y. (2019). Epithelial-mesenchymal transition contributes to pulmonary fibrosis via aberrant epithelial/fibroblastic cross-talk. *Journal of lung health and diseases*, 3(2), 31–35.
51. Hinton, T. J., Jallerat, Q., Palchesko, R. N., Park, J. H., Grodzicki, M. S., Shue, H.-J., Ramadan, M. H., Hudson, A. R., & Feinberg, A. W. (2015). Three-dimensional printing of complex biological structures by freeform reversible embedding of suspended hydrogels. *Biomedical Engineering*, 1(9), e1500758. https://doi.org/10.1126/sciadv.1500758.
52. Hinton, T. J., Hudson, A., Pusch, K., Lee, A., & Feinberg, A. W. (2016). 3D printing PDMS elastomer in a hydrophilic support bath via freeform reversible embedding. *ACS Biomaterials Science & Engineering*, 2, 1781–1786. https://doi.org/10.1021/acsbiomaterials.6b00170.
53. Horváth, L., Umehara, Y., Jud, C., Blank, F., Petri-Fink, A., & Rothen-Rutishauser, B. (2015). Engineering an in vitro air-blood barrier by 3D bioprinting. *Sci Rep*, 5, 7974. https://doi.org/10.1038/srep07974.
54. Htwe, S. S., Harrington, H., Knox, A., Rose, F., Aylott, J., Haycock, J. W., et al. (2015). Investigating NF-κB signaling in lung fibroblasts in 2D and 3D culture systems. *Respiratory research*, 16. https://doi.org/10.1186/s12931-015-0302-7.

55. Hu, X. (2021). Protein-based composite materials | Elsevier Enhanced Reader. https://doi.org/10.1016/S1369-7021(12)70091-3.
56. Huang, X., Li, L., Ammar, R., Zhang, Y., Wang, Y., Ravi, K., et al. (2019). Molecular characterization of a precision-cut rat lung slice model for the evaluation of antifibrotic drugs. *American journal of physiology. Lung cellular and molecular physiology*, 316(2). https://doi.org/10.1152/ajplung.00339.2018.
57. Hubbell, J. (1995). Biomaterials in tissue engineering. *Bio/technology (Nature Publishing Company)*, 13(6). https://doi.org/10.1038/nbt0695-565.
58. Huh, D., Matthews, B. D., Mammoto, A., Montoya-Zavala, M., Hsin, H. Y., & Ingber, D. E. (2010). Reconstituting organ-level lung functions on a chip. *Science (New York, N.Y.)*, 328(5986). https://doi.org/10.1126/science.1188302.
59. Humayun, M., Chow, C.-W., & Young, E. W. K. (2018). Microfluidic lung airway-on-a-chip with arrayable suspended gels for studying epithelial and smooth muscle cell interactions. *Lab on a chip*, 18(9). https://doi.org/10.1039/c7lc01357d.
60. Isakson, B. E., Lubman, R. L., Seedorf, G. J., & Boitano, S. (2001). Modulation of pulmonary alveolar type II cell phenotype and communication by extracellular matrix and KGF (research-article). https://doi.org/10.1152/ajpcell.2001.281.4.C1291.
61. Jacob, A., Morley, M., Hawkins, F., McCauley, K. B., Jean, J. C., Heins, H., et al. (2017). Differentiation of Human Pluripotent Stem Cells into Functional Lung Alveolar Epithelial Cells. *Cell stem cell*, 21(4). https://doi.org/10.1016/j.stem.2017.08.014.
62. Janssen, R., Piscaer, I., Franssen, F. M. E., & Wouters, E. F. M. (2019). Emphysema: looking beyond alpha-1 antitrypsin deficiency. *Expert review of respiratory medicine*, 13(4). https://doi.org/10.1080/17476348.2019.1580575.
63. Joshi, R., Batie, M. R., Fan, Q., & Varisco, B. M. (2022). Mouse lung organoid responses to reduced, increased, and cyclic stretch (research-article). https://doi.org/10.1152/ajplung.00310.2020.
64. Jung, M., Han, Y., Woo, C., & Ki, C. S. (2021). Pulmonary tissue-mimetic hydrogel niches for small cell lung cancer cell culture. *Journal of materials chemistry. B*, 9(7). https://doi.org/10.1039/d0tb02609c.
65. Kandow, C. E., Georges, P. C., Janmey, P. A., & Beningo, K. A. (2007). Polyacrylamide hydrogels for cell mechanics: steps toward optimization and alternative uses. *Methods in cell biology*, 83. https://doi.org/10.1016/S0091-679X(07)83002-0.
66. Khan, M. M., Poeckel, D., Halavatyi, A., Zukowska-Kasprzyk, J., Stein, F., Vappiani, J., et al. (2020). An integrated multiomic and quantitative label-free microscopy-based approach to study pro-fibrotic signalling in ex vivo human precision-cut lung slices. *The European respiratory journal*. https://doi.org/10.1183/13993003.00221-2020.
67. Kirschner, C. M., & Anseth, K. S. (2013). Hydrogels in Healthcare: From Static to Dynamic Material Microenvironments. *Acta materialia*, 61(3). https://doi.org/10.1016/j.actamat.2012.10.037.
68. Kitano, K., Ohata, K., Economopoulos, K. P., Gorman, D. E., Gilpin, S. E., Becerra, D. C., et al. (2021). Orthotopic Transplantation of Human Bioartificial Lung Grafts in a Porcine Model: A Feasibility Study. *Seminars in thoracic and cardiovascular surgery*. https://doi.org/10.1053/j.semtcvs.2021.03.006.
69. Kloxin, A. M., Benton, J. A., & Anseth, K. S. (2010). In situ elasticity modulation with dynamic substrates to direct cell phenotype. *Biomaterials*, 31(1). https://doi.org/10.1016/j.biomaterials.2009.09.025.
70. Kloxin, A. M., Lewis, K. J. R., DeForest, C. A., Seedorf, G., Tibbitt, M. W., Balasubramaniam, V., et al. (2012). Responsive culture platform to examine the influence of microenvironmental geometry on cell function in 3D. *Integrative biology: quantitative biosciences from nano to macro*, 4(12). https://doi.org/10.1039/c2ib20212c.
71. Knudsen, L., & Ochs, M. (2018). The micromechanics of lung alveoli: structure and function of surfactant and tissue components. *Histochemistry and cell biology*, 150(6). https://doi.org/10.1007/s00418-018-1747-9.

72. Kosmala, A., Fitzgerald, M., Moore, E., & Stam, F. (2016). Evaluation of a Gelatin-Modified Poly(ε-Caprolactone) Film as a Scaffold for Lung Disease (research-article). https://doi.org/10.1080/00032719.2016.1163363.
73. Kruk, D. M. L. W., Wisman, M., Bruin, H. G.D., Lodewijk, M. E., Hof, D. J., Borghuis, T., et al. (2021). Abnormalities in reparative function of lung-derived mesenchymal stromal cells in emphysema. *American journal of physiology. Lung cellular and molecular physiology*, 320(5). https://doi.org/10.1152/ajplung.00147.2020.
74. Lee, A., Hudson, A. R., Shiwarski, D. J., Tashman, J. W., Hinton, T. J., Yerneni, S., et al. (2019). 3D bioprinting of collagen to rebuild components of the human heart. *Science (New York, N.Y.)*, 365(6452). https://doi.org/10.1126/science.aav9051.
75. Lee, J.-H., Bhang, D. H., Beede, A., Huang, T. L., Stripp, B. R., Bloch, K. D., et al. (2014). Lung stem cell differentiation in mice directed by endothelial cells via a BMP4-NFATc1-thrombospondin-1 axis. *Cell*, 156(3). https://doi.org/10.1016/j.cell.2013.12.039.
76. Lehmann, M., Buhl, L., Alsafadi, H. N., Klee, S., Hermann, S., Mutze, K., et al. (2018). Differential effects of Nintedanib and Pirfenidone on lung alveolar epithelial cell function in ex vivo murine and human lung tissue cultures of pulmonary fibrosis. *Respiratory research*, 19(1). https://doi.org/10.1186/s12931-018-0876-y.
77. Lewis, K. J. R., Hall, J. K., Kiyotake, E. A., Christensen, T., Balasubramaniam, V., & Anseth, K. S. (2018). Epithelial-mesenchymal crosstalk influences cellular behavior in a 3D alveolus-fibroblast model system. *Biomaterials*, 155. https://doi.org/10.1016/j.biomaterials.2017.11.008.
78. Lewis, K. J. R., Tibbitt, M. W., Zhao, Y., Branchfield, K., Sun, X., Balasubramaniam, V., et al. (2015). In vitro model alveoli from photodegradable microsphere templates. *Biomaterials science*, 3(6). https://doi.org/10.1039/c5bm00034c.
79. Li, J., Chen, M., Fan, X., & Zhou, H. (2016). Recent advances in bioprinting techniques: approaches, applications and future prospects. *Journal of translational medicine*, 14. https://doi.org/10.1186/s12967-016-1028-0.
80. Li, Y., Wu, Q., Li, L., Chen, F., Bao, J., & Li, W. (2021). Decellularization of porcine whole lung to obtain a clinical-scale bioengineered scaffold. *Journal of biomedical materials research. Part A*, 109(9). https://doi.org/10.1002/jbm.a.37158.
81. Link, P. A., Pouliot, R. A., Mikhaiel, N. S., Young, B. M., & Heise, R. L. (2017). Tunable Hydrogels from Pulmonary Extracellular Matrix for 3D Cell Culture. *Journal of visualized experiments: JoVE*, (119). https://doi.org/10.3791/55094.
82. Liu, F., Mih, J. D., Shea, B. S., Kho, A. T., Sharif, A. S., Tager, A. M., et al. (2010). Feedback amplification of fibrosis through matrix stiffening and COX-2 suppression. *The Journal of cell biology*, 190(4). https://doi.org/10.1083/jcb.201004082.
83. Mabry, K. M., Payne, S. Z., & Anseth, K. S. (2016). Microarray analyses to quantify advantages of 2D and 3D hydrogel culture systems in maintaining the native valvular interstitial cell phenotype. *Biomaterials*, 74, 31–41. https://doi.org/10.1016/j.biomaterials.2015.09.035.
84. Mandrycky, C., Wang, Z., Kim, K., & Kim, D.-H. (2016). 3D bioprinting for engineering complex tissues. *Biotechnology advances*, 34(4). https://doi.org/10.1016/j.biotechadv.2015.12.011.
85. Mandrycky, C., Wang, Z., Kim, K., & Kim, D.-H. (2021). 3D Bioprinted Multicellular Vascular Models. *Advanced healthcare materials*, 10(21). https://doi.org/10.1002/adhm.202101141.
86. Marinković, A., Liu, F., & Tschumperlin, D. J. (2013). Matrices of physiologic stiffness potently inactivate idiopathic pulmonary fibrosis fibroblasts. *American journal of respiratory cell and molecular biology*, 48(4). https://doi.org/10.1165/rcmb.2012-0335OC.
87. Massa, C. B., Allen, G. B., & Bates, J. H. T. (2008). Modeling the dynamics of recruitment and derecruitment in mice with acute lung injury. *Journal of Applied Physiology*, 105, 1813–1821. https://doi.org/10.1152/japplphysiol.90806.2008.
88. Matera, D. L., DiLillo, K. M., Smith, M. R., Davidson, C. D., Parikh, R., Said, M., et al. (2020). Microengineered 3D pulmonary interstitial mimetics highlight a critical role for

matrix degradation in myofibroblast differentiation. *Science Advances*, 6(37), eabb5069. https://doi.org/10.1126/sciadv.abb5069.
89. McCrary, M. W., Bousalis, D., Mobini, S., Song, Y. H., & Schmidt, C. E. (2020). Decellularized tissues as platforms for in vitro modeling of healthy and diseased tissues. *Acta biomaterialia*, 111. https://doi.org/10.1016/j.actbio.2020.05.031.
90. Mejías, J. C., Nelson, M. R., Liseth, O., & Roy, K. (2020). A 96-well format microvascularized human lung-on-a-chip platform for microphysiological modeling of fibrotic diseases. *Lab on a chip*, 20(19). https://doi.org/10.1039/d0lc00644k.
91. Memic, A., Navaei, A., Mirani, B., Cordova, J. A. V., Aldhahri, M., Dolatshahi-Pirouz, A., et al. (2017). Bioprinting technologies for disease modeling. *Biotechnology letters*, 39(9). https://doi.org/10.1007/s10529-017-2360-z.
92. Mishra, D. K., Sakamoto, J. H., Thrall, M. J., Baird, B. N., Blackmon, S. H., Ferrari, M., et al. (2012). Human lung cancer cells grown in an ex vivo 3D lung model produce matrix metalloproteinases not produced in 2D culture. *PloS one*, 7(9). https://doi.org/10.1371/journal.pone.0045308.
93. Mondoñedo, J. R., Bartolák-Suki, E., Jawde, S. B., Nelson, K., Cao, K., Sonnenberg, A., et al. (2020). A High-Throughput System for Cyclic Stretching of Precision-Cut Lung Slices During Acute Cigarette Smoke Extract Exposure. *Frontiers in physiology*, 11. https://doi.org/10.3389/fphys.2020.00566.
94. Murphy, S. V., & Atala, A. (2014). 3D bioprinting of tissues and organs. *Nature biotechnology*, 32(8). https://doi.org/10.1038/nbt.2958.
95. Murray, J. F. (2010). The structure and function of the lung. *The international journal of tuberculosis and lung disease: the official journal of the International Union against Tuberculosis and Lung Disease*, 14(4).
96. Naba, A., Clauser, K. R., Hoersch, S., Liu, H., Carr, S. A., & Hynes, R. O. (2012). The matrisome: in silico definition and in vivo characterization by proteomics of normal and tumor extracellular matrices. *Molecular & cellular proteomics: MCP*, 11(4). https://doi.org/10.1074/mcp.M111.014647.
97. Nair, L. S. (2021). https://doi.org/10.1016/j.progpolymsci.2007.05.017 | Elsevier Enhanced Reader.
98. Nakamura, H., Sugano, M., Miyashita, T., Hashimoto, H., Ochiai, A., Suzuki, K., et al. (2019). Organoid culture containing cancer cells and stromal cells reveals that podoplanin-positive cancer-associated fibroblasts enhance proliferation of lung cancer cells. *Lung cancer (Amsterdam, Netherlands)*, 134. https://doi.org/10.1016/j.lungcan.2019.04.007.
99. Narciso, M., Otero, J., Navajas, D., Farré, R., Almendros, I., & Gavara, N. (2021). Image-Based Method to Quantify Decellularization of Tissue Sections. *International journal of molecular sciences*, 22(16). https://doi.org/10.3390/ijms22168399.
100. Neelakantan, S., Xin, Y., Gaver, D. P., Cereda, M., Rizi, R., Smith, B. J., & Avazmohammadi, R. (2022). Computational lung modelling in respiratory medicine. *Journal of The Royal Society Interface*, 19, 20220062. https://doi.org/10.1098/rsif.2022.0062.
101. Nelson, C. M., Jean, R. P., Tan, J. L., Liu, W. F., Sniadecki, N. J., Spector, A. A., et al. (2005). Emergent patterns of growth controlled by multicellular form and mechanics. *Proceedings of the National Academy of Sciences of the United States of America*, 102(33). https://doi.org/10.1073/pnas.0502575102.
102. Ng-Blichfeldt, J.-P., Jong, T. de, Kortekaas, R. K., Wu, X., Lindner, M., Guryev, V., et al. (2019). TGF-β activation impairs fibroblast ability to support adult lung epithelial progenitor cell organoid formation. *American journal of physiology. Lung cellular and molecular physiology*, 317(1). https://doi.org/10.1152/ajplung.00400.2018.
103. Novak, R., Ingram, M., Marquez, S., Das, D., Delahanty, A., Herland, A., et al. (2020). Robotic fluidic coupling and interrogation of multiple vascularized organ chips. *Nature biomedical engineering*, 4(4). https://doi.org/10.1038/s41551-019-0497-x.

104. O'Neill, J. D., Anfang, R., Anandappa, A., Costa, J., Javidfar, J., Wobma, H. M., et al. (2013). Decellularization of human and porcine lung tissues for pulmonary tissue engineering. *The Annals of thoracic surgery*, 96(3). https://doi.org/10.1016/j.athoracsur.2013.04.022.
105. Obata, T., Tsuchiya, T., Akita, S., Kawahara, T., Matsumoto, K., Miyazaki, T., et al. (2019). Utilization of Natural Detergent Potassium Laurate for Decellularization in Lung Bioengineering. *Tissue engineering. Part C, Methods*, 25(8). https://doi.org/10.1089/ten.TEC.2019.0016.
106. Ochs, M., & Mühlfeld, C. (2013). Quantitative microscopy of the lung: a problem-based approach. Part 1: basic principles of lung stereology (review-article). https://doi.org/10.1152/ajplung.00429.2012.
107. Ochs, M., Nyengaard, J. R., Jung, A., Knudsen, L., Voigt, M., Wahlers, T., et al. (2004). The number of alveoli in the human lung. *American journal of respiratory and critical care medicine*, 169(1). https://doi.org/10.1164/rccm.200308-1107OC.
108. Oldenkamp, H. F., Ramirez, J. E. V., & Peppas, N. A. (2019). Re-evaluating the importance of carbohydrates as regenerative biomaterials. *Regenerative biomaterials*, 6(1). https://doi.org/10.1093/rb/rby023.
109. Paddenberg, R., Mermer, P., Goldenberg, A., & Kummer, W. (2014). Videomorphometric analysis of hypoxic pulmonary vasoconstriction of intra-pulmonary arteries using murine precision cut lung slices. *Journal of visualized experiments: JoVE*, (83). https://doi.org/10.3791/50970.
110. Parker, M. W., Rossi, D., Peterson, M., Smith, K., Sikström, K., White, E. S., et al. (2014). Fibrotic extracellular matrix activates a profibrotic positive feedback loop. https://doi.org/10.1172/JCI71386.
111. Paré, P. D., & Mitzner, W. (2012). Airway-parenchymal interdependence. *Comprehensive Physiology*, 2(3). https://doi.org/10.1002/cphy.c110039.
112. Petrou, C. L., D'Ovidio, T. J., Bölükbas, D. A., Tas, S., Brown, R. D., Allawzi, A., et al. (2020). Clickable decellularized extracellular matrix as a new tool for building hybrid-hydrogels to model chronic fibrotic diseases in vitro. *Journal of materials chemistry. B*, 8(31). https://doi.org/10.1039/d0tb00613k.
113. Placke, M. E., & Fisher, G. L. (1987). Adult peripheral lung organ culture--a model for respiratory tract toxicology. *Toxicology and applied pharmacology*, 90(2). https://doi.org/10.1016/0041-008x(87)90336-x.
114. Platz, J., Bonenfant, N. R., Uhl, F. E., Coffey, A. L., McKnight, T., Parsons, C., et al. (2016). Comparative Decellularization and Recellularization of Wild-Type and Alpha 1,3 Galactosyltransferase Knockout Pig Lungs: A Model for Ex Vivo Xenogeneic Lung Bioengineering and Transplantation. *Tissue engineering. Part C, Methods*, 22(8). https://doi.org/10.1089/ten.TEC.2016.0109.
115. Pouliot, R. A., Link, P. A., Mikhaiel, N. S., Schneck, M. B., Valentine, M. S., Gninzeko, F. J. K., et al. (2016). Development and characterization of a naturally derived lung extracellular matrix hydrogel. *Journal of biomedical materials research. Part A*, 104(8). https://doi.org/10.1002/jbm.a.35726.
116. Pouliot, R. A., Young, B. M., Link, P. A., Park, H. E., Kahn, A. R., Shankar, K., et al. (2020). Porcine Lung-Derived Extracellular Matrix Hydrogel Properties Are Dependent on Pepsin Digestion Time. *Tissue engineering. Part C, Methods*, 26(6). https://doi.org/10.1089/ten.TEC.2020.0042.
117. Price, A. P., England, K. A., Matson, A. M., Blazar, B. R., & Panoskaltsis-Mortari, A. (2010). Development of a decellularized lung bioreactor system for bioengineering the lung: the matrix reloaded. *Tissue engineering. Part A*, 16(8). https://doi.org/10.1089/ten.TEA.2009.0659.
118. Pugin, J. (2003). Molecular mechanisms of lung cell activation induced by cyclic stretch. *Critical care medicine*, 31(4 Suppl). https://doi.org/10.1097/01.CCM.0000057844.31307.ED.
119. Raeber, G. P., Lutolf, M. P., & Hubbell, J. A. (2005). Molecularly engineered PEG hydrogels: a novel model system for proteolytically mediated cell migration. *Biophysical journal*, 89(2). https://doi.org/10.1529/biophysj.104.050682.

120. Rahman, L., Williams, A., Gelda, K., Nikota, J., Wu, D., Vogel, U., et al. (2020). 21st Century Tools for Nanotoxicology: Transcriptomic Biomarker Panel and Precision-Cut Lung Slice Organ Mimic System for the Assessment of Nanomaterial-Induced Lung Fibrosis. *Small (Weinheim an der Bergstrasse, Germany)*, 16(36). https://doi.org/10.1002/smll.202000272.
121. Reyfman, P. A., Walter, J. M., Joshi, N., Anekalla, K. R., McQuattie-Pimentel, A. C., Chiu, S., et al. (2018). Single-Cell Transcriptomic Analysis of Human Lung Provides Insights into the Pathobiology of Pulmonary Fibrosis. *American Journal of Respiratory and Critical Care Medicine*, 199(12), 1517–1536. https://doi.org/10.1164/rccm.201712-2410OC.
122. Roan, E., & Waters, C. M. (2011). What do we know about mechanical strain in lung alveoli? (review-article). https://doi.org/10.1152/ajplung.00105.2011.
123. Saldin, L. T., Cramer, M. C., Velankar, S. S., White, L. J., & Badylak, S. F. (2017). Extracellular matrix hydrogels from decellularized tissues: Structure and function. *Acta biomaterialia*, 49. https://doi.org/10.1016/j.actbio.2016.11.068.
124. Saleh, K. S., Hewawasam, R., Šerbedžija, P., Blomberg, R., Noreldeen, S. E., Edelman, B., et al. (2022). Engineering Hybrid-Hydrogels Comprised of Healthy or Diseased Decellularized Extracellular Matrix to Study Pulmonary Fibrosis (OriginalPaper). *Cellular and Molecular Bioengineering*, 1–15. https://doi.org/10.1007/s12195-022-00726-y.
125. Sava, P., Ramanathan, A., Dobronyi, A., Peng, X., Sun, H., Ledesma-Mendoza, A., et al. (2017). Human pericytes adopt myofibroblast properties in the microenvironment of the IPF lung. *JCI insight*, 2(24). https://doi.org/10.1172/jci.insight.96352.
126. Scarritt, M. E., Bonvillain, R. W., Burkett, B. J., Wang, G., Glotser, E. Y., Zhang, Q., et al. (2014). Hypertensive rat lungs retain hallmarks of vascular disease upon decellularization but support the growth of mesenchymal stem cells. *Tissue engineering. Part A*, 20(9–10). https://doi.org/10.1089/ten.TEA.2013.0438.
127. Seiji, Y., Ito, T., Nakamura, Y., Nakaishi-Fukuchi, Y., Matsuo, A., Sato, N., et al. (2019). Alveolus-like organoid from isolated tip epithelium of embryonic mouse lung. *Human cell*, 32(2). https://doi.org/10.1007/s13577-019-00236-6.
128. Shelke, N. B. (2021). Polysaccharide biomaterials for drug delivery and regenerative engineering – Shelke – 2014 – Polymers for Advanced Technologies – Wiley Online Library. https://doi.org/10.1002/pat.3266.
129. Shiraishi, K., Nakajima, T., Shichino, S., Deshimaru, S., Matsushima, K., & Ueha, S. (2019). In vitro expansion of endogenous human alveolar epithelial type II cells in fibroblast-free spheroid culture. *Biochemical and biophysical research communications*, 515(4). https://doi.org/10.1016/j.bbrc.2019.05.187.
130. Shirani, A., Ganji, F., Golmohammadi, M., Hashemi, S. M., Mozafari, M., Amoabediny, G., et al. (2021). Cross-linked acellular lung for application in tissue engineering: Effects on biocompatibility, mechanical properties and immunological responses. *Materials science & engineering. C, Materials for biological applications*, 122. https://doi.org/10.1016/j.msec.2021.111938.
131. Shiwarski, D. J., Hudson, A. R., Tashman, J. W., & Feinberg, A. W. (2021). Emergence of FRESH 3D printing as a platform for advanced tissue biofabrication. *APL Bioengineering*, 5(1), 010904. https://doi.org/10.1063/5.0032777 (Accessed 2023/03/20).
132. Skardal, A., Murphy, S. V., Devarasetty, M., Mead, I., Kang, H.-W., Seol, Y.-J., et al. (2017). Multi-tissue interactions in an integrated three-tissue organ-on-a-chip platform. *Scientific reports*, 7(1). https://doi.org/10.1038/s41598-017-08879-x.
133. Smith, B. J., Bartolak-Suki, E., Suki, B., Roy, G. S., Hamlington, K. L., Charlebois, C. M., et al. (2017). Linking Ventilator Injury-Induced Leak across the Blood-Gas Barrier to Derangements in Murine Lung Function (Text). https://doi.org/10.3389/fphys.2017.00466.
134. Smith, B. J., Grant, K. A., & Bates, J. H. T. (2013). Linking the Development of Ventilator-Induced Injury to Mechanical Function in the Lung (Text). https://doi.org/10.1007/s10439-012-0693-2.
135. Soleas, J. P., D'Arcangelo, E., Huang, L., Karoubi, G., Nostro, M. C., McGuigan, A. P., et al. (2020). Assembly of lung progenitors into developmentally-inspired geom-

etry drives differentiation via cellular tension. *Biomaterials*, 254. https://doi.org/10.1016/j.biomaterials.2020.120128.
136. Stucki, A. O., Stucki, J. D., Hall, S. R. R., Felder, M., Mermoud, Y., Schmid, R. A., et al. (2015). A lung-on-a-chip array with an integrated bio-inspired respiration mechanism. *Lab on a chip*, 15(5). https://doi.org/10.1039/c4lc01252f.
137. Sucre, J. M. S., Wilkinson, D., Vijayaraj, P., Paul, M., Dunn, B., Alva-Ornelas, J. A., et al. (2016). A three-dimensional human model of the fibroblast activation that accompanies bronchopulmonary dysplasia identifies Notch-mediated pathophysiology. *American journal of physiology. Lung cellular and molecular physiology*, 310(10). https://doi.org/10.1152/ajplung.00446.2015.
138. Suki, B. (2014). Assessing the functional mechanical properties of bioengineered organs with emphasis on the lung. *Journal of cellular physiology*, 229(9). https://doi.org/10.1002/jcp.24600.
139. Suki, B., Ito, S., Stamenovic, D., Lutchen, K. R., & Ingenito, E. P. (2005). Biomechanics of the lung parenchyma: critical roles of collagen and mechanical forces. *Journal of applied physiology (Bethesda, Md.: 1985)*, 98(5). https://doi.org/10.1152/japplphysiol.01087.2004.
140. Suki, B., Stamenović, D., & Hubmayr, R. (2011). Lung parenchymal mechanics. *Comprehensive Physiology*, 1(3). https://doi.org/10.1002/cphy.c100033.
141. Sundarakrishnan, A., Zukas, H., Coburn, J., Bertini, B. T., Liu, Z., Georgakoudi, I., et al. (2019). Bioengineered in Vitro Tissue Model of Fibroblast Activation for Modeling Pulmonary Fibrosis. *ACS biomaterials science & engineering*, 5(5). https://doi.org/10.1021/acsbiomaterials.8b01262.
142. Tan, Q., Choi, K. M., Sicard, D., & Tschumperlin, D. J. (2017). Human airway organoid engineering as a step toward lung regeneration and disease modeling. *Biomaterials*, 113. https://doi.org/10.1016/j.biomaterials.2016.10.046.
143. Tan, Q., Ma, X. Y., Liu, W., Meridew, J. A., Jones, D. L., Haak, A. J., et al. (2019). Nascent Lung Organoids Reveal Epithelium- and Bone Morphogenetic Protein-mediated Suppression of Fibroblast Activation. *American journal of respiratory cell and molecular biology*, 61(5). https://doi.org/10.1165/rcmb.2018-0390OC.
144. Travaglini, K. J., Nabhan, A. N., Penland, L., Sinha, R., Gillich, A., Sit, R. V., et al. (2020). A molecular cell atlas of the human lung from single-cell RNA sequencing (OriginalPaper). *Nature*, 587(7835), 619–625. https://doi.org/10.1038/s41586-020-2922-4.
145. Tuan, R. S., Boland, G., & Tuli, R. (2003). Adult mesenchymal stem cells and cell-based tissue engineering. *Arthritis research & therapy*, 5(1). https://doi.org/10.1186/ar614.
146. Uhl, F. E., Wagner, D. E., & Weiss, D. J. (2017). Preparation of Decellularized Lung Matrices for Cell Culture and Protein Analysis. *Methods in molecular biology (Clifton, N.J.)*, 1627. https://doi.org/10.1007/978-1-4939-7113-8_18.
147. Uriarte, J. J., Uhl, F. E., Enes, S. E. R., Pouliot, R. A., & Weiss, D. J. (2018). Lung bioengineering: advances and challenges in lung decellularization and recellularization. *Current opinion in organ transplantation*, 23(6). https://doi.org/10.1097/MOT.0000000000000584.
148. Vadasz, S., Jensen, T., Moncada, C., Girard, E., Zhang, F., Blanchette, A., et al. (2014). Second and third trimester amniotic fluid mesenchymal stem cells can repopulate a decellularized lung scaffold and express lung markers. *Journal of pediatric surgery*, 49(11). https://doi.org/10.1016/j.jpedsurg.2014.04.006.
149. Vazquez-Armendariz, A. I., Heiner, M., Agha, E. E., Salwig, I., Hoek, A., Hessler, M. C., et al. (2020). Multilineage murine stem cells generate complex organoids to model distal lung development and disease. *The EMBO journal*, 39(21). https://doi.org/10.15252/embj.2019103476.
150. Wanczyk, H., Jensen, T., Weiss, D. J., & Finck, C. (2021). Advanced single-cell technologies to guide the development of bioengineered lungs. *American journal of physiology. Lung cellular and molecular physiology*, 320(6). https://doi.org/10.1152/ajplung.00089.2021.

151. Wang, H., Haeger, S. M., Kloxin, A. M., Leinwand, L. A., & Anseth, K. S. (2012). Redirecting valvular myofibroblasts into dormant fibroblasts through light-mediated reduction in substrate modulus. *PloS one*, 7(7). https://doi.org/10.1371/journal.pone.0039969.
152. Weibel, E. R. (2017). Lung morphometry: the link between structure and function. *Cell and tissue research*, 367(3). https://doi.org/10.1007/s00441-016-2541-4.
153. Weymann, A., Patil, N. P., Sabashnikov, A., Korkmaz, S., Li, S., Soos, P., et al. (2015). Perfusion-Decellularization of Porcine Lung and Trachea for Respiratory Bioengineering. *Artificial organs*, 39(12). https://doi.org/10.1111/aor.12481.
154. Wilkinson, D. C., Mellody, M., Meneses, L. K., Hope, A. C., Dunn, B., & Gomperts, B. N. (2018). Development of a Three-Dimensional Bioengineering Technology to Generate Lung Tissue for Personalized Disease Modeling. *Current protocols in stem cell biology*, 46(1). https://doi.org/10.1002/cpsc.56.
155. Williams, D. (1999) 'The Williams Dictionary of Biomaterials'. Liverpool: Liverpool University Press.
156. Wolters, P. J., Collard, H. R., & Jones, K. D. (2014). Pathogenesis of idiopathic pulmonary fibrosis. *Annual review of pathology*, 9. https://doi.org/10.1146/annurev-pathol-012513-104706.
157. Wu, X., Dijk, E. M. van, Ng-Blichfeldt, J.-P., Bos, I. S. T., Ciminieri, C., Königshoff, M., et al. (2019). Mesenchymal WNT-5A/5B Signaling Represses Lung Alveolar Epithelial Progenitors. *Cells*, 8(10). https://doi.org/10.3390/cells8101147.
158. Yang, X., Li, K., Zhang, X., Liu, C., Guo, B., Wen, W., et al. (2018). Nanofiber membrane supported lung-on-a-chip microdevice for anti-cancer drug testing. *Lab on a chip*, 18(3). https://doi.org/10.1039/c7lc01224a.
159. Zamprogno, P., Wüthrich, S., Achenbach, S., Thoma, G., Stucki, J. D., Hobi, N., et al. (2021). Second-generation lung-on-a-chip with an array of stretchable alveoli made with a biological membrane (Text). https://doi.org/10.1038/s42003-021-01695-0.
160. Zscheppang, K., Berg, J., Hedtrich, S., Verheyen, L., Wagner, D. E., Suttorp, N., et al. (2018). Human Pulmonary 3D Models For Translational Research. *Biotechnology journal*, 13(1). https://doi.org/10.1002/biot.201700341.
161. Zvarova, B., Uhl, F. E., Uriarte, J. J., Borg, Z. D., Coffey, A. L., Bonenfant, N. R., et al. (2016). Residual Detergent Detection Method for Nondestructive Cytocompatibility Evaluation of Decellularized Whole Lung Scaffolds. *Tissue engineering. Part C, Methods*, 22(5). https://doi.org/10.1089/ten.TEC.2015.0439.

# Chapter 10
# Lung-on-a-Chip Models of the Lung Parenchyma

**Pauline Zamprogno, Jan Schulte, Dario Ferrari, Karin Rechberger, Arunima Sengupta, Lisette van Os, Tobias Weber, Soheila Zeinali, Thomas Geiser, and Olivier T. Guenat**

## 10.1 Introduction

The lung parenchyma, located at the end of the respiratory tree, comprises millions of alveoli, highly fragile three-dimensional structures. Yet, the parenchyma is robust enough to be constantly subjected to various mechanical forces, including surface tension forces and cyclic respiratory motion, while exposed to pathogens and foreign particles. Several cell types, each with a specific phenotype, are located in the lung parenchyma and form the ultrathin air-blood barrier. The latter has many functions, such as ensuring gas exchange, forming the first line of defence against invaders, building tiny vessels that transport blood cells, healing lesions, and more.

The in-vitro modelling of the alveolar barrier has proven to be a significant challenge and has been the subject of intense research worldwide. The first

Pauline Zamprogno and Jan Schulte contributed equally.

P. Zamprogno · J. Schulte · D. Ferrari · K. Rechberger · A. Sengupta
L. van Os · T. Weber · S. Zeinali
Organs-on-Chip Technologies Laboratory, ARTORG Center, University of Bern, Bern, Switzerland

T. Geiser
Department of Pulmonary Medicine, University Hospital of Bern, Bern, Switzerland

O. T. Guenat (✉)
Organs-on-Chip Technologies Laboratory, ARTORG Center, University of Bern, Bern, Switzerland

Department of Pulmonary Medicine, University Hospital of Bern, Bern, Switzerland

Department of General Thoracic Surgery, University Hospital of Bern, Bern, Switzerland
e-mail: olivier.guenat@unibe.ch

 
C. M. Magin (ed.), *Engineering Translational Models of Lung Homeostasis and Disease*, Advances in Experimental Medicine and Biology 1413,
https://doi.org/10.1007/978-3-031-26625-6_10

lung-on-a-chip (LOC) model was reported in 2010, following which the lung alveolar barrier models were suddenly propelled to the top of the most commented and discussed in-vitro models [1]. With this ground-breaking work, the organs-on-chip (OOC) technology was born. OOC are microengineered systems that provide cells with an environment that closely resembles their native in vivo milieu [2]. The primary LOC was the first to report a flexible and ultra-thin cell culture substrate, mimicking the alveolar barrier, that could be stretched cyclically to mimic respiratory movements. This setup also allows lung alveolar epithelial cells to be exposed to air, while endothelial cells remain in contact with a blood-like medium. This first system revealed the effects of specific parameters of the cellular environment. For example, the toxic effect of interleukin-2 (IL-2), an anticancer drug, was shown to be enhanced by cyclic mechanical stress, leading to vascular damage and consequent intra-alveolar fluid accumulation resembling pulmonary edema [3]. Since its introduction, LOC technology has made tremendous progress, and many more complex biological functions can now be reproduced in vitro.

This chapter first presents the specific cellular landscape of the lung parenchyma. Second, the progress made over the past ten years reveals how LOC technology has become relevant for the drug discovery industry and the study of fundamental physiological phenomena. Next, the chapter provides an overall picture of the complex biological processes this technology can mimic, including lung disease modelling. Finally, challenges and perspectives in lung alveolar barrier modelling are discussed.

## 10.2 Lung Alveolar Cells and the Alveolar Environment

### *10.2.1 Lung Alveolar Cells and Their Environment*

Respiration, essentially the exchange of oxygen and carbon dioxide, takes place in the 300 million [4] lung alveoli. The alveoli are tiny thin-walled air sacs, which are connected to the terminal part of the bronchi (Fig. 10.1). A single alveolus is considered the smallest structural and functional unit of the lung. The cells populating the alveoli are exposed to specific, often unique, biochemical and physical conditions that define their phenotype. The lung alveolar epithelium, composed of type 1 (AEC1) and type 2 (AEC2) lung alveolar epithelial cells, provides an extensive surface for gas exchange. AEC1 have a very thin cell body but cover a large area. This particular cell type facilitates respiratory exchange between the blood vessel plexus and the alveolar lumen. AEC2 are small and cuboidal and are responsible for epithelial homeostasis, the regulation of ions and water transport and surfactant production. Although AEC1 cover about 95% of the alveolar surface, their number is relatively small compared to AEC2 [5, 6]. AEC2 have been described as alveolar epithelium progenitor/stem cells that can differentiate into AEC1 during development and regeneration [7]. Further research revealed that a Wnt-responsive AEC2 subpopulation is the primary progenitor population of AEC1 [8]. Moreover, after

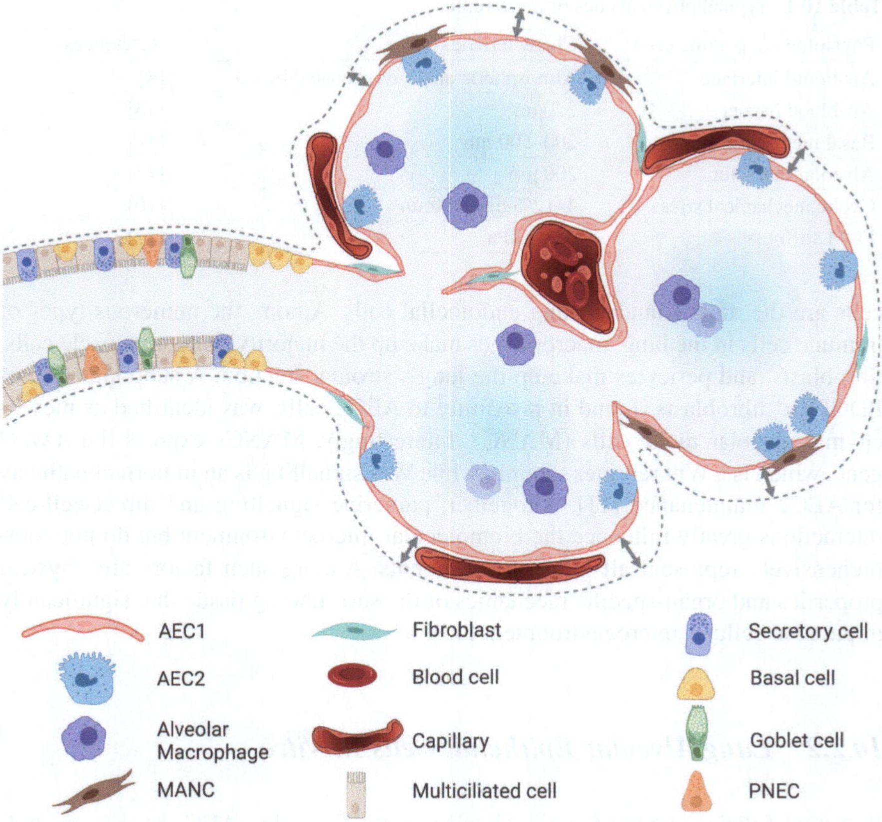


**Fig. 10.1** Illustration of the cells populating the distal airways. Alveoli, the lungs functional units, are composed of several cell types, all together forming and maintaining the cyclically stretched (arrows) air-blood-barrier. Air is conveyed via airways towards the alveoli, where the gas exchange takes place. At the end of the respiratory tree, the airway epithelium with its basal, multiciliated, goblet, secretory and pulmonary neuroendocrine (PNEC) cells, transitions into the alveolar epithelium with thin and large alveolar epithelial cells type 1 (AEC1) and small cuboidal alveolar epithelial cells type 2 (AEC2). The mesenchymal cell fraction in the alveoli consists of various fibroblasts, including mesenchymal alveolar niche cells (MANC). Closely aligned with the alveolar epithelium are capillaries that allow the transport of oxygenated blood to the heart. In addition to the immune cells circulating in within the blood, tissue resident alveolar macrophages serve as important responders to pathogens and injuries. (Created with BioRender.com)

injury, high Krt8+ expressing AEC2 act as alveolar differentiation intermediate cells, a transitional cell state between AEC2 and AEC1 [9].

In addition to the epithelium, the alveolar microenvironment consists of other cell types and various biochemical and physical conditions (Fig. 10.1 and Table 10.1). To date, 17 different cell types of epithelial, endothelial, stromal, and immune origin have been identified in the alveoli [10]. In the lung, cells of the endothelial compartment are ubiquitous and part of the larger vessels down to the smallest capillaries. Given the lung's respiratory function, it is not surprising that capillary

**Table 10.1** Typical physical cues of the alveolus

| Physiological parameters | Characteristics | References |
|---|---|---|
| Air-liquid interface | Atmospheric air/Deoxygenated blood | [4] |
| Air-blood barrier | 2.2 μm | [13] |
| Basal membrane | 200–500 nm | [14] |
| Alveolar diameter | 200 μm | [15] |
| Cyclic mechanical stress | 4–12% linear strain | [16] |
| ECM stiffness | 0.2–2 kPa | [17] |

cells are the most abundant lung endothelial cells. Among the numerous types of immune cells in the lung, macrophages make up the majority. Smooth muscle cells, fibroblasts, and pericytes make up the lung's stromal fraction. A subpopulation of PDGFR$\alpha^+$ fibroblasts, found in proximity to AEC2 cells, was identified as mesenchymal alveolar niche cells (MANC). Interestingly, MANCs express the *AXIN2* gene, which is a Wnt reporter subunit, while Wnt signalling is an important pathway for AEC2 maintenance [11]. Altogether, paracrine signalling and direct cell-cell interactions greatly influence the biomolecular microenvironment but do not comprehensively represent all influential elements. Among such factors are physical properties and organ-specific mechanics of the surrounding tissue that significantly impact the cellular microenvironment [12].

## 10.2.2 Lung Alveolar Epithelial Cells In Vitro

Because of their essential functional and regenerative roles, AECs have been studied extensively. The A549 cell line has been used as the AEC standard for decades [18]. However, this cell line does not represent healthy cells but rather non-small cell lung cancer since this cell line originates from a tumour. Recently, Tièche and colleagues described that A549 cells consist of three subpopulations: holo-, para- and meroclone, which exhibit epithelial, intermediate, and mesenchymal properties, respectively [19].

In recent years, efforts have been made to generate healthy AEC cell lines for research. To this end, genetic reprogramming has been used to create AEC from induced pluripotent stem cells (iPSCs) [20–23]. Several protocols have already been reported, including using patient-specific cells [24]. However, iPSCs have their own limitations, often unable to differentiate into AEC1 and their similarity to the embryonic/early developmental state [25].

When available, primary human alveolar epithelial cells are still the preferred cell source. However, given the high variability between donors' cells, the difficulty to culture AEC remains challenging. Many researchers chose to culture AEC2 in hydrogels, often co-cultured with mesenchymal cells to allow for cellular crosstalk [11, 26–29]. Today, due to our increased knowledge regarding their biomolecular processes, AEC2 can be maintained for extended periods of time in hydrogel without adding other cell types [25].

## 10.3 Reproducing the Alveolar Barrier with a Lung-on-a-Chip

Lung alveolar models hold great interest for a broad range of applications, such as drug toxicity and efficacy testing, lung disease modelling, chemical risk assessments of aerosols, and answering fundamental research questions.

### *10.3.1 Reproducing the Lung Alveolar Environment on Chip*

The development of lung alveolar models has evolved rapidly over the past decade, mainly due to the advent of OOC technology. Lung biologists have traditionally used Transwell cell culture inserts to reproduce the air-blood barrier. These inserts, made of a 10 μm-thin, rigid, porous membrane, onto which cells are cultured on both sides, are gradually being replaced by LOC technology. The latter can mimic the thin basement membrane, similarly to Transwells, but also more accurately simulate other parameters of the alveolar environment, such as the mechanical stress of respiration, lung extracellular matrix (ECM) softness, capillary flow, and even the size and three-dimensional morphology of the alveolar sacs. Recent advances in mimicking specific aspects of the alveolar environment with OOC technology are detailed in Fig. 10.2.

#### Scaffolds for the Alveolar Barrier: Engineering a Thin, Flexible and Soft Basement Membrane

An ideal cell culture substrate for the alveolar barrier should be thin, soft (with a typical stiffness of 0.2–2 kPa), elastic, permeable to gas and fluids, and comprise lung extracellular matrix proteins. This last aspect is crucial since the lung ECM plays a critical role in regulating lung homeostasis in health and disease [17, 33, 34]. In addition, from a technical point of view, the scaffold should be resilient to enable reproducible results and simple to manufacture.

The first-generation LOCs used a thin and porous membrane made of a flexible polymer polydimethylsiloxane (PDMS) as a cell culture substrate. Due to the intrinsic hydrophobicity of PDMS, the PDMS membrane must be coated with ECM proteins such as collagen, fibronectin, laminin for the cells to adhere. Membrane thickness ranges from about 3 μm [34] to 10 μm [1], with pore sizes varying from 3 to 10 μm. Porous PDMS membranes are usually fabricated using micromolding techniques [35]. Although PDMS membranes have extremely interesting elastic and optical properties, they poorly mimic the biochemical and physical properties of the alveolar wall ECM scaffold. Furthermore, PDMS typically has a stiffness of 500 kPa or more and is known to absorb and adsorb small molecules, a significant limitation when the PDMS membrane is exposed to drug compounds [36].

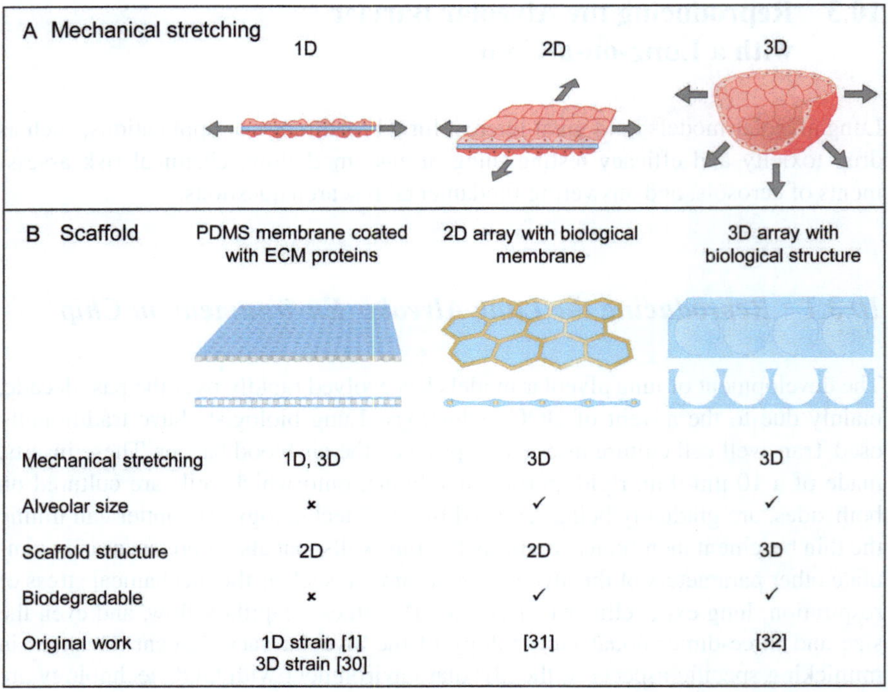


**Fig. 10.2** Parameters of the alveolar environment reproduced in LOC. Respiratory movements and the cell culture alveolar scaffold are two of the aspects reproduced by LOCs. (**a**) Three types of mechanical stretch can be distinguished: 1D, 2D or 3D. (**b**) Development of the scaffolds used to mimic the alveolar basement membrane. Thin polydimethylsiloxane (PDMS) membranes with microscopic pores coated with ECM proteins were incorporated into LOCs to be stretched in 1D or 3D. Biological membranes made of hydrogels are replacing PDMS membranes. Arrays of alveoli have been mimicked by fabricating such a membrane in a hexagonal grid with dimensions similar to those of the alveoli. More recently, a microstructured biological hydrogel that forms alveolar sacs has been developed to mimic the 3D architecture of the alveoli

Hydrogel materials have attracted considerable interest as an alternative to PDMS because of their biodegradability and adaptable chemical and mechanical properties [37]. Synthetic hydrogels, such as polycaprolactone (PCL) [38–40], poly(lactic-co-glycolic acid) (PLGA) [41], and poly(L-lactic acid) (PLLA) [42, 43] have been used to mimic basement membranes, mainly for LOC applications. For example, 2 μm-thin PCL-nanofibre mesh membrane produced by a one-step PCL electrospinning could be stretched cyclically [38]. Similarly, a 20 μm-thin PCL membrane was used to mimic ventilated induced lung injury with a pathophysiological stretch [44]. Furthermore, Pensabene and colleagues succeeded in fabricating an ultrathin 0.1 μm PLLA membrane with 2 μm pores by spin-coating [42]. However, these synthetic hydrogel membranes are relatively stiff with a Young's modulus of 360 kPa up to several MPa [39].

In contrast, ultra-soft membranes, made of natural hydrogels, provide stiffnesses in the physiological (0.2–2 kPa) and fibrotic (10–50 kPa) range [45–47].

Therefore, it is not surprising that the recently presented second-generation LOCs are based on hydrogel alveolar barrier scaffolds. Our group developed a thin, stretchable membrane made of collagen and elastin, two proteins found in the lung ECM [31]. In contrast to first-generation LOCs, the second-generation reproduce an array of alveoli with in vivo-like dimensions. The straightforward fabrication process involves drop-casting a solution of collagen and elastin onto a hexagonal mesh, where the solution spreads and is held in place by surface tension forces. Vitrified membranes have a stiffness of about 170 kPa, while those without vitrification have a Young's modulus of about 1–2 kPa [46]. One of the most important properties of this new membrane is its biodegradability, which was demonstrated using MMP-8, a known neutrophil collagenase. Recently, Huang et al. succeeded in reproducing the three-dimensional structure of alveolar sacs by using a porous hydrogel of gelatin methacryloyl with an inverse opal structure. This structure mimics an array of interconnected alveolar sacs populated with primary alveolar epithelial cells [32].

**Mechanical Stress Induced by the Respiratory Movements**

The engineering aspect that made the first LOC so famous is undoubtedly its ability to reproduce the exposure of the alveolar barrier to respiratory-like movements. During normal breathing, the mechanical stress affecting lung alveolar cells corresponds to a linear elongation of 4%, while the linear elongation is 12% for deep respiration. As illustrated in Fig. 10.2a, unidirectional and three-dimensional strains have been used to mimic the respiratory movement in LOCs [12]. Huh and colleagues [1] used a thin, porous PDMS membrane to centrally divide a microfluidic channel into two compartments, an air-filled apical chamber and a basal chamber filled with cell culture medium. Lung alveolar epithelial cells were cultured on the membrane's apical side, thereby in contact with air, while endothelial cells were cultured on the basal side, contacting the physiological medium. The researchers cleverly designed two adjacent channels, placed parallel to the main channel, from which thin walls separated them. By applying cyclic negative pressure in the adjacent channels, the thin walls attaching the porous membrane rhythmically deformed and stretched the PDMS membrane laterally [1]. To improve on this design of a unidirectional strain, our group engineered a LOC in which the alveolar barrier was stretched in three directions, similarly to in-vivo conditions [30, 48]. To achieve this, a second membrane was integrated into the basal compartment. A cyclic vacuum is applied to the cavity below this second membrane, resulting in a cyclic deflection of the alveolar membrane, inducing three-dimensional mechanical stress. Most LOCs reported so far have used the unidirectional breathing principle designed by Huh, by either using the original design [49–51] or by adapting its geometry to a circular channel around a round open-top alveolar barrier [52]. Recently, a LOC aimed at mimicking mechanotrauma in lung epithelial cells induced by pathophysiological stretch (25%) was reported [44].

### 10.3.2 *Effects of Biochemical and Physical Cues on the Lung Alveolar Barrier*

**Effects of Mechanical Forces on Alveolar Epithelial Cells**

When we inhale and exhale, the lung tissue is cyclically stretched and relaxed. The induced mechanical stress affects several signaling pathways, particularly the Hippo signaling pathway and YAP/TAZ in lung development and disease, as well as TGF-β, which significantly influences ECM remodeling and cell differentiation processes [53]. The interplay of parameters, such as the duration of stretch, cycle frequency, and stretch variability the release of cytokines. Via the MAPK ERK1/2 and SAPK/JNK signaling pathways, stretched cells release cytokines, triggering a cellular inflammatory response [54]. In addition to cytokines, the release of surfactants in vivo is also associated with cell elongation. For example, when AEC1 caveolae are stretched, paracrine signals trigger the release of stored surfactant from AEC2 into the alveolar lumen [55]. Additionally, cyclic stretch affects epithelial permeability and tight junction formation. Particularly, pathophysiological stretch can lead to ventilation-induced microinjuries, resulting in cellular processes, such as apoptosis and autophagy [56].

**Effects of Mechanical Forces on Lung Endothelial Cells**

Another mechanical stress on the lungs is formed by blood cells acting on the tiny alveolar capillary walls. Mechanical forces induce several cell signalling pathways, affecting vascular function and remodelling. Among these, reactive oxygen species (ROS) produced by vascular cells play a crucial role in signal transduction and physiological regulation of vascular function [57]. Most LOCs reproduce this alveolar-capillary interface using a monolayer of endothelial cells cultured on the basal side of a porous membrane [1, 50, 51]. However, in those systems, the basal chamber is significantly larger than the typical capillary 7–9 μm diameter [4], resulting in the exposure of endothelial cells to flow-induced shear stress [58]. Therefore, this reductionist approach does not reflect the size of the capillaries nor their three-dimensionality. Furthermore, the effect of the combined mechanical forces, shear stress, and respiratory movements on the alveolar capillaries is only starting to emerge. Indeed, our group recently demonstrated that small vessels made of pulmonary microvascular endothelial cells respond differently to mechanical stimuli in 2D and 3D [59]. The most striking finding was that under 3D stress, these vessels could be stretched to a very high level without barrier dysfunction, which would typically lead to an endothelial barrier disruption when exposed to 2D stress. Similarly, other groups have developed gelatin-based three-dimensional vessels subjected to both shear stress and cyclic strain [60]. Such systems will undoubtedly allow further study into the effects of mechanotransduction on vascular function and remodelling.

**Lung Alveolar Extracellular Matrix (ECM)**

The lung alveolar ECM lung transforms during lung development and in health and disease. The ECM comprises various proteins such as collagen, elastin, tenascin-C, and several growth factors [34]. The combination of mechanical forces exerted on cells via focal adhesion and local growth factors has been shown to synergistically regulate alveolar epithelial differentiation. Recently, different basement membranes made of natural fibres were tested in a LOC environment. Epithelial and endothelial cells respond to changes in fibre stiffness and architecture. Epithelial cells are more spread out and form tight junctions on softer and less dense fibre networks [39]. Hydrogel membranes have only recently been reported, and the implications of their stiffness undoubtedly need to be evaluated. Several hydrogels composed of collagen and Matrigel were recently tested to investigate the migration potential of lung cancer cells. Matrigel was shown to facilitate migration at low concentrations, most likely by providing a supportive and growth factor-retaining environment, in contrast to high concentrations [61].

**Effects Induced by the Air-Liquid Interface**

In vivo, a thin layer of surfactant in direct contact with air, called an air-liquid interface (ALI), covers the lung alveolar epithelium. The surfactant is synthesised by AEC2 and released from intracellular lamellar bodies. To date, few studies are reporting the effects of ALI in a LOC setting. However, the porous membrane incorporated in most LOC systems makes them a great tool to study this phenomenon, as lung alveolar epithelial cells receive sufficient nutrients through the pores of the membrane [1, 30, 31]. Culturing AEC2 at ALI conditions significantly decreases hypophase surface tension compared to cells cultured in submerged conditions [62]. Furthermore, a recent study on inserts shows that hiPSC-derived AEC2 cultured at ALI induces alveolar properties [63], suggesting that these culture conditions could further improve LOC output.

### *10.3.3 Read-Outs: Extracting Information from a Lung-on-a-Chip*

Extraction of cellular and tissue information from LOC systems is critical to capture cellular and tissue responses. Table 10.2 summarises the main read-outs and methods used to monitor these changes in the three lung alveolar barrier compartments: the alveolar space, barrier, and blood capillaries. It should be noted that unlike standard cell culture systems (e.g., inserts), LOC systems can provide information about the tissue response [2]. For example, remodelling by alveolar barrier stiffening [64] or degradation [31] can be detected. Other read-outs, such as edema formation [3] or vasoactive responses can also be recorded [65].

**Table 10.2** Main read-outs from LOC. Main read-outs and methods used for monitoring changes in the three lung alveolar barrier compartments: alveolar space, alveolar barrier and blood capillaries

| Compartment | Components | Main functions | In-vitro read-outs |
|---|---|---|---|
| Alveolar space | Gas | Gas exchange ($O_2$, $CO_2$) | Gas concentration (gas sensor) |
| | Surfactant lining | Surface tension reduction | Surface tension quantification (contact angle measurement, surfactometer) |
| | Alveolar macrophages | Innate immunity and defense | Surfactant release |
| | | Cytokines/chemokines secretion | |
| | | Immune cells recruitment | |
| Alveolar barrier | Epithelium | Barrier formation | Tight junction formation (immunostaining, TEM) |
| | Interstitium (ECM, fibroblasts) | | Barrier tightness (TEER) |
| | Endothelium | | Barrier permeability (transport) |
| | Immune cells | | |
| | | Barrier remodelling | ECM accumulation |
| | | | ECM stiffness (bulge test, AFM) |
| | | | Cellular differentiation |
| | | | Cellular morphology |
| Blood capillaries | Dissolved gas | Transport (blood, soluble factors, cells,…) | Flow rate measurement (flow sensor) |
| | Blood cells | | Cytokines, chemokines secretion |
| | Plasma (soluble factors, incl. cytokines, chemokines, growth factors,…) | | Dissolved gas concentration (gas sensor) |
| | | | Migration of immune cells |
| | | | Vasoactivity |
| | | | Vessel permeability |

## 10.4 Lung Disease-on-a-Chip Models

Models of lung diseases are of great importance for preclinical (drug discovery, pathophysiology) and clinical investigations and may be developed as tools for precision medicine applications. These models are created using either healthy cells induced to produce specific disease patterns or patient cells. A mixture of both is often preferred to limit the heterogeneity of patient cells. The following is a brief overview of the major lung parenchyma diseases: idiopathic pulmonary fibrosis (IPF), emphysema, acute lung infection, and lung cancer. Figure 10.3 illustrates the changes in the alveolar environment induced by these diseases.

### *10.4.1 Idiopathic Pulmonary Fibrosis (IPF)*

IPF is a fatal lung disease believed to be triggered by repetitive injury of the lung alveolar epithelium followed by an abnormal wound-healing response, leading to the accumulation of excess fibrous connective tissue (Fig. 10.3a). The soft, healthy lung matrix, with a normal stiffness of 0.2–2 kPa, stiffens in a fibrotic lung to an average of 16 kPa with localised stiffer areas [66]. The strength of LOC's systems lies in their ability to mimic specific aspects of this complex disease sequentially, allowing the identification of the effects of individual cell culture parameters. For example, epithelial microinjury induced either chemically [67] or mechanically [68] can be followed by fibroblast activation by TGF-β and by antifibrotic drug treatment [67, 69, 70]. Meijas and colleagues created specific fluidic compartments to mimic the interface between the human lung's interstitium and airways. Using this model, it was possible to study the interactions between epithelial, endothelial, and fibroblasts (especially IPF fibroblasts) cultured in separate compartments [69]. Sundarakrishnan et al. developed a sophisticated perfusable model based on a cell-laden silk collagen hydrogel placed on a stretchable polymeric membrane [67]. After two weeks of culture, during which cells were exposed to TGF-β1 and bleomycin sulfate, the hydrogel was significantly stiffer (34–76 kPa) than an unexposed one (2 kPa). Furthermore, the antifibrotic treatment significantly improved the TGF-β1-induced myofibroblast phenotype. Asmani et al. showed a similar antifibrotic drug response obtained with free-standing membranous lung microtissue [70]. Additionally, Felder et al. used a breathing LOC to provide a proof-of-concept of the wound healing response that was partially inhibited by the cyclic mechanical stress [71].

### *10.4.2 Emphysema*

Pulmonary emphysema is a severe complication of chronic obstructive pulmonary disease (COPD) triggered by inhalation of air pollutants, chemical fumes, and cigarette smoke, causing an increased inflammatory response in the lungs, with progressive and irreversible alveolar destruction [72, 73]. The emphysematous enlargement of the airspace results in the necrosis of the cells lining the alveoli and the destruction of the underlying ECM. The degradation of the alveolar ECM by MMP8 [74], a neutrophil collagenase, was recently mimicked in a LOC setting [31]. Huang and

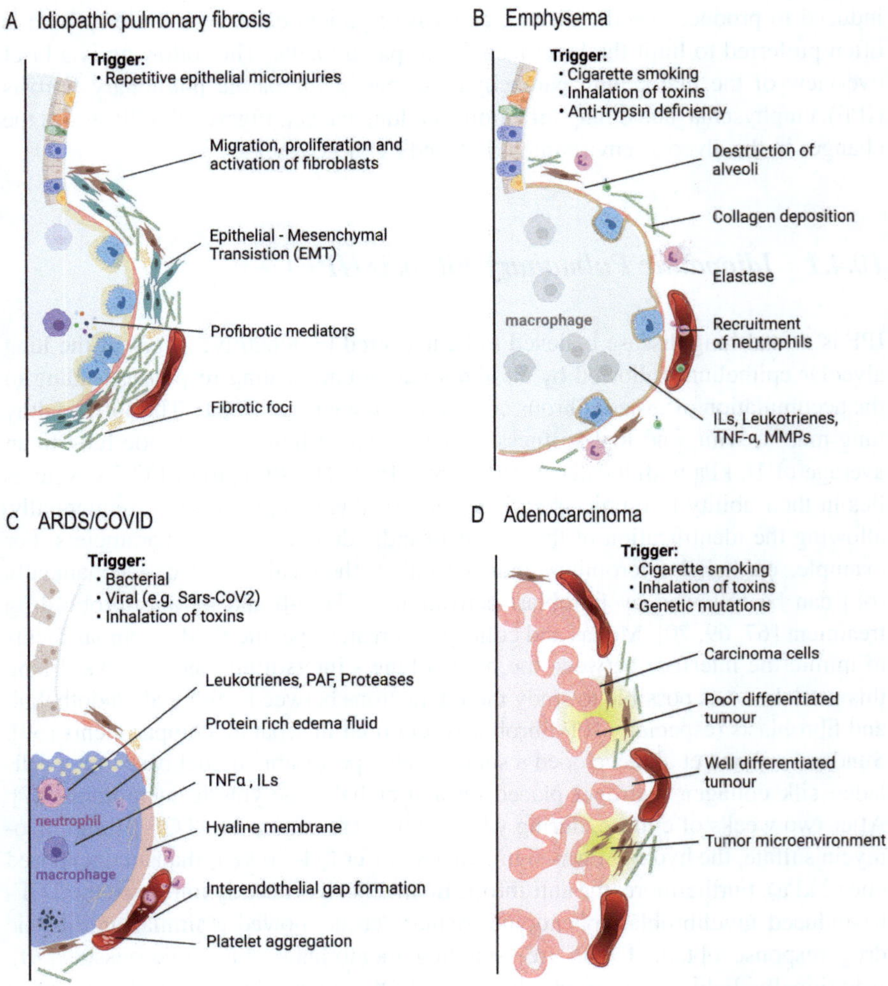


**Fig. 10.3** Illustrations of four lung diseases affecting the alveoli. (**a**) Idiopathic pulmonary fibrosis results most likely from recurrent epithelial microinjuries. Risk factors include gastroesophageal reflux, cigarette smoking and genetics. The tissue remodeling occurring in IPF is due to excessive secretion of ECM proteins, which significantly impairs normal gas exchange and compromises the elasticity of the lung parenchyma. (**b**) Alveolar damage in emphysema is usually due to inhalation of toxins (e.g., cigarette smoking) or anti-trypsin deficiency. The breakdown of the alveoli results in impaired lung function due to loss of elastic recoil of the lung. (**c**) Bacterial, viral (e.g., Sars-CoV2), parasitic, or mycotic pneumonia, chemical irritation from aspiration or inhalation, embolism, or extrapulmonary sepsis can cause acute respiratory distress syndrome. Accumulation of fluid in the alveoli, due particularly to excessive alveolocapillary permeability, reduces oxygenation of the blood. (**d**) Risk factors for developing adenocarcinoma include cigarette smoking or inhalation of other toxins such as radon or asbestos, and genetic mutations. The native ECM composition is altered, leading to modifications of the cellular microenvironment (e.g., hypoxia, interactions between tumour – non-tumour cells). (Created with BioRender.com)

colleagues observed profound alveolar epithelial cell damage along with elevated levels of the proinflammatory cytokines IL-8, IL-6, and IL-1β, upon exposure to cigarette smoke. For this experimental setup, they used a 3D LOC platform comprising porous hydrogel that mimics human alveolar sacs [32]. In addition to cigarette smoke, inhalation of other exogenous oxidants, such as toxic airborne irritants, leads to severe alveolar damage and subsequent emphysema [75]. Several researchers have used LOC systems to mimic nanoparticle-induced pulmonary toxicity in a physiologically relevant strain-based in vitro environment to mimic occupational exposure to ZnO, $TiO_2$, and silica particles [1, 76]. The development of these sophisticated alveolar injury models is critical to understanding the pathomechanism of emphysema and conducting drug safety and toxicity studies.

### *10.4.3 Acute Respiratory Distress Syndrome (ARDS)*

Acute respiratory distress syndrome (ARDS), also called acute lung injury (ALI), is hallmarked by severe lung injury and inflammation [77]. The injury can be induced by various causes, such as mechanical ventilation, sepsis, and respiratory infections, which can be reproduced in LOC ARDS models. Ventilation-induced lung injury (VILI) was modelled in a pressure-based system, where dexamethasone treatment had a protective effect against VILI [78]. In another VILI model, mouse alveolar epithelial cells were exposed to pathophysiological mechanical stress [44], resulting in increased epithelial injury and cell detachment. Nanoparticle-induced lung injury was mimicked without mechanical strain but with circulating flow [76]. Additionally, fine particulate matter (PM2.5) induced alveolar barrier dysfunction, specifically increased ROS generation and barrier permeability, cell apoptosis, and monocyte attachment [79]. Modelling of lung infection-on-a-chip was first performed in the seminal work of Huh, in which neutrophil migration was induced by the proinflammatory cytokines IL-8 and TNFα [1]. Using a similar LOC setting, a mouse model of tuberculosis-on-a-chip was used to study the dynamics of host-Mycobacterium tuberculosis interactions at an air-liquid interface [50]. Another immunocompetent model, coinfection with S. Aureus and influenza, showed that dual infection was far more damaging to the lung than a single infection [80]. Overall, the diversity of lung injury models reflects the variety of ARDS causes. The importance of the immune system in ARDS is represented in many LOCs by adding an immune component.

### *10.4.4 COVID*

The emergence of the SARS-CoV-2 pandemic has led to the investigation of COVID19-related ARDS in LOC systems. Initial clinical studies have shown that SARS-CoV-2 replicates primarily in the lung and can lead to pneumonia, pulmonary edema, ARDS, pulmonary fibrosis, tissue necrosis, and even multiple organ failure in severe cases [81–83]. Activated innate immune cells trigger a robust immune response leading to a cytokine storm, resulting in ARDS. A SARS-CoV-2 infection model, based on a LOC system lined with human lung epithelial, endothelial, and immune cells was used to study the cytokine storm and cellular barrier disruption triggered by the virus [84], as well as host cell response [85], and vascular damage [86]. In addition, Huang and colleagues demonstrated that infection rates decrease with treatment with antiviral drugs such as remdesivir, hydroxychloroquine, and amodiaquine [32].

### *10.4.5 Lung Adenocarcinoma*

The most common type of lung neoplasm is adenocarcinoma, a subtype of non-small cell lung cancer [87]. Over the past decade, two schools of thought on the tumour microenvironment (TME) have been established. One focuses on the interaction between tumour and non-tumour cells, while the other on physicochemical gradients ($O_2$, pH, etc.) of the tumour interstitium [88]. Although still in its infancy, lung cancer on-chip technology is best suited to engineer these TME aspects. Hassel et al. developed an orthotopic lung cancer-on-a-chip model by seeding H1975 human lung adenocarcinoma cells on healthy primary alveolar cells while exposing the LOC system to cyclic mechanical stress. Their results show that implementing respiratory motion leads to a heterogeneous growth pattern of cancer cells and the development of drug resistance, similar to the observed clinical responses [49]. Furthermore, Park and colleagues added vascularised cancer spheroids embedded in a perfusable vascular network to better understand tumour angiogenesis, a hallmark of cancer [89].

## 10.5 Challenges of Lung-on-a-Chip Technologies

Over the past decade, LOC technologies have evolved from proof-of-concept to commercial solutions. However, the situation is not as clear-cut as it seems at first glance. First, a distinction must be made between home-built LOC devices used in academic laboratories and the first-generation commercial devices mainly implemented in the pharmaceutical industry due to their high costs. Unlike commercial products, academic devices often **lack the resilience** one would expect from engineered systems. Particularly, fluid management remains a challenge that often determines whether the experiment will be successful. Although this challenge applies to all OOC technologies, the LOC setting is even more critical due to the air-filled apical compartment and stretchable membrane that is sensitive to pressure variations induced by flow fluctuations—induced for instance by a peristaltic pump—in the basal compartment.

The **lack of standards**, especially in terms of technical and biological performance, is another challenge hindering the widespread adoption of LOC technologies [90]. Appropriate reference values, such as defined TEER settings and epithelial barrier thresholds, and standard operating protocols, would greatly facilitate benchmarking of new LOC systems.

Another challenge that is currently the focus of significant research efforts is the **lack of available lung alveolar cells**, particularly AEC2, with their original phenotype. Primary cells, obtained from several donors, are a relevant cell source, but donor-to-donor variability remains a problem as well as availability. Human iPSC-derived AEC2 are promising candidates, but their differentiation and maturation remains a challenge [91]. Another aspect that needs to be improved is the development of easy-to-culture hydrogels with **suitable stiffness**, which are required to provide an in vivo-like niche for AEC2.

Finally, **the fabrication of microfluidic chips** in academic settings remains labor-intensive, limiting the time available for biological experimentation. This is due to the predominant use of PDMS as the primary LOC construction material, using soft-lithographic techniques. Soft lithography relies on the prior fabrication of molds with high aspect ratio structures, often micrometers in size. This accuracy requires cleanroom facilities as PDMS cannot yet be produced by non-lithographic, low-cost techniques such as stereolithography, 3D printing, cutting plotters, or micromilling. Replacing PDMS with rigid polymers produced by injection molding imposes further limitations, such as cost or lower tunability. In addition, the integration of a flexible membrane to mimic the air-blood barrier is another limitation of this technology.

## 10.6 Perspectives for Lung-on-a-Chip Technologies

Although many biological processes can already be mimicked in lung-on-chips, this technology holds even greater potential just waiting to unfold. Aside from the above challenges that need to be addressed, we anticipate that their ability to decipher **more complex biological questions** will increase in the coming years. On the one hand, the engineered alveolar microenvironment will increasingly resemble the natural environment, allowing cells to differentiate and mature into specific cellular subtypes [10]. On the other hand, multi-organ-on-a-chip (MOOC) systems will be developed to mimic **crosstalk and interplay between different tissues**. The integration of the pulmonary capillary network with the alveolar barrier is the first obvious combination, which will enable investigating pulmonary vascular diseases and lung diseases of the distal airways (emphysema, fibrosis). The introduction of microvasculature is not only physiologically relevant but will also allow studying the immune response to various bacterial or viral infections [92]. In this context, developing a universal medium, possibly including parts of the patients' blood (plasma), will become necessary to provide the various organ-specific cells with nutrients and other required soluble molecules [93]. Another two organs-on-chip system may explore the gut-lung axis, particularly the interplay with their specific microbiota [94].

**Automation of LOC systems** requires more resilient solutions that should be user-centric to ease LOC handling, ultimately improving their widespread implementation. This development should be coupled with integrating sensors to **obtain quantifiable data** to feed machine learning algorithms aimed at providing yet to be discovered biological information [95]. These sensors can measure the mechanical stress to which the cells are exposed, the dissolved oxygen or other parameters of the cell culture media, or the tightness of the lung alveolar barrier [96, 97]. For example, Khalid et al. developed a lung cancer on-a-chip platform including an optical pH sensor to monitor the circulating culture medium, a trans-epithelial electrical resistance sensor (TEER), and a portable digital microscope for real-time monitoring [98]. In another study, we reported a microimpedance tomography system for monitoring membrane movements in a LOC, overcoming the limitations of the classical TEER technique with opposing electrodes, for inclusion in three-dimensional designs by using three coplanar electrodes.

With the advent of **precision medicine**, cells from patients, possibly iPSC-derived alveolar cells, will eventually be tested in LOC to determine the best possible treatment for each patient. For example, one can foresee that the antifibrotic effects of IPF drugs on patient cells can be predicted using such systems [70]. In any case, the correlation of in vitro data with clinical data needs to be rigorously established, which could pave the way to clinical trials with on LOC populated with patient cells.

**Acknowledgements** This work was supported by funding from the Swiss National Science Foundation (project: 185365), the Swiss 3R Competence Centre (project: 3RCC-OC-2019-025), the H2020 Eurostars (project: AIM4DoC) and MSCA-ITN (EUROoC, project: 812954), Innosuisse (project: 48818.1 IP-LS) and the Novartis Foundation for medical-biological Research (project: #20B146). The authors thank Dr. Anne Morbach for the illustration of Fig. 10.2.

## References

1. Huh D, Matthews BD, Mammoto A, Montoya-Zavala M, Hsin HY, Ingber DE. Reconstituting organ-level lung functions on a chip. Science. 2010;328:1662–8.
2. Guenat OT, Geiser T, Berthiaume F. Clinically Relevant Tissue Scale Responses as New Readouts from Organs-on-a-Chip for Precision Medicine. Annu Rev Anal Chem (Palo Alto Calif). 2020;13:111–33.
3. Huh D, Leslie DC, Matthews BD, Fraser JP, Jurek S, Hamilton GA, et al. A human disease model of drug toxicity-induced pulmonary edema in a lung-on-a-chip microdevice. Sci Transl Med. 2012;4:159ra147.
4. Weibel ER, Gomez DM. Architecture of the human lung. Use of quantitative methods establishes fundamental relations between size and number of lung structures. Science. 1962;137:577–85.
5. Basil MC, Katzen J, Engler AE, Guo M, Herriges MJ, Kathiriya JJ, et al. The Cellular and Physiological Basis for Lung Repair and Regeneration: Past, Present, and Future. Cell Stem Cell. Elsevier Inc.; 2020;26:482–502.

6. Zepp JA, Morrisey EE. Cellular crosstalk in the development and regeneration of the respiratory system. Nature Reviews Molecular Cell Biology [Internet]. Springer US; 2019; Available from: https://doi.org/10.1038/s41580-019-0141-3
7. Barkauskas CE, Cronce MJ, Rackley CR, Bowie EJ, Keene DR, Stripp BR, et al. Type 2 alveolar cells are stem cells in adult lung. Journal of Clinical Investigation. 2013;123:3025–36.
8. Zacharias WJ, Frank DB, Zepp JA, Morley MP, Alkhaleel FA, Kong J, et al. Regeneration of the lung alveolus by an evolutionarily conserved epithelial progenitor. Nature. Nature Publishing Group; 2018;555:251–5.
9. Strunz M, Simon LM, Ansari M, Kathiriya JJ, Angelidis I, Mayr CH, et al. Alveolar regeneration through a Krt8+ transitional stem cell state that persists in human lung fibrosis. Nature Communications. Springer US; 2020;11:3559.
10. Travaglini KJ, Nabhan AN, Penland L, Sinha R, Gillich A, Sit RV, et al. A molecular cell atlas of the human lung from single-cell RNA sequencing. Nature. 2020;587:619–25.
11. Zepp JA, Zacharias WJ, Frank DB, Cavanaugh CA, Zhou S, Morley MP, et al. Distinct Mesenchymal Lineages and Niches Promote Epithelial Self-Renewal and Myofibrogenesis in the Lung. Cell. 2017;170:1134–1148.e10.
12. Guenat OT, Berthiaume F. Incorporating mechanical strain in organs-on-a-chip: Lung and skin. Biomicrofluidics. 2018;12.
13. Gehr P, Bachofen M, Weibel ER. The normal human lung: ultrastructure and morphometric estimation of diffusion capacity. Respiration Physiology. 1978;32:121–40.
14. Kopf M, Schneider C, Nobs SP. The development and function of lung-resident macrophages and dendritic cells. Nat Immunol. 2015;16:36–44.
15. Ochs M, Nyengaard JR, Jung A, Knudsen L, Voigt M, Wahlers T, et al. The number of alveoli in the human lung. Am J Respir Crit Care Med. 2004;169:120–4.
16. Waters CM, Roan E, Navajas D. Mechanobiology in lung epithelial cells: measurements, perturbations, and responses. Compr Physiol. 2012;2:1–29.
17. Burgstaller G, Oehrle B, Gerckens M, White ES, Schiller HB, Eickelberg O. The instructive extracellular matrix of the lung: basic composition and alterations in chronic lung disease. Eur Respir J. 2017;50:1601805.
18. Giard DJ, Aaronson SA, Todaro GJ, Arnstein P, Kersey JH, Dosik H, et al. In vitro cultivation of human tumors: establishment of cell lines derived from a series of solid tumors. J Natl Cancer Inst. 1973;51:1417–23.
19. Tièche CC, Gao Y, Bührer ED, Hobi N, Berezowska SA, Wyler K, et al. Tumor Initiation Capacity and Therapy Resistance Are Differential Features of EMT-Related Subpopulations in the NSCLC Cell Line A549. Neoplasia. 2019;21:185–96.
20. Miller AJ, Dye BR, Ferrer-Torres D, Hill DR, Overeem AW, Shea LD, et al. Generation of lung organoids from human pluripotent stem cells in vitro. Nat Protoc. 2019;14:518–40.
21. Jacob A, Morley M, Hawkins F, McCauley KB, Jean JC, Heins H, et al. Differentiation of Human Pluripotent Stem Cells into Functional Lung Alveolar Epithelial Cells. Cell Stem Cell. Elsevier Inc.; 2017;21:472–488.e10.
22. Ghaedi M, Mendez JJ, Bove PF, Sivarapatna A, Raredon MSB, Niklason LE. Alveolar epithelial differentiation of human induced pluripotent stem cells in a rotating bioreactor. Biomaterials. 2014;35:699–710.
23. Tamò L, Hibaoui Y, Kallol S, Alves MP, Albrecht C, Hostettler KE, et al. Generation of an alveolar epithelial type II cell line from induced pluripotent stem cells. Am J Physiol Lung Cell Mol Physiol. 2018;315:L921–32.
24. Alysandratos K-D, Russo SJ, Petcherski A, Taddeo EP, Acín-Pérez R, Villacorta-Martin C, et al. Patient-specific iPSCs carrying an SFTPC mutation reveal the intrinsic alveolar epithelial dysfunction at the inception of interstitial lung disease. Cell Rep. 2021;36:109636.
25. Youk J, Kim T, Evans KV, Jeong Y-I, Hur Y, Hong SP, et al. Three-Dimensional Human Alveolar Stem Cell Culture Models Reveal Infection Response to SARS-CoV-2. Cell Stem Cell. Elsevier Inc.; 2020;27:905–919.e10.

26. Weiner AI, Jackson SR, Zhao G, Quansah KK, Farshchian JN, Neupauer KM, et al. Mesenchyme-free expansion and transplantation of adult alveolar progenitor cells: steps toward cell-based regenerative therapies. NPJ Regen Med. 2019;4:17.
27. Nikolić MZ, Rawlins EL. Lung Organoids and Their Use To Study Cell-Cell Interaction. Curr Pathobiol Rep. 2017;5:223–31.
28. Leeman KT, Pessina P, Lee J-H, Kim CF. Mesenchymal Stem Cells Increase Alveolar Differentiation in Lung Progenitor Organoid Cultures. Scientific Reports. Springer US; 2019;9:6479.
29. Evans KV, Lee J-H. Alveolar wars: The rise of in vitro models to understand human lung alveolar maintenance, regeneration, and disease. Stem Cells Transl Med. 2020;9:867–81.
30. Stucki AO, Stucki JD, Hall SRR, Felder M, Mermoud Y, Schmid RA, et al. A lung-on-a-chip array with an integrated bio-inspired respiration mechanism. Lab Chip. 2015;15:1302–10.
31. Zamprogno P, Wüthrich S, Achenbach S, Thoma G, Stucki JD, Hobi N, et al. Second-generation lung-on-a-chip with an array of stretchable alveoli made with a biological membrane. Commun Biol. 2021;4:168.
32. Huang D, Liu T, Liao J, Maharjan S, Xie X, Pérez M, et al. Reversed-engineered human alveolar lung-on-a-chip model. Proc Natl Acad Sci U S A. 2021;118:e2016146118.
33. Burgess JK, Mauad T, Tjin G, Karlsson JC, Westergren-Thorsson G. The extracellular matrix - the under-recognized element in lung disease? J Pathol. 2016;240:397–409.
34. Zhou Y, Horowitz JC, Naba A, Ambalavanan N, Atabai K, Balestrini J, et al. Extracellular matrix in lung development, homeostasis and disease. Matrix Biology [Internet]. Elsevier B.V.; 2018; Available from: https://doi.org/10.1016/j.matbio.2018.03.005
35. Tan X, Rodrigue D. A Review on Porous Polymeric Membrane Preparation. Part II: Production Techniques with Polyethylene, Polydimethylsiloxane, Polypropylene, Polyimide, and Polytetrafluoroethylene. Polymers (Basel). 2019;11:E1310.
36. van Meer BJ, de Vries H, Firth KSA, van Weerd J, Tertoolen LGJ, Karperien HBJ, et al. Small molecule absorption by PDMS in the context of drug response bioassays. Biochem Biophys Res Commun. 2017;482:323–8.
37. Petrou CL, D'Ovidio TJ, Bölükbas DA, Tas S, Brown RD, Allawzi A, et al. Clickable decellularized extracellular matrix as a new tool for building hybrid-hydrogels to model chronic fibrotic diseases in vitro. J Mater Chem B. 2020;8:6814–26.
38. Nishiguchi A, Singh S, Wessling M, Kirkpatrick CJ, Möller M. Basement Membrane Mimics of Biofunctionalized Nanofibers for a Bipolar-Cultured Human Primary Alveolar-Capillary Barrier Model. Biomacromolecules. 2017;18:719–27.
39. Higuita-Castro N, Nelson MT, Shukla V, Agudelo-Garcia PA, Zhang W, Duarte-Sanmiguel SM, et al. Using a Novel Microfabricated Model of the Alveolar-Capillary Barrier to Investigate the Effect of Matrix Structure on Atelectrauma. Sci Rep. 2017;7:11623.
40. Dohle E, Singh S, Nishigushi A, Fischer T, Wessling M, Möller M, et al. Human Co- and Triple-Culture Model of the Alveolar-Capillary Barrier on a Basement Membrane Mimic. Tissue Eng Part C Methods. 2018;24:495–503.
41. Yang X, Li K, Zhang X, Liu C, Guo B, Wen W, et al. Nanofiber membrane supported lung-on-a-chip microdevice for anti-cancer drug testing. Lab Chip. 2018;18:486–95.
42. Pensabene V, Costa L, Terekhov AY, Gnecco JS, Wikswo JP, Hofmeister WH. Ultrathin Polymer Membranes with Patterned, Micrometric Pores for Organs-on-Chips. ACS Appl Mater Interfaces. 2016;8:22629–36.
43. Montesanto S, Smithers NP, Bucchieri F, Brucato V, La Carrubba V, Davies DE, et al. Establishment of a pulmonary epithelial barrier on biodegradable poly-L-lactic-acid membranes. PLoS One. 2019;14:e0210830.
44. Tas S, Rehnberg E, Bölükbas DA, Beech JP, Kazado LN, Svenningsson I, et al. 3D printed lung on a chip device with a stretchable nanofibrous membrane for modeling ventilator induced lung injury [Internet]. Bioengineering; 2021 Jul. Available from: http://biorxiv.org/lookup/doi/10.1101/2021.07.02.450873

45. Dunphy SE, Bratt JAJ, Akram KM, Forsyth NR, El Haj AJ. Hydrogels for lung tissue engineering: Biomechanical properties of thin collagen-elastin constructs. Journal of the Mechanical Behavior of Biomedical Materials. 2014;38:251–9.
46. Zamprogno P, Thoma G, Cencen V, Ferrari D, Putz B, Michler J, et al. Mechanical Properties of Soft Biological Membranes for Organ-on-a-Chip Assessed by Bulge Test and AFM. ACS Biomater Sci Eng. 2021;7:2990–7.
47. Barkal LJ, Procknow CL, Álvarez-García YR, Niu M, Jiménez-Torres JA, Brockman-Schneider RA, et al. Microbial volatile communication in human organotypic lung models. Nat Commun. 2017;8:1770.
48. Stucki JD, Hobi N, Galimov A, Stucki AO, Schneider-Daum N, Lehr C-M, et al. Medium throughput breathing human primary cell alveolus-on-chip model. Sci Rep. 2018;8:14359.
49. Hassell BA, Goyal G, Lee E, Sontheimer-Phelps A, Levy O, Chen CS, et al. Human Organ Chip Models Recapitulate Orthotopic Lung Cancer Growth, Therapeutic Responses, and Tumor Dormancy In Vitro. Cell Reports. ElsevierCompany.; 2017;21:508–16.
50. Thacker VV, Dhar N, Sharma K, Barrile R, Karalis K, McKinney JD. A lung-on-chip model of early Mycobacterium tuberculosis infection reveals an essential role for alveolar epithelial cells in controlling bacterial growth. Elife. 2020;9:e59961.
51. Seo J, Conegliano D, Farrell M, Cho M, Ding X, Seykora T, et al. A microengineered model of RBC transfusion-induced pulmonary vascular injury. Sci Rep. 2017;7:3413.
52. Varone A, Nguyen JK, Leng L, Barrile R, Sliz J, Lucchesi C, et al. A novel organ-chip system emulates three-dimensional architecture of the human epithelia and the mechanical forces acting on it. Biomaterials. 2021;275:120957.
53. Saito A, Nagase T. Hippo and TGF-β interplay in the lung field. Am J Physiol Lung Cell Mol Physiol. 2015;309:L756–767.
54. Rentzsch I, Santos CL, Huhle R, Ferreira JMC, Koch T, Schnabel C, et al. Variable stretch reduces the pro-inflammatory response of alveolar epithelial cells. PLoS One. 2017;12:e0182369.
55. Diem K, Fauler M, Fois G, Hellmann A, Winokurow N, Schumacher S, et al. Mechanical stretch activates piezo1 in caveolae of alveolar type I cells to trigger ATP release and paracrine stimulation of surfactant secretion from alveolar type II cells. FASEB J. 2020;34:12785–804.
56. Cong X, Hubmayr RD, Li C, Zhao X. Plasma membrane wounding and repair in pulmonary diseases. Am J Physiol Lung Cell Mol Physiol. 2017;312:L371–91.
57. Birukov KG. Cyclic stretch, reactive oxygen species, and vascular remodeling. Antioxid Redox Signal. 2009;11:1651–67.
58. Ishahak M, Hill J, Amin Q, Wubker L, Hernandez A, Mitrofanova A, et al. Modular Microphysiological System for Modeling of Biologic Barrier Function. Front Bioeng Biotechnol. 2020;8:581163.
59. Zeinali S, Thompson EK, Gerhardt H, Geiser T, Guenat OT. Remodeling of an in vitro microvessel exposed to cyclic mechanical stretch. APL Bioeng. 2021;5:026102.
60. Shimizu A, Goh WH, Itai S, Hashimoto M, Miura S, Onoe H. ECM-based microchannel for culturing in vitro vascular tissues with simultaneous perfusion and stretch. Lab Chip. 2020;20:1917–27.
61. Anguiano M, Castilla C, Maška M, Ederra C, Peláez R, Morales X, et al. Characterization of three-dimensional cancer cell migration in mixed collagen-Matrigel scaffolds using microfluidics and image analysis. PLoS One. 2017;12:e0171417.
62. Nalayanda DD, Puleo C, Fulton WB, Sharpe LM, Wang T-H, Abdullah F. An open-access microfluidic model for lung-specific functional studies at an air-liquid interface. Biomed Microdevices. 2009;11:1081–9.
63. van Riet S, Ninaber DK, Mikkers HMM, Tetley TD, Jost CR, Mulder AA, et al. In vitro modelling of alveolar repair at the air-liquid interface using alveolar epithelial cells derived from human induced pluripotent stem cells. Sci Rep. 2020;10:5499.
64. Mermoud Y, Felder M, Stucki JD, Stucki AO, Guenat OT. Microimpedance tomography system to monitor cell activity and membrane movements in a breathing lung-on-chip. Sensors and Actuators B: Chemical. 2018;255:3647–53.

65. Bichsel CA, Hall SRR, Schmid RA, Guenat OT, Geiser T. Primary Human Lung Pericytes Support and Stabilize In Vitro Perfusable Microvessels. Tissue Eng Part A. 2015;21:2166–76.
66. Haak AJ, Tan Q, Tschumperlin DJ. Matrix biomechanics and dynamics in pulmonary fibrosis. Matrix Biol. 2018;73:64–76.
67. Sundarakrishnan A, Zukas H, Coburn J, Bertini BT, Liu Z, Georgakoudi I, et al. Bioengineered in Vitro Tissue Model of Fibroblast Activation for Modeling Pulmonary Fibrosis. ACS Biomater Sci Eng. 2019;5:2417–29.
68. Felder M, Trueeb B, Stucki AO, Borcard S, Stucki JD, Schnyder B, et al. Impaired Wound Healing of Alveolar Lung Epithelial Cells in a Breathing Lung-On-A-Chip. Front Bioeng Biotechnol. 2019;7:3.
69. Mejías JC, Nelson MR, Liseth O, Roy K. A 96-well format microvascularized human lung-on-a-chip platform for microphysiological modeling of fibrotic diseases. Lab Chip. 2020;20:3601–11.
70. Asmani M, Velumani S, Li Y, Wawrzyniak N, Hsia I, Chen Z, et al. Fibrotic microtissue array to predict anti-fibrosis drug efficacy. Nat Commun. 2018;9:2066.
71. Felder M, Trueeb B, Stucki AO, Borcard S, Stucki JD, Schnyder B, et al. Impaired Wound Healing of Alveolar Lung Epithelial Cells in a Breathing Lung-On-A-Chip. Frontiers in bioengineering and biotechnology. 2019;7:3.
72. Foreman KJ, Marquez N, Dolgert A, Fukutaki K, Fullman N, McGaughey M, et al. Forecasting life expectancy, years of life lost, and all-cause and cause-specific mortality for 250 causes of death: reference and alternative scenarios for 2016-40 for 195 countries and territories. Lancet. 2018;392:2052–90.
73. Tuder RM, Yoshida T, Arap W, Pasqualini R, Petrache I. State of the art. Cellular and molecular mechanisms of alveolar destruction in emphysema: an evolutionary perspective. Proc Am Thorac Soc. 2006;3:503–10.
74. Krimmer DI, Oliver BGG. What can in vitro models of COPD tell us? Pulm Pharmacol Ther. 2011;24:471–7.
75. Madl AK, Plummer LE, Carosino C, Pinkerton KE. Nanoparticles, lung injury, and the role of oxidant stress. Annu Rev Physiol. 2014;76:447–65.
76. Zhang M, Xu C, Jiang L, Qin J. A 3D human lung-on-a-chip model for nanotoxicity testing. Toxicol Res (Camb). 2018;7:1048–60.
77. ARDS Definition Task Force, Ranieri VM, Rubenfeld GD, Thompson BT, Ferguson ND, Caldwell E, et al. Acute respiratory distress syndrome: the Berlin Definition. JAMA. 2012;307:2526–33.
78. Nalayanda DD, Fulton WB, Colombani PM, Wang T-H, Abdullah F. Pressure induced lung injury in a novel in vitro model of the alveolar interface: protective effect of dexamethasone. J Pediatr Surg. 2014;49:61–5; discussion 65.
79. Xu C, Zhang M, Chen W, Jiang L, Chen C, Qin J. Assessment of Air Pollutant PM2.5 Pulmonary Exposure Using a 3D Lung-on-Chip Model. ACS Biomater Sci Eng. 2020;6:3081–90.
80. Deinhardt-Emmer S, Rennert K, Schicke E, Cseresnyés Z, Windolph M, Nietzsche S, et al. Co-infection with Staphylococcus aureus after primary influenza virus infection leads to damage of the endothelium in a human alveolus-on-a-chip model. Biofabrication. 2020;12:025012.
81. Carsana L, Sonzogni A, Nasr A, Rossi RS, Pellegrinelli A, Zerbi P, et al. Pulmonary postmortem findings in a series of COVID-19 cases from northern Italy: a two-centre descriptive study. Lancet Infect Dis. 2020;20:1135–40.
82. Zaim S, Chong JH, Sankaranarayanan V, Harky A. COVID-19 and Multiorgan Response. Curr Probl Cardiol. 2020;45:100618.
83. Kiener M, Roldan N, Machahua C, Sengupta A, Geiser T, Guenat OT, et al. Human-Based Advanced in vitro Approaches to Investigate Lung Fibrosis and Pulmonary Effects of COVID-19. Front Med (Lausanne). 2021;8:644678.
84. Deinhardt-Emmer S, Böttcher S, Häring C, Giebeler L, Henke A, Zell R, et al. SARS-CoV-2 causes severe epithelial inflammation and barrier dysfunction. J Virol. 2021;JVI.00110-21.

85. Zhang M, Wang P, Luo R, Wang Y, Li Z, Guo Y, et al. Biomimetic Human Disease Model of SARS-CoV-2 Induced Lung Injury and Immune Responses on Organ Chip System. Adv Sci (Weinh). 2020;2002928.
86. Thacker VV, Sharma K, Dhar N, Mancini G-F, Sordet-Dessimoz J, McKinney JD. Rapid endotheliitis and vascular damage characterize SARS-CoV-2 infection in a human lung-on-chip model. EMBO Rep. 2021;22:e52744.
87. Denisenko TV, Budkevich IN, Zhivotovsky B. Cell death-based treatment of lung adenocarcinoma. Cell Death Dis. 2018;9:117.
88. Singleton DC, Macann A, Wilson WR. Therapeutic targeting of the hypoxic tumour microenvironment. Nat Rev Clin Oncol. 2021;
89. Park S, Kim TH, Kim SH, You S, Jung Y. Three-Dimensional Vascularized Lung Cancer-on-a-Chip with Lung Extracellular Matrix Hydrogels for In Vitro Screening. Cancers (Basel). 2021;13:3930.
90. Piergiovanni M, Leite SB, Corvi R, Whelan M. Standardisation needs for organ on chip devices. Lab Chip. 2021;21:2857–68.
91. Beers MF, Moodley Y. When Is an Alveolar Type 2 Cell an Alveolar Type 2 Cell? A Conundrum for Lung Stem Cell Biology and Regenerative Medicine. Am J Respir Cell Mol Biol. 2017;57:18–27.
92. McMinn PH, Hind LE, Huttenlocher A, Beebe DJ. Neutrophil trafficking on-a-chip: an in vitro, organotypic model for investigating neutrophil priming, extravasation, and migration with spatiotemporal control. Lab Chip. 2019;19:3697–705.
93. Oleaga C, Bernabini C, Smith AST, Srinivasan B, Jackson M, McLamb W, et al. Multi-Organ toxicity demonstration in a functional human in vitro system composed of four organs. Sci Rep. 2016;6:20030.
94. Marsland BJ, Trompette A, Gollwitzer ES. The Gut-Lung Axis in Respiratory Disease. Ann Am Thorac Soc. 2015;12 Suppl 2:S150–156.
95. Mencattini A, Mattei F, Schiavoni G, Gerardino A, Businaro L, Di Natale C, et al. From Petri Dishes to Organ on Chip Platform: The Increasing Importance of Machine Learning and Image Analysis. Front Pharmacol. 2019;10:100.
96. Zhang B, Korolj A, Lai BFL, Radisic M. Advances in organ-on-a-chip engineering. Nat Rev Mater. 2018;3:257–78.
97. Kilic T, Navaee F, Stradolini F, Renaud P, Carrara S. Organs-on-chip monitoring: sensors and other strategies. Microphysiol Syst. 2018;1:1–1.
98. Khalid MAU, Kim YS, Ali M, Lee BG, Cho Y-J, Choi KH. A lung cancer-on-chip platform with integrated biosensors for physiological monitoring and toxicity assessment. Biochemical Engineering Journal. 2020;155:107469.

# Chapter 11
# Assessment of Collagen in Translational Models of Lung Research

**Claudia A. Staab-Weijnitz, Ceylan Onursal, Deepika Nambiar, and Roberto Vanacore**

## 11.1 Introduction

The extracellular matrix (ECM) is a complex protein network which not only provides structural support for adherent and migrating cells, but also important mechanical and biochemical cues for cell phenotypes and functions including stem cell fate [114]. The ECM is classified in two major structural compartments: (1) Basement membranes (BM) are a specialized form of ECM that appear as thin and dense acellular sheets underneath epithelial and endothelial cell layer providing structural integrity and mechanical support to cell layers and conferring cell polarity [56, 114]. They are mainly composed of the network-forming type IV collagen, laminins, nidogens and the proteoglycan perlecan. In contrast, the (2) interstitial ECM typically surrounds cells, e.g., fibroblasts residing in the lung interstitium, completely and its main components are the fibrillar type I and type III collagens, fibronectin, decorin, and hyaluronan [87] (Fig. 11.1a). Both ECM compartments undergo fundamental changes in lung disease which directly cause loss of lung function, increase susceptibility to inhaled toxicants and respiratory pathogens, and may strongly affect therapeutic efficacy [17, 123].

C. A. Staab-Weijnitz (✉) · C. Onursal
Institute of Lung Health and Immunity and Comprehensive Pneumology Center with the CPC-M BioArchive, Member of the German Center for Lung Research (DZL), Ludwig-Maximilians-Universität and Helmholtz Zentrum München, Munich, Germany
e-mail: staab-weijnitz@helmholtz-muenchen.de

D. Nambiar · R. Vanacore (✉)
Center for Matrix Biology, Department of Medicine, Division of Nephrology and Hypertension, Vanderbilt University Medical Center, Nashville, TN, USA
e-mail: roberto.vanacore@vumc.org

 
C. M. Magin (ed.), *Engineering Translational Models of Lung Homeostasis and Disease*, Advances in Experimental Medicine and Biology 1413,
https://doi.org/10.1007/978-3-031-26625-6_11

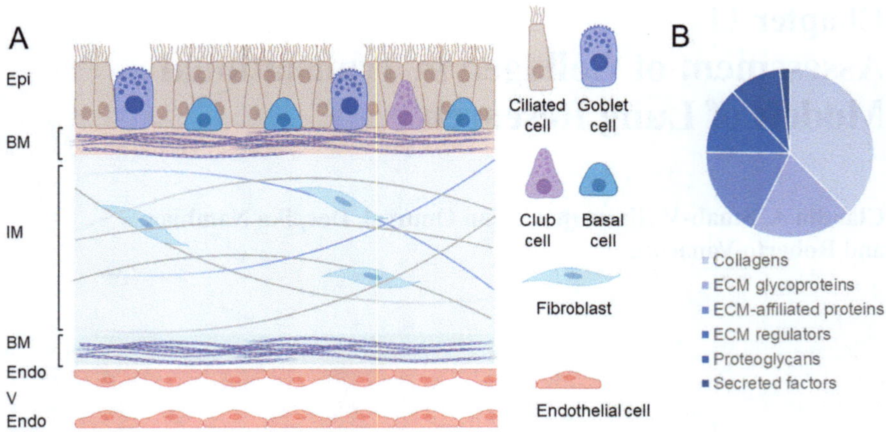


**Fig. 11.1** Types and composition of the extracellular matrix (ECM) in the lung. (**a**) Schematic overview on types of ECM in the lung in relation to the epithelial and endothelial cell layer, focusing on the conducting airways. Epi, bronchial epithelium displaying all major cell types; BM basement membrane; IM, interstitial matrix harboring scarce fibroblasts; Endo, endothelial cell layer; V, vessel lumen. Smooth muscle cells and cartilage are omitted for increased clarity. (Figure was created with BioRender.com). (**b**) ECM composition in the lung according to the matrisome categorization established by Naba et al. [77]. (Adapted from Beachley et al. [9])

Collagen is a large protein superfamily and the main ECM protein component in almost every tissue type including the lung [9] (Fig. 11.1b). The unifying feature of collagens is the triple-helical collagenous domain, which is assembled in the endoplasmic reticulum (ER) from three α-chains (Fig. 11.2) consisting of regular amino acid repeats of (Gly-X-Y)$_n$, where Y often is 4-hydroxyproline. There are 28 different human collagen types which form homo- or heterotrimeric triple helices during folding in the ER and which are categorized in seven different classes, based on their final extracellular supramolecular assembly: (1) Fibril-forming collagens (I, II, III, V, XI, XXIV, XXVII), (2) fibril-associated collagens with interrupted triple helices (FACITs, IX, XII, XIV, XVI, XIX, XX, XXI, XXII), (3) network-forming collagens (IV, VIII, X), (4) transmembrane collagens (XIII, XVII, XXIII, XXV), (5) endostatin-producing collagens (also termed multiplexins, XV, XVIII), (6) anchoring fibrils (VII), and (7) beaded-filament-forming collagen (VI). The collagen types XXVI and XXVIII do not fit well in any of the above-listed categories [83].

Because of the central role of the ECM and collagen in lung disease, quantification, determination of molecular properties, and visualization of three-dimensional structure of collagen is important for the development and characterization of translational models of lung research. For in vitro, ex vivo, and in vivo models for lung fibrosis, collagen quantity and crosslinking remain the most important readouts for evaluation of protection from disease [1, 43, 74]. In this chapter, we aim to give a comprehensive overview on the various methodologies for quantification and characterization of collagen currently available including their advantages and disadvantages.

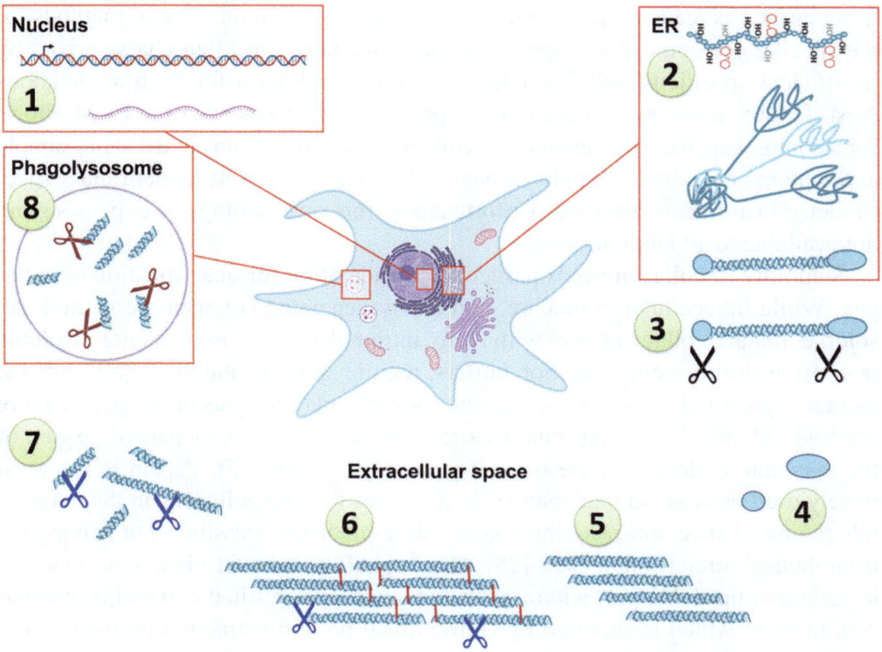


**Fig. 11.2** Overview of the most important intracellular and extracellular collagen biosynthesis, maturation, and degradation pathways. Using the example of an activated myofibroblast and biosynthesis of fibrillar collagen, the figure depicts the following major steps of this pathway: (**1**) Collagen gene transcription to mRNA in the nucleus; (**2**) Translation of mRNA to protein, co- and post-translational modifications and triple helix formation in the rough endoplasmic reticulum (rER); after secretion to the extracellular space via trafficking through the Golgi network, (**3**) propeptide cleavage by specific N- and C-terminal collagen proteinases, yielding (**4**) propeptides which may serve as peripheral markers for collagen formation. Propeptide cleavage triggers (**5**) fibril formation, followed by (**6**) extracellular crosslinking and stabilization and further maturation of the resulting fibers. (**7**) Extracellular degradation is performed by collagenolytic matrix metalloproteases (MMPs) and generated fragments, but also larger fibrils, can be internalized and (**8**) degraded within the cell in the phagolysosome. (Figure was created with BioRender.com)

## 11.2 Quantification of Collagen

Methods for the identification and quantification of collagen and its subtypes have been continuously developed for more than one century. These methods differ considerably in specificity and sensitivity [12, 24, 81]. Many current methods for quantification of collagen take advantage of collagen-specific properties such as hydroxyproline content and affinity of dyes to the triple helical domains, and therefore determine total collagen content. Quantification of specific collagen types is currently only possible by transcript analysis, immuno-based or mass spectrometry-based approaches. Considering the numerous post-transcriptional regulation events involved in collagen biosynthesis and maturation (Fig. 11.2), transcript analysis, although informative on the regulatory level of collagen expression changes, can never suffice as readout on its own. On the other hand, immuno-based methods for the specific detection of collagen types, unfortunately, are often characterized by insufficient specificity and few reliable specific antibodies for such applications exist. Finally, mass spectrometry-based proteomic assessment of the ECM allows for a comprehensive assessment of all collagen types and chains in the same sample and requires only small sample amounts when state-of-the-art tandem mass spectrometry instruments are used. Unfortunately, this methodology is expensive and not available to all laboratories.

Solubility of collagen needs particular consideration for quantification of collagen. While intracellular, immature, and newly generated collagen are neutral-salt soluble, mature collagen fibers with many intramolecular cross-links are insoluble in conventional protein extraction buffers, require special solubilization protocols, and are typically dissolved in acetic acid for molecular analysis or usage as culture scaffold [54, 89]. Following protein extraction, chemical or enzymatic digestion of the insoluble pellet will increase collagen coverage [39, 75]. Pepsin is the most widely used protease in that context, as it cleaves fibrillar collagens in the telopeptide regions, hence removing the lysyl oxidase-mediated crosslinks but leaving the triple-helical stretch untouched [28]. Notably, efforts should always be made to include this insoluble part and/or assess secreted or deposited extracellular collagen, none of which is captured by conventional protein extraction protocols. It is important to acknowledge that an increase in intracellular collagen does not necessarily reflect an increase in extracellular collagen—actually, the opposite may be the case if collagen secretion is impaired, and collagen accumulates in the ER.

### 11.2.1 The Sircol Assay

The Sircol assay is a fast and simple colorimetric method based on binding of Sirius Red F3B to collagen [59]. The binding specificity of Sirius Red relies on its elongated structure which associates with triple helical collagen along the linear axis and exposes numerous acid sulfonate groups which interact with basic residues in the collagen sequence [60]. Following appropriate solubilization, this method can in principle be used to determine all pools of collagen in complex protein solution in the context of in vitro and in vivo experiments. We find it particularly suitable for determination of newly synthesized collagen content in the cell culture supernatant as a readout for in vitro experiments [102]. However, caution must be taken to use serum-starved culture settings, as serum components are known to interfere with the assay [21, 59]. Application of Sirius Red for collagen visualization in situ will be described in Sect. 11.4.2.

### 11.2.2 Hydroxyproline Quantification

The amino acid 4-hydroxyproline (4-Hyp) occurs in high abundance in triple-helical collagenous domains where it frequently occupies Y positions of the above mentioned $(Gly\text{-}X\text{-}Y)_n$ repeats. 4-Hydroxylation of proline is catalyzed in a co- or post-translational fashion by ER-resident prolyl-4-hydroxylases which act on the unfolded polypeptide chain [83]. Presence of 4-Hyp is known to increase the thermodynamic stability of collagen [90]. Although widely considered specific for collagen, it should be mentioned that 4-Hyp also occurs in other proteins. For instance, it has been estimated that up to 33% of the about 90 proline residues in elastin can be hydroxylated [96]. Furthermore, a single 4-Hyp in the hypoxia-inducible factor (HIF) α subunit (HIF-1α) acts as oxygen sensor and plays an important role in the regulation of gene expression by hypoxia [42]. More such examples may exist; nevertheless, considering that collagen is much more abundant than other 4-Hyp containing proteins and the exceptionally high abundance of 4-Hyp in collagen, 4-Hyp quantity can still be considered a reasonable measure for total collagen content. In addition, 3-Hyp is also present in collagens but it is found in much less abundance than 4-Hyp [41].

Collagen quantification by the hydroxyproline assay is particularly suitable for insoluble and solid samples such as mature collagen in tissue or insoluble pellets following protein extraction from complex samples. Samples are completely hydrolyzed by boiling in 6 M hydrochloric acid for several hours and subsequently subjected to amino acid analysis including the quantification of 4-hydroxyproline using high-performance liquid chromatography (HPLC) [54]. This method is still considered the gold standard for measuring hydroxyproline content, although it is not particularly sensitive, requires large sample sizes, and an HPLC set-up is not available in every laboratory [24]. LC-MS/MS quantitation of 4-Hyp methods are more sensitive and given the popularity of LC-MS instrumentation, the technique is becoming more accessible in research centers. Notably, colorimetric alternatives are increasingly being offered by suppliers and appear to yield similar results [88].

### 11.2.3 Immuno-Based Methods

Immune-based methods such as enzyme-linked immunosorbent assays (ELISAs) and Western blotting have the potential to allow for specific detection of collagen types and their absolute or relative quantification. Given that it is becoming increasingly clear that collagens fulfill very different roles in health and disease, collagen-type-specific approaches must be considered much more often. For instance, a mass spectrometry-based analysis of TGF-β-induced changes in lung fibroblast-deposited ECM has shown the expected upregulation of the fibrillar type I, but at the same time decreases in type VI collagen, and more variably altered or unchanged levels of other collagen types. Hence, already in such a simple in vitro model of lung fibrosis changes in collagen go far beyond a simple increase in levels overall, but specifically affect collagen types differently [74]. Indeed, while in lung fibrosis fibrillar collagen types I and III are consistently increased, levels of other collagen types remain unchanged or are downregulated during fibrogenesis; others again are variably regulated dependent on disease stage or anatomic location [52]. These observations undoubtedly call for more collagen type-specific assessments to increase our understanding of the distinct roles of different collagen types in disease. However, immune-based approaches are limited by the availability of specific antibodies and, unfortunately, inadequate validation of antibodies and poor specificity remain a major issue in the biomedical research community [30]. Only well-validated antibodies should be used, and proper controls included to ensure specificity. Special attention must be paid to the immunogen used for raising the antibody—some antibodies are specifically raised to target propeptide sequences and hence will only detect intracellular and immature collagen. Others are directed against specific chains of a collagen type or against three-dimensional epitopes of fully assembled extracellular mature collagen types. Hence, some antibodies will require sample denaturation while others will most specifically detect the native fold—sample selection and processing must be adjusted accordingly.

Finally, tandem mass spectrometry (MS/MS) allows for simultaneous assessment of all collagen chains and types in a single run with comparatively little sample amount given that a state-of-the-art instrument is used. In addition, the data generated also enables determination of collagen chain stoichiometries and site-specific identification of post-translational modifications in distinct collagen types [8, 74]. Clearly, it represents an expensive technology, but at the same time it is by far the most powerful technique, offering information about collagen in unprecedented molecular detail. Therefore, Sect. (11.3) will follow to provide guidance for the usage of LC-MS/MS-based proteomics approaches for the assessment of collagen (Table 11.1).

## 11.3 Mass Spectrometry Characterization of Collagen

### *11.3.1 Assessment of Collagens in Proteomics Analyses of Pulmonary ECM*

The study of ECM proteins may provide important insights about the molecular mechanisms underlying disease including lung fibrosis. The advent of proteomics has propelled the study of the so-called matrisome, a term coined by Naba et al., which catalogued both "core" and "associated" proteins present in the ECM of many tissues [77]. Although in the beginning proteomics used other biochemical techniques such as two-dimensional electrophoresis, nowadays it relies almost exclusively on liquid chromatography tandem mass spectrometry (LC-MS/MS), an analytical technique that combines the power of liquid chromatography to separate

**Table 11.1** Methodologies for the quantification of collagen in crude samples from in vitro, ex vivo, and in vivo models of lung research

| Method | Principle | Advantages | Disadvantages |
|---|---|---|---|
| Sircol assay | Sirius red dye binds specifically to elongated triple helical collagenous domains | Rapid colorimetric assay<br>No special equipment necessary | May require additional solubilization steps<br>Serum components interfere with the assay<br>Does not differentiate between collagen types |
| Hydroxyproline quantification | In collagens, 4-hydroxyproline is highly abundant and reasonably specific. | Suitable for insoluble samples and crude tissue<br>Colorimetric alternative is available | Requires HPLC set-up if not colorimetric assay is used<br>Requires large sample size<br>Does not differentiate between collagen types<br>Hydroxyproline content in elastin may falsify results |
| Immuno-based methods (ELISA, Western Blot) | Antibodies are raised against specific collagen types or chains. | In principle suitable for all kinds of samples<br>No special equipment necessary<br>Can differentiate between collagen types | Strongly dependent on antibody specificity—may require additional validation<br>Antibodies are expensive |
| Proteomics | Mass-spectrometry-based identification and quantification of the matrisome including all detectable collagen types in the sample | Can differentiate between collagen types<br>Little sample necessary for comprehensive analysis of all collagen types in the same run | Requires a facility with state-of-the-art tandem mass spectrometry<br>Expensive methodology |

complex peptide mixtures and the high sensitivity of modern high-resolution mass spectrometers. Most ECM proteomics studies have used a "bottom-up" approach in which proteins are first digested into peptides before LC-MS/MS analyses. The peptide's amino acid sequences are identified by database search engines where tandem mass spectra predicted from a protein database are compared to MS/MS spectra obtained experimentally. Identified peptide sequences are assembled into proteins using bioinformatics tools. Thus, unlike collagen-specific methods mentioned above, mass spectrometry-based proteomics is a powerful technological platform that allows not only the assessment of hundreds of proteins, including the different collagen types, at once, but also the characterization of detailed biochemical features such as posttranslational modifications and/or chemical changes of proteins.

Matrisome studies involved the sequential extraction of several cellular fractions until a fraction enriched in insoluble ECM proteins was obtained and analyzed by LC-MS/MS. An example of a general workflow is given in Fig. 11.3. Notably, the study also developed an in-silico human matrisome database composed of a "core" and "associated" matrisome genes that facilitated the identification and classification of matrisome proteins present in different tissues. The "core matrisome" is comprised of 278 genes (274 in the mouse) encoding ECM glycoproteins, collagens, and proteoglycans. In addition, 778 genes are cataloged as "matrisome-associated proteins" comprised of ECM regulators and modifiers and secreted factors [77]. In these studies, the protein composition of the lung ECM was investigated using LC-MS/MS. Following sequential extraction, insoluble ECM-rich

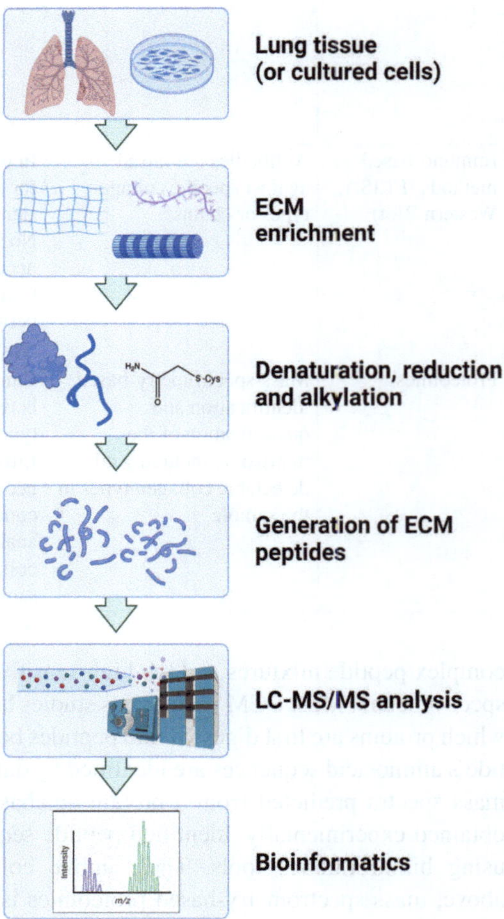


**Fig. 11.3** General workflow of sample preparation for extracellular matrix proteomics. Decellularization of the tissues or cultured cells enhances detection of ECM proteins. Many strategies have been developed to take advantage of the insoluble nature and enrich for ECM proteins using detergent mixtures that dissolve lipid membranes and allow the removal of soluble proteins. The enriched ECM proteins are denatured, reduced and alkylated to facilitate digestion into peptides by enzymatic or chemical protocols. The peptide mixture is fractionated by different means, typically liquid chromatography, before detection and quantification by high-resolution mass spectrometry. (Figure is created with BioRender.com)

samples were obtained and solubilized in high urea followed by reduction, alkylation, and deglycosylation. ECM proteins were digested with trypsin and the resulting peptides were further separated and purified on off-gel electrophoresis. ECM samples were analyzed by LC-MS/MS and the resulting spectra were searched against the mouse database. The study revealed that the murine lung ECM comprises 143 total matrisome proteins: 92 core matrisome proteins, and 51 matrisome-associated proteins [77]. Notably, 43 collagen gene products were identified in the lung samples.

In addition to learning about the composition of normal tissue ECM, LC-MS/MS proteomics studies provide an effective platform for the characterization of ECM proteins from diseased lungs, which may give cues into molecular pathways underlying fibrotic disease. For instance, a comprehensive dynamic proteomics effort to characterize changes in ECM protein biosynthesis during bleomycin-induced lung fibrosis was undertaken [27]. Lung tissues were collected from bleomycin- and vehicle-treated mice labelled with deuterated water prior to injury. Labeled tissue samples were subjected to sequential extraction with salt, detergent, and guanidine. Analysis of soluble and insoluble fractions revealed that while ECM proteins were not significantly solubilized with salt and detergent, they were highly enriched in guanidine-soluble and insoluble fractions. Furthermore, unlike traditional "static" proteomics methods, the use of stable isotope labelling combined with the sequential extractions allowed estimation of fractional synthesis rate (FSR) of single ECM proteins. The study revealed the dynamic changes in synthesis of certain collagen types along with elevated levels of fibrillar collagen, confirming that bleomycin injury induced deposition of insoluble matrix that eventually leads to fibrosis [27].

In a subsequent study, quantitative detergent solubility profiling (QDSP) of lung tissue homogenates was used to evaluate tissue composition from the onset of inflammation and fibrosis to its full recovery [98]. Unlike the previous study, the authors separated whole tissue proteins based on their differential solubility in response to increased detergent stringency, which allowed monitoring interactions of secreted proteins with the insoluble ECM. A label-free mass spectrometry-based approach was used to estimate the relative levels of proteins in the different fractions from all time-points from fibrosis to recovery. The results revealed that some core matrisome proteins, including collagens, were altered upon bleomycin injury with it moving from the soluble to the insoluble fraction. In contrast, many of these fell back to baseline during repair and remodeling of the lungs after injury. Such QDSP method has also been adopted to develop a proteomics workflow for human lung fibrosis biopsies and to study protein involved in lung aging [2, 99].

The studies described above identified peptides with few posttranslational modifications from the most abundant collagen types. leaving many unidentified collagen types identified in a later study concentrating on pulmonary ECM [74]. While the MS methods used in previous general studies included hydroxylation of proline in database search parameters, allowing the identification of collagen peptides, they are not optimized to exhaustively identify highly modified collagen peptides. Thus, this limitation is more relevant for basement membrane collagens such as collagen IV and IV which are highly hydroxylated and glycosylated. In the following Sect.

11.3.2 we provide more details of database search strategies that have been designed to overcome these limitations.

### *11.3.2 Analysis of Posttranslational Modifications of Collagen*

During biosynthesis, a number of co- and post-translational modifications (PTMs) are added onto the collagen molecule, most notably hydroxylation and glycosylation, both of which are essential for maintaining the architecture and function of tissues [8, 90]. For instance, 4-hydroxylation of proline is important for folding and stabilization of triple-helical molecules and fibril assembly [117]. Similarly, hydroxylation and glycosylation of lysine are important for crosslinking and stabilization of collagen fibrils as well as regulation of cell-matrix interactions [61, 122]. In addition, N-glycosylation of asparagine is thought to play an important role in collagen degradation [47]. Mutations on either collagens and/or the enzymes that modify them can have a detrimental effect on collagen structure, alter tissue function and thus result in disease. Perhaps the best characterized collagen-related disease resulting from genetical alterations is *osteogenesis imperfecta* [71]. Although the extent of modification may vary, quantitative mapping efforts to catalog modifications along the collagen molecule are very few. However, in recent years improvement of mass spectrometry technologies may allow a better characterization of collagen PTM changes [74].

Although several protocols have been developed for the generation of suitable samples for analysis of these highly modified molecules, the majority of them take advantage of the insoluble nature of collagens for their enrichment. As can be observed in Fig. 11.3, after collagen peptides are generated by chemical degradation or enzymatic digestion, they may be fractionated by different means, but typically liquid chromatography, followed by tandem mass spectrometry analysis.

Because collagen peptides are highly hydroxylated and glycosylated, they pose a challenge for characterization by conventional mass spectrometry-based proteomics approaches. To generate tandem mass spectra, peptides are subjected to gas-phase fragmentation techniques such as collision-induced dissociation (CID) which breaks peptide bonds generating so-called b- and y-fragment ions from which the peptide sequence can be deduced. Hydroxylation of proline and lysine residues is identified by adding 16 mass units to such amino acids during the database search. For this it is key that hydroxylation is sufficiently stable and does not undergo cleavage under CID conditions. Although O-glycosylation was thought to be labile, Perdivara et al. showed that o-glycosidic bonds between glucose, galactose and hydroxylysine are unusually stable to CID in collagen tryptic peptides [85]. Because of this characteristic of collagen peptides, CID and higher-energy C-trap dissociation (HCD), a technology available in high-resolution Orbitrap instruments, have successfully been used alone or in conjunction with PTM-friendly chemical ionization technologies such as electron-transfer dissociation (ETD) for the MS analysis of highly hydroxylated and glycosylated collagen peptides [8].

For instance, a label-free mass spectrometry approach was used to understand hydroxylation and glycosylation on collagen in response to profibrotic cytokine transforming growth factor β1 (TGF-β1). As a proof of concept, fibroblasts derived from IPF human samples were treated with TGF-β1 to induce changes to their ECM within an environment that mimics fibrosis [74]. After decellularization, the enriched ECM was digested into peptides by trypsin/LysC mix and peptides were analyzed by LC-MS/MS. MS-based label-free quantification revealed that upon exposure to TGF-β1 in-vitro, lung fibroblasts ECM experienced changes commonly associated with lung fibrosis such as increased expression of fibrillar collagens such as collagen I, collagen II, collagen III & collagen V. Furthermore, a new bioinformatic platform was developed to allow for the comprehensive mapping and site-specific quantitation of collagen PTMs in these crude ECM preparations. For the identification of PTM on collagen peptides, mass spectrometry data files were searched using the unique motif search feature MyriMatch that allowed identification of sites of hydroxylation and glycosylation [8, 70, 106]. The analyses yielded a comprehensive map of prolyl and lysyl hydroxylations as well as lysyl glycosylations for 15 collagen chains. PTM analysis revealed novel sites of prolyl-3-hydroxylation and lysyl glycosylation in type I collagen. In addition, the same data-dependent acquisition MS data were subjected to an MS1 analysis using Skyline software to assess changes in collagen PTM [74]. Skyline MS1 is a label-free quantification technique in which the areas of each peptide chromatographic peak (a.k.a. extracted ion chromatogram – XIC) are recorded, averaged, and compared between the different sample groups [https://pubmed.ncbi.nlm.nih.gov/20147306/]. The Skyline MS1 workflow was able to identify significant changes in prolyl-3-hydroxylation and O-glycosylation at specific sites within type I collagen molecules present in ECM samples taken from human lung fibroblasts stimulated with TGF-β1 concentrations mimicking a fibrotic environment.

### *11.3.3 Assessment of Enzymatic Crosslinks in Collagen*

Crosslinks play an important role in maintaining and strengthening the intricate structure of collagens in the ECM. In fibrotic diseases and cancer, abnormal ECM dynamics and crosslinks disturb the homeostatic state of cells and promote organ failure. Collagen-crosslinking lysyl oxidases (LOX) are upregulated in many forms of lung fibrotic disorders such as idiopathic pulmonary fibrosis (IPF) [16, 45]. Increased crosslinking leads to pathologic deposition of collagen which may disrupt elasticity, promote stiffness, and reduce lung function. Thus, when evaluating collagen in translational models of lung disease, it is important to quantify collagen crosslinks.

The precursors to the crosslinks are formed by the oxidative deamination of lysine/hydroxylysine in collagen by the members of the lysyl oxidase family (LOX), resulting in lysine aldehydes also known as "allysines." The latter may spontaneously condense with a lysine or hydroxylysine in a neighboring collagen α-chain to form immature divalent crosslinks dehydro-dihydroxylysinonorleucine (deH-DHLNL) and dehydro-hydroxylysinonorleucine (deH-HLNL). These immature crosslinks can undergo Amadori rearrangement to form the keto forms hydroxylysinoketonorleucine and lysinoketonorleucine which can condense with a hydroxylysine in third alpha collagen chain to form trivalent crosslink lysylpyridinoline or deoxypyridinoline (dePyr) and hydroxylysylpyridinoline or pyridinoline (Pyr) [12]. Apart from these, there are crosslinks involving the condensation of immature crosslinks with histidine residue forming histidinohydroxylysinonorleucine (HHL) and dehydrohistidinohydroxymeridesmosine (HHMD). Pyrrole is another kind of mature trivalent crosslink present in collagen I formed by condensation of a hydroxylysinoketonorleucine with hydroxylysinonorleucine [34]. Glycosylation is observed on certain helical lysine residues involved in crosslinking and are thus present on immature divalent crosslink DHLNL & HLNL as well as on mature trivalent crosslinks. Collagen crosslink biosynthesis has been extensible reviewed elsewhere [34]. In addition to LOX-mediated crosslinks, non-enzymatic action of reducing sugars on amino groups of proteins (via Maillard reaction) leads to formation of advanced glycation end products (AGEs). Pentosidine is one such AGE highly characterized in aged lungs [11]. The presence of mature crosslinks is thought to be related to matrix stiffness which increases in age related lung diseases such as pulmonary fibrosis [72].

With such structural diversity and complexity, selection of methods for the detection and quantification of collagen crosslinks has historically required careful experimental considerations. A possible workflow is outlined in Fig. 11.4. For instance, the unstable nature of the immature crosslinks to strong acids used for hydrolysis requires prior reduction of collagen samples with sodium borohydride. HHMD crosslinks are stable in acid but can also be detected in the same $NaBH_4$-reduced samples. A small number of laboratories still use tritiated sodium borohydride ($NaB^3H_4$) to radiolabel immature crosslinks because this method has high sensitivity allowing the use of smaller sample size. If $NaB^3H_4$ is used for reduction, separation of immature crosslinks can be achieved on either a cation -exchange column or $C_{18}$ reverse-phase column with detection on a liquid scintillation counter [66, 101]. Although radioactive methods present many advantages, it is not readily available to most collagen researchers around the world who are obligated to look for alternative methods.

More accessible non-radioactive methods to quantify immature reducible crosslinks have been developed. In this case, ultra-performance liquid chromatography ESI-MS/MS has been developed and used for a variety of tissues. Since immature crosslinks are small and have a more polar character, the mobile phase of reverse-phase $C_{18}$ column includes heptafluorobutyric acid (HFBA) which allows a better retention and fractionation of these crosslinks on the column. However, because HFBA reduces columns shelf life and it is difficult to remove from the mass spectrometer source, other columns and mobile phase agents have been implemented. One such column that has emerged in the field of crosslink quantification is hydrophilic interaction chromatography (HILIC) columns coupled with mass spectrometry detection. Unlike $C_{18}$ reverse-phase columns, HILIC columns are hydrophilic

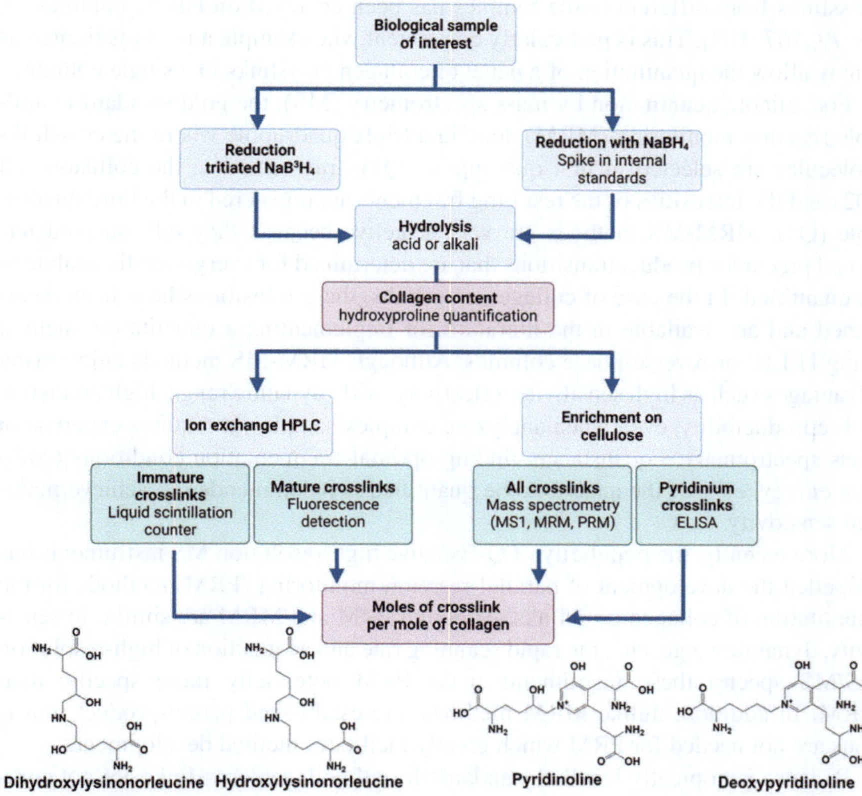


**Fig. 11.4** Schematic workflow for detection of collagen crosslinks in biological samples. The biological sample is crushed in liquid nitrogen to generate an insoluble powder that is lyophilized before determination of its dry-weight. Stabilization of acid-labile immature crosslinks can be achieved with either radiolabeled (left) or non-radiolabeled sodium borohydride (right). After hydrolysis amino acids and lysyl-derived crosslinks can be separated on different HPLC columns. In case of immature crosslinks, those that have been radiolabeled are detected with a liquid scintillation counter, whereas non-labeled crosslinks are detected by mass spectrometry. For mature crosslinks, quantitation can be achieved using either mass spectrometry using MS1 extracted-ion chromatogram (MS1), multiple reaction monitoring (MRM) or parallel-reaction monitoring (PRM). In the particular case of pyridinolines, they can also be quantified using their intrinsic fluorescence or using an ELISA detection kit. A portion of each hydrolyzed sample may also be used to quantify hydroxyproline, which is used to determine the number of crosslinks per molecule of collagen in each sample. The addition of a known amount of an internal standard at the beginning of the procedure accounts of sample loss and greatly improves results. The crosslinks are represented in moles per mole of collagen in the sample. Commonly analyzed collagen crosslinks are shown at the bottom. (Figure is created with BioRender.com)

where polar compounds such as immature collagen crosslinks are typically fractionated by a gradient starting with a high concentration of organic solvent (e.g., acetonitrile) and increasing the polarity of the mobile phase with water. Notably, in addition to immature crosslinks, fractionation of a mixture of immature and mature crosslinks from different tissue samples has been achieved on HILIC columns [3, 78, 79, 107, 108]. This is particularly convenient when sample amount is limited as it may allow the quantitation of a panel of collagen crosslinks in a single column.

For reliable quantitation by mass spectrometry (MS), the gold standard is multiple reaction monitoring (MRM) done in a triple quadrupole where the crosslinks molecules are selected in first quadrupole (Q1), fragmented in the collision cell (Q2) and the intensities of the resulting fragments are registered in the third quadrupole (Q3). MRM-MS methods are very selective because they rely on predetermined precursor-product transitions that are determined for every specific analyte to be quantified. In the case of collagen crosslinks, these transitions have been determined and are available in the literature for implementing a quantitation method using HILIC or reverse-phase columns. Although MRM-MS methods enjoy many advantages such as high sensitivity, selectivity, wide dynamic range, high precision, and reproducibility, even when analyzing complex samples, it requires expertise in mass spectrometry. For instance, finding optimal fragmentation conditions (collision energy, etc.) of the analyte to be quantified is recommended to achieve maximal sensitivity.

More recently, the popularity of Q-Exactive high-resolution MS instruments has propelled the development of parallel-reaction monitoring (PRM) methods for the quantitation of collagen crosslinks. Although PRM and MRM are similar in sensitivity, dynamic range, etc., the rapid scanning rate and acquisition of high-resolution MS/MS spectra these instruments make PRM potentially more specific than MRM. In addition, unlike MRM methods, pre-established parent-product transitions are not needed for PRM which greatly facilitates method development.

Because isotopically labelled standards for all collagen crosslinks are not commercially available, a typical quantitation method relies on an external calibration curved constructed with crosslink standards. The inclusion of a related molecule such as pyridoxamine as internal standard [4] could help account for losses during sample preparation and thus significantly improve accuracy.

Due to the fluorescent nature of the pyridinolines, HPLC with fluorescent detection can be used to detect these trivalent mature crosslinks. Notably, fluorescent detection of pyridinolines in human urine is a well characterized method as it has been used to understand the collagen turnover or as biomarker in many forms of bone disease [31–33, 35] & lung fibrosis [116]. Although pyrrole crosslinks are also likely to be present in tissue samples, they are not typically quantified because unlike pyridinolines they are not fluorescent and require Ehrlich's chromogen for detection (Table 11.2).

## 11.4 Assessment of Collagen Architecture In Situ

Collagen architecture, including fibril length, density, and alignment, directly and profoundly influences adherent cell behavior and function [58, 86, 100, 120]. Visualization of collagen architecture in translational models of lung disease may

**Table 11.2** Examples of collagen crosslink quantification in lung diseases

| Nature/Origin of tissue | Method of quantification | Crosslink detected & interpretation |
|---|---|---|
| Rat bleomycin induced injury [92] | $NaB^3H_4$ reduction & labelling of immature crosslink, separation on C18 RP-HPLC and detected on liquid scintillation counter. Crosslinks normalized to hydroxyproline content | DHLNL reported as dpm/50ug of hydroxyproline. Increase in DHLNL in initial few weeks<br>Pyridinolines reported as pmol per 5ug of hydroxyproline. Increased amounts of pyridinolines in injured lungs |
| Bleomycin induced injury [7] | Sirius red staining | Increased insoluble crosslinked collagen found in injured tissue. Use of AB0023 LOXL2 inhibitor decreased the crosslinked collagen amount |
| Idiopathic pulmonary fibrosis patient urine samples [116] | Fluorescence based detection of mature crosslinks | Increase in pyridinolines reported in IPF patient urine, expressed in nmoles/mmoles of creatinine |
| 344SQ tumors [116] | $NaB^3H_4$ reduction & labelling of immature crosslink & fluorescence detection of mature crosslinks | Immature and mature crosslink amounts are reported in mol/mol of collagen<br>Treatment with trihydroxyphenolic compounds decrease in immature DHLNL and mature pyridinoline crosslinks and attenuated lung injury as well as metastasis |
| Urine from bleomycin treated mice [116] | Fluorescence based detection of mature crosslinks | PyD & DPD were measured as nmole/ mmole of creatinine<br>Increase in pyridinolines were reported in injured mice and treatment with trihydroxyphenolic compounds reduced the mature crosslink levels in urine |
| Lung stromal tumor cells [23] | $NaB^3H_4$ reduction & labelling of immature crosslink, C18 separation & UHPLC-ESI-MS/ MS.<br>ELISA for detection of pyridinolines | DHLNL, HLNL, Pyr & HHMD crosslinks detected and reported in mol/mol of collagen from tumor tissues<br>Increase in DHLNL and pyridinoline in comparison to HHMD crosslinks. LH2 was proposed to be regulatory switch converting lysines into hydroxylysines which in turn participates in formation of DHLNL & pyridinolines |
| Lung fibroblasts [45] | $KBH_4$ reduction of immature crosslink & fluorescence detection of mature crosslinks | Crosslinks were normalized to total collagen content in fibroblasts. Increase of immature and mature crosslinks in fibrosis which was markedly reduced with use of LOXL2/3 dual inhibitor |
| Human lung fibroblasts CCS19-Lu [94] | ELISA kit for detection of pyridinolines | Increase in pyridinolines (nM/$10^6$ cells) reported in cells treated to LOX & BMP-1 |
| Rat lungs [11] | Separation on C18 and fluorescence detection of AGE | Pentosidine AGE detected in increasing amount in aged lungs and reported in pmol/mg of collagen |

therefore provide important clues for underlying disease-driving mechanisms beyond the mere increase in deposition of fibrillar collagen, which can be quantified using the methodology outlined above. Several staining and microscopy techniques allow for the assessment of collagen or ECM architecture and are not only used to visualize collagen in tissue sections, but also in more complex three-dimensional samples. These methods range from traditional histological staining and immunohistochemistry coupled with light or fluorescence microscopy to polarization-based microscopy methods, second-harmonic generation (SHG) microscopy, and scanning and transmission electron microscopy (Fig. 11.5). While most of these

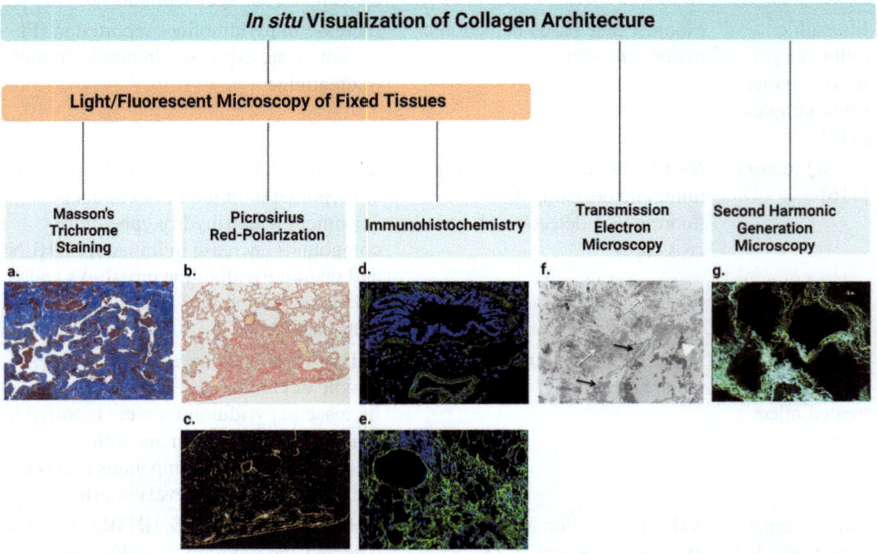


**Fig. 11.5** Examples for in situ visualization of collagen architecture. (**a**) Masson Trichrome staining of an IPF lung tissue section. (Reproduced from Harris WT et al. [124] (https://journals.plos.org/plosone/article?id=10.1371/journal.pone.0070196)). (**b**) Picrosirius Red staining of a mouse lung with bleomycin-induced lung fibrosis, visualized using conventional light microscopy, in comparison to (**c**) the same section visualized using polarized light microscopy. ((**b**) and (**c**) are reproduced from Egger C et al. [125] (https://journals.plos.org/plosone/article?id=10.1371/journal.pone.0063432)). (**d**) Immunofluorescent staining of a normal human lung section for type IV collagen (green). (**e**) Immunofluorescent staining of a mouse lung section depicting bleomycin-induced lung fibrosis for type I collagen (green). Staab-Weijnitz CA, unpublished results. (**f**) Transmission electron microscopy of idiopathic pulmonary fibrosis (IPF) acellular lung matrix. (Reproduced from Booth AJ et al. [126], Figure 6, right-hand panel, with permission of the American Thoracic Society. Copyright © 2021 American Thoracic Society. All rights reserved. The American Journal of Respiratory and Critical Care Medicine is an official journal of the American Thoracic Society. Readers are encouraged to read the entire article for the correct context at https://www.atsjournals.org/doi/full/10.1164/rccm.201204-0754OC. The authors, editors, and The American Thoracic Society are not responsible for errors or omissions in adaptations). (**g**) Second Harmonic Generation image of lung parenchyma from an IPF patient's lung section. (Reproduced from Tjin G et al. [127] (https://doi.org/10.1242/dmm.030114), with permission of Disease Models & Mechanisms (https://journals.biologists.com/dmm))

methods allow for quantification of collagen content and alignment in situ, they differ considerably in specificity and resolution [53, 60, 68, 115]. Their advantages and disadvantages will be outlined below.

### 11.4.1 Masson's Trichrome Staining

Masson's trichrome staining is a widely used method to for the visualization of collagen in the context of tissue structure which allows for detection of morphological alterations and the qualitative and quantitative assessment of the extent of collagen deposition [110]. The latter makes it particularly useful for routine use in pathological diagnosis of fibrotic diseases. Numerous variants of this method and combinations with other staining protocols exist, but in principle they all rely on the sequential use of three dyes with different molecular weights and acid-base chemistry in a precisely controlled timely manner. Weigert's iron hematoxylin is used first to stain the nuclei. This dye is resistant to subsequent acid decolorization procedures. Then, a red acid dye (the so-called plasma stain, e.g., Biebrich scarlet or acid fuchsine) is applied which binds all acidophilic structures including cytoplasm, muscle, and collagen. A solution containing large acid heteropolymetalates such as phosphomolybdic or phosphotungstic acid removes the plasma stain from collagen, but not from muscle fibers or cytoplasm. Finally, a green or blue dye (e.g., light green SF or aniline blue) is used to stain collagen fibers (Fig. 11.5).

### 11.4.2 Picrosirius Red Staining

The underlying principle for the specific binding of Sirius Red to collagen has been described above (Sect. 11.2.1). When used for tissue staining, Sirius Red is often called Picrosirius Red (PSR), as the stain is dissolved in an aqueous solution of picric acid in the protocol. In bright-field microscopy, collagen appears red on a pale-yellow background. Also, fluorescence microscopy has been used for visualization of PSR-stained sections using excitation/emission settings for rhodamine—stained fibers also then appear red and can be combined with green autofluorescence of elastic fibers or live cells [14, 29, 112]. Importantly, specificity and sensitivity for visualization of collagen fibers can be considerably enhanced when polarized light microscopy is used for analysis [119]. Parallel alignment of collagen fibers causes a strong natural birefringence, which is further enhanced by the association of elongated Sirius Red molecules along the linear fiber axis. Thus, collagen can be visualized under linear polarized light and will then present as red, orange, yellow or green fibers [46, 60, 67, 115] (Fig. 11.5). The different colors were initially thought to reflect distinct collagen types [46, 67], but are more likely to be a measure of fiber thickness and/or degree of parallel orientation [60, 69].

### 11.4.3 Second Harmonic Generation Microscopy

Second harmonic generation (SHG) is a non-linear optical process where photons from a strong laser pass through a so-called non-centrosymmetric environment, i.e., an environment without an inversion center as symmetry element, and interact with aligned harmonophores that possess a permanent dipole moment. This leads to emission of second-harmonic light at half the wavelength of the light that originally entered the material. Only very few biological materials/proteins are such harmonophores, i.e., meet the physical requirements for efficient second harmonic generation. These materials include type I and II collagens and myosin in actin-myosin complexes [22]. Hence, specificity for fibrillar type I and II collagens in biological samples is excellent. In addition, this technique allows for assessment of the three-dimensional fibrillar collagen network in high resolution [5]. Another advantage is that, as SHG microscopy entirely relies on the above-described intrinsic physical properties of fibrillar collagen, it does not require staining procedures or any type of labelling [53, 105, 121].

### 11.4.4 Immunohistochemistry

While all techniques described above are best suited for or even restricted to the visualization of fibrillar collagens, immunohistochemistry (IHC) provides the opportunity to stain for distinct collagen types, given that specific antibodies are available. Hence, the same limitations as already outlined above for immune-based methods for quantification of collagen types (Sect. 11.2.3) apply. Notably, Rickelt and Hynes have published a highly useful summary on IHC methods and suitable antibodies for the detection of ECM proteins including numerous collagens [93]. Immunohistochemical techniques are typically highly sensitive. Dependent on the antibodies used, efficient detection of collagen may require the unmasking of epitopes by limited predigestion of tissue sections prior to the staining procedure [26].

### 11.4.5 Transmission Electron Microscopy

In transmission electron microscopy (TEM), electrons instead of light are sent through the sample specimen, resulting in superior magnification by a factor of 1000 and much higher resolving power. This facilitates the visualization of subcellular compartments and single collagen fibrils, diameters of which may range from 12 to 500 nm, far below or at best at the limit of resolution of standard light microscopy [103]. This property makes TEM the only currently available technique to quantify fibril diameter and length and to directly visualize collagen fibrillogenesis at the interface fibroblast-extracellular space [104]. Methods for 3D reconstruction have been developed [76, 104]. Exemplifying the power of this technique, reconstructing 3D images from serial TEM images ultimately lead to the discovery that, in tendon fibroblasts, Golgi to plasma membrane carriers (GPCs) carry collagen fibril cargo and target them to special plasma membrane protrusions, fibripositors, for secretion [18] (Table 11.3).

## *11.4.6 Selected Complementary and Emerging Techniques*

### Confocal Reflection Microscopy (CRM)

In confocal reflection microscopy (CRM), a confocal microscope is used to take reflection images of a sample specimen at sequential focal planes along the $z$ axis followed by three-dimensional reconstruction. While reflection as an intrinsic optical property is not particularly specific for collagen, in contrast to SHG as described above (Sect. 11.4.3), CRM is a comparatively simple alternative for the detailed assessment of collagen microarchitecture in artificially prepared scaffolds from purified collagen or collagen-enriched material [15, 100].

### Atomic Force Microscopy (AFM)

Collagen architecture and the extent of collagen crosslinking affect tissue stiffness which, in turn, regulates adherent cell behavior via biomechanical signaling [37, 45]. Therefore, for some research questions, it may be of interest to assess mechanical properties of a biological sample with altered collagen properties. Atomic force microscopy (AFM) is a type of scanning probe microscopy, where interactions between a sharp tip at the end of a cantilever and the sample surface are recorded, allowing for the visualization of surface topography at nanoscale [82]. Importantly, alongside topographic analysis, AFM can be used for nanoindentation, where specified forces are applied to the surface of the sample and the extent of indentation is recorded by the instrument. This yields a measure of the sample's stiffness/elasticity and allows for the generation of Young's modulus maps of the sample's surface [37, 45, 82].

### Imaging Probes for Magnetic Resonance Imaging (MRI)

Magnetic resonance imaging (MRI) is an imaging modality routinely used in the clinic when diagnosis requires high-resolution images of soft tissues or when ionizing radiation should be avoided. MRI relies on strong magnetic fields, magnetic field gradients, and radiofrequency pulses to generate images of anatomical structures. Molecular MRI is an emerging field where targeted probes are designed and, for instance, coupled to MRI contrast agents such as clinically approved

**Table 11.3** Methodologies for the visualization of collagen architecture in tissue sections and three-dimensional samples (for tissue staining examples, refer to Fig. 11.5)

| Method | Principle | Advantages | Disadvantages |
|---|---|---|---|
| Masson Trichrome Staining | Dyes of different molecular weights and acid-base chemistry differentially bind to collagen, muscle fibers, nuclei, and cytoplasm | Cheap and rapid procedure, routine use in pathology<br>Allows for visualization of collagen distribution and architecture | Does not differentiate between collagen types as differential binding mostly relies on acid-base interactions |
| Picrosirius Red Polarization | Sirius red dye binds specifically to elongated triple helical collagenous domains. Its association along collagen fibers enhances the natural birefringency of collagen and allows for specific visualization in polarized light microscopy | Highly specific for collagen fibers<br>Accurate visualization of fibrillar collagen networks | Requires a linear polarized light microscope with a rotating stage, or a circular polarized light microscope<br>Strongly dependent on sample orientation<br>Does not differentiate between collagen types<br>Only suitable for collagen fibers |
| Second harmonic generation microscopy | Relies on the intrinsic property of type I and II fibers to act as harmonophores and emit a strong second harmonic generation signal | No staining or labelling required suitable for three-dimensional visualization of collagen architecture<br>high specificity for fibrillar type I and II collagens | Requires a laser scanning microscope with special modifications<br>Only suitable for type I and II collagen fibers |
| Immunohisto-chemistry | | No special equipment necessary<br>Can differentiate between collagen types | Strongly dependent on antibody specificity—may require additional validation<br>Antibodies are expensive |

(continued)

**Table 11.3** (continued)

| Method | Principle | Advantages | Disadvantages |
|---|---|---|---|
| Electron microscopy | | Very high resolution allows for visualization of single collagen fibers<br>Facilitates determination of individual fibril length and diameter in situ suitable for the study of fibrillogenesis at the cell-ECM interface | Requires an electron microscope, special expertise, and training |

gadolinium-based structures [95]. ECM-targeting moieties have much been explored in the context of cardiovascular disease [91]. For in vivo models of lung fibrosis, successful targeting of type I collagen by a gadolinium-coupled probe has been reported by Caravan et al. [19] and, more recently, the same group demonstrated that an allysine-binding gadolinium chelate can be used to monitor fibrogenesis in the mouse model of bleomycin-induced lung fibrosis [113]. Also, the latter targets mostly collagen, as fibrogenesis typically associates with increased LOX-mediated oxidations of lysine to the allysine aldehyde intermediate. Clearly, ECM-targeted MRI may provide the unique opportunity for non-invasive and non-destructive analysis of ECM changes in lung disease, ultimately allowing for sequential assessment of disease development in the same animal.

## 11.5 Monitoring Fibril Formation in Real Time Using Purified Collagen

Fibril formation and crosslinking stabilizes fibrillar collagen and largely protects it from proteolytic degradation. Inhibition of fibril formation can therefore be viewed as a promising therapeutic strategy for all types of pulmonary disease where excessive secretion and deposition of ECM is a pathological feature. Notably, this is not only the case for fibrotic disease, but also for cancer, where the formation of tumor-encapsulating ECM shields the tumor cells from therapy and is typically associated with poor prognosis [38].

A well-established and straightforward assay for the assessment of collagen fibril formation in vitro relies on the principle that purified and acid-dissolved type I collagen spontaneously forms fibrils upon neutralization in an entropy-driven process. Fibrillogenesis can be monitored in real time by light scattering and the resulting fibrils, which are similar to those formed in vivo, further examined by electron microscopy [48, 118]. We have previously used this assay to show that two approved therapeutics for the treatment of idiopathic pulmonary fibrosis, nintedanib and pirfenidone, both delay collagen fibril formation, identifying a potential novel mechanism of action [55].

As simple and attractive as this assay appears, several things need to be considered. First, it should be acknowledged that it reflects a highly artificial environment and that assay conditions must be strictly controlled as temperature, buffer, salt composition, and pH strongly affect fibril formation [118]. Second, the source of the collagen used must be carefully chosen. In many studies, pepsin-digested collagen is used which is devoid of telopeptides and therefore harbors few LOX-crosslinking sites if any [34, 97]. However, absence of telopeptides has been shown to slow fibril formation and to alter the morphology of resulting collagen fibrils [97]. Also, with telopeptides present, this assay will be better suited to directly study the influence of enzymatic and non-enzymatic collagen crosslinking on fibril formation and on resulting fibril architecture. Therefore, collagen preparation methods should be considered that leave telopeptides intact [20, 97, 118].

## 11.6 Assessment of Collagen Turnover by Peripheral Markers

Normal collagen turnover maintains a healthy balance between collagen synthesis and degradation processes. Impairment of this balance and uncontrolled ECM remodeling also is a major pathological hallmark of many chronic lung diseases [13, 17, 123]. Importantly, these processes lead to cleaved-off collagen propeptides as well as MMP-dependent cleavage products which are detectable in peripheral blood. Although a relatively new concept, several studies already support the concept that assessment of peripheral collagen propeptides and matrikines may be beneficial for prognosis and diagnosis of chronic lung disease [10, 44, 49–51, 65, 84, 109]. Markers of collagen formation can also be used as a read-out when efficiency of novel therapeutic strategies is assessed in preclinical models [73]. In addition, these cleavage products allow for inferences on the responsible protease for collagen turnover, provide information about the tissue of origin, and thus contribute to our understanding of disease pathogenesis and comorbidities [25, 44, 49–51, 109]. Table 11.4 provides a list of the most frequently currently used peripheral markers of collagen research.

**Table 11.4** Peripheral markers of collagen turnover with potential applicability as biomarkers for lung disease

| Substrate | Protease | Protein fragment | Investigations in lung | References |
|---|---|---|---|---|
| Collagen type I degradation | MMP-2/9/13 | C1M | Elevated in COPD and IPF | [44, 65, 62] |
| Collagen type III degradation | MMP-9 | C3M | Elevated in COPD and IPF | [6, 44, 65] |
| Collagen type IV degradation | MMP | C4M | Elevated in COPD | [64, 65, 111] |
| Collagen type V degradation | MMP-2/9 | C5M | Elevated in COPD and IPF | [44, 65, 111] |
| Collagen type VI degradation | MMP-2/9 | C6M | Elevated in COPD and IPF | [44, 65, 111] |
| Collagen type I formation | ADAMTS2, ADAMTS14 | PINP | Not elevated in progressing IPF disease | [10, 57, 63, 84] |
| | BMP1 | PICP | Elevated in bronchoalveolar lavage fluid of IPF patients | [36, 40, 57] |
| Collagen type III formation | ADAMTS2 | PIIINP | Elevated in progressing IPF disease | [10, 57, 80, 84] |
| | BMP1 | PRO-C3 | Elevated in progressing IPF disease | [40, 84] |
| Collagen type VI formation | | PRO-C6 | Elevated in progressing IPF disease | [40, 84] |

## 11.7 Conclusion

In most mammalian tissues, collagens represent the main ECM component and play a major role in the maintenance of tissue integrity and function. Hence, it is not surprising that alterations of collagen quantity and molecular properties contribute considerably to chronic lung disease development and thus need to be considered in translational models of lung disease. The high structural and functional diversity of collagens, however, entail a number of challenges for their characterization.

While methods for the detection and visualization of fibrillar collagens in situ are well-established, the analysis of collagen types in molecular detail, including PTMs and extracellular crosslinks, requires very specific approaches, many of which are still under development. Proteomics studies, in combination with detailed functional analysis of observed changes, may reveal novel molecular mechanisms underlying fibrosis and thus help identify potential future targets for treatment of chronic lung disease. Identification and quantification of collagen PTMs, crosslinks, collagen-derived propeptides and matrikines may help elucidate their functions in lung health and disease, and also serve as a platform for the development of future diagnostic and therapeutic strategies.

In conclusion, here, we provided a summary on the current state-of-the-art of collagen quantification, detection, and assessment of properties in molecular detail. The methodology reviewed includes traditional and well-established approaches that take advantage of well-known intrinsic and unique properties of collagen fibers, but also cutting-edge methodologies such as tandem mass spectrometry, which allows not only for assessment of collagen composition but also for elucidation of site-specific PTMs and crosslinks.

**Acknowledgements** Work in the authors' laboratories is supported in part by the Helmholtz Association, the German Center for Lung Research (DZL), the Deutsche Forschungsgemeinschaft (DFG) within the Research Training Group GRK2338 (grant to CASW), the Federal Institute for Risk Assessment (Bundesinstitut für Risikobewertung, BfR) (#1328-570, grant to CASW), R01 DK099467 (RMV) and R01 NS096173 (Subaward 10529sc to RMV).

## References

1. Alsafadi, H. N., Staab-Weijnitz, C. A., Lehmann, M., Lindner, M., Peschel, B., Konigshoff, M. & Wagner, D. E. 2017. An Ex Vivo Model To Induce Early Fibrosis-Like Changes In Human Precision-Cut Lung Slices. *American Journal Of Physiology: Lung Cellular And Molecular Physiology,* 312, L896–L902.
2. Angelidis, I., Simon, L. M., Fernandez, I. E., Strunz, M., Mayr, C. H., Greiffo, F. R., Tsitsiridis, G., Ansari, M., Graf, E., Strom, T. M., Nagendran, M., Desai, T., Eickelberg, O., Mann, M., Theis, F. J. & Schiller, H. B. 2019. An Atlas Of The Aging Lung Mapped By Single Cell Transcriptomics And Deep Tissue Proteomics. *Nat Commun,* 10, 963.
3. Arakawa, S., Suzuki, R., Kurosaka, D., Ikeda, R., Hayashi, H., Kayama, T., Ohno, R. I., Nagai, R., Marumo, K. & Saito, M. 2020. Mass Spectrometric Quantitation Of Ages And Enzymatic Crosslinks In Human Cancellous Bone. *Sci Rep,* 10, 18774.
4. Avery, N. C., Sims, T. J. & Bailey, A. J. 2009. Quantitative Determination Of Collagen Cross-Links. *Methods Mol Biol,* 522, 103–21.
5. Bancelin, S., Aime, C., Coradin, T. & Schanne-Klein, M. C. 2012. In Situ Three-Dimensional Monitoring Of Collagen Fibrillogenesis Using Shg Microscopy. *Biomedical Optics Express,* 3, 1446–54.
6. Barascuk, N., Veidal, S. S., Larsen, L., Larsen, D. V., Larsen, M. R., Wang, J., Zheng, Q., Xing, R., Cao, Y., Rasmussen, L. M. & Karsdal, M. A. 2010. A Novel Assay For Extracellular Matrix Remodeling Associated With Liver Fibrosis: An Enzyme-Linked Immunosorbent Assay (Elisa) For A Mmp-9 Proteolytically Revealed Neo-Epitope Of Type Iii Collagen. *Clinical Biochemistry,* 43, 899–904.
7. Barry-Hamilton, V., Spangler, R., Marshall, D., Mccauley, S., Rodriguez, H. M., Oyasu, M., Mikels, A., Vaysberg, M., Ghermazien, H., Wai, C., Garcia, C. A., Velayo, A. C., Jorgensen, B., Biermann, D., Tsai, D., Green, J., Zaffryar-Eilot, S., Holzer, A., Ogg, S., Thai, D., Neufeld, G., Van Vlasselaer, P. & Smith, V. 2010. Allosteric Inhibition Of Lysyl Oxidase-Like-2 Impedes The Development Of A Pathologic Microenvironment. *Nat Med,* 16, 1009–17.
8. Basak, T., Vega-Montoto, L., Zimmerman, L. J., Tabb, D. L., Hudson, B. G. & Vanacore, R. M. 2016. Comprehensive Characterization Of Glycosylation And Hydroxylation Of Basement Membrane Collagen Iv By High-Resolution Mass Spectrometry. *Journal Of Proteome Research,* 15, 245–58.
9. Beachley, V. Z., Wolf, M. T., Sadtler, K., Manda, S. S., Jacobs, H., Blatchley, M. R., Bader, J. S., Pandey, A., Pardoll, D. & Elisseeff, J. H. 2015. Tissue Matrix Arrays For High-Throughput Screening And Systems Analysis Of Cell Function. *Nature Methods,* 12, 1197–204.
10. Bekhouche, M. & Colige, A. 2015. The Procollagen N-Proteinases Adamts2, 3 And 14 In Pathophysiology. *Matrix Biology,* 44–46, 46–53.

11. Bellmunt, M. J., Portero, M., Pamplona, R., Muntaner, M. & Prat, J. 1995. Age-Related Fluorescence In Rat Lung Collagen. *Lung,* 173, 177–85.
12. Bielajew, B. J., Hu, J. C. & Athanasiou, K. A. 2020. Collagen: Quantification, Biomechanics, And Role Of Minor Subtypes In Cartilage. *Nature Reviews: Materials,* 5, 730–747.
13. Bonnans, C., Chou, J. & Werb, Z. 2014. Remodelling The Extracellular Matrix In Development And Disease. *Nature Reviews: Molecular Cell Biology,* 15, 786–801.
14. Borges, L. F., Taboga, S. R. & Gutierrez, P. S. 2005. Simultaneous Observation Of Collagen And Elastin In Normal And Pathological Tissues: Analysis Of Sirius-Red-Stained Sections By Fluorescence Microscopy. *Cell And Tissue Research,* 320, 551–2.
15. Brightman, A. O., Rajwa, B. P., Sturgis, J. E., Mccallister, M. E., Robinson, J. P. & Voytik-Harbin, S. L. 2000. Time-Lapse Confocal Reflection Microscopy Of Collagen Fibrillogenesis And Extracellular Matrix Assembly In Vitro. *Biopolymers,* 54, 222–34.
16. Burgstaller, G., Oehrle, B., Gerckens, M., White, E. S., Schiller, H. B. & Eickelberg, O. 2017a. The Instructive Extracellular Matrix Of The Lung: Basic Composition And Alterations In Chronic Lung Disease. *Eur Respir J,* 50.
17. Burgstaller, G., Oehrle, B., Gerckens, M., White, E. S., Schiller, H. B. & Eickelberg, O. 2017b. The Instructive Extracellular Matrix Of The Lung: Basic Composition And Alterations In Chronic Lung Disease. *European Respiratory Journal,* 50.
18. Canty, E. G., Lu, Y., Meadows, R. S., Shaw, M. K., Holmes, D. F. & Kadler, K. E. 2004. Coalignment Of Plasma Membrane Channels And Protrusions (Fibripositors) Specifies The Parallelism Of Tendon. *Journal Of Cell Biology,* 165, 553–63.
19. Caravan, P., Yang, Y., Zachariah, R., Schmitt, A., Mino-Kenudson, M., Chen, H. H., Sosnovik, D. E., Dai, G., Fuchs, B. C. & Lanuti, M. 2013. Molecular Magnetic Resonance Imaging Of Pulmonary Fibrosis In Mice. *American Journal Of Respiratory Cell And Molecular Biology,* 49, 1120–6.
20. Chandrakasan, G., Torchia, D. A. & Piez, K. A. 1976. Preparation Of Intact Monomeric Collagen From Rat Tail Tendon And Skin And The Structure Of The Nonhelical Ends In Solution. *Journal Of Biological Chemistry,* 251, 6062–7.
21. Chen, C. Z. & Raghunath, M. 2009. Focus On Collagen: In Vitro Systems To Study Fibrogenesis And Antifibrosis State Of The Art. *Fibrogenesis And Tissue Repair,* 2, 7.
22. Chen, X., Nadiarynkh, O., Plotnikov, S. & Campagnola, P. J. 2012. Second Harmonic Generation Microscopy For Quantitative Analysis Of Collagen Fibrillar Structure. *Nature Protocols,* 7, 654–69.
23. Chen, Y., Terajima, M., Yang, Y., Sun, L., Ahn, Y. H., Pankova, D., Puperi, D. S., Watanabe, T., Kim, M. P., Blackmon, S. H., Rodriguez, J., Liu, H., Behrens, C., Wistuba, Ii, Minelli, R., Scott, K. L., Sanchez-Adams, J., Guilak, F., Pati, D., Thilaganathan, N., Burns, A. R., Creighton, C. J., Martinez, E. D., Zal, T., Grande-Allen, K. J., Yamauchi, M. & Kurie, J. M. 2015. Lysyl Hydroxylase 2 Induces A Collagen Cross-Link Switch In Tumor Stroma. *J Clin Invest,* 125, 1147–62.
24. Cissell, D. D., Link, J. M., Hu, J. C. & Athanasiou, K. A. 2017. A Modified Hydroxyproline Assay Based On Hydrochloric Acid In Ehrlich's Solution Accurately Measures Tissue Collagen Content. *Tissue Engineering. Part C, Methods,* 23, 243–250.
25. Conrozier, T., Poole, A. R., Ferrand, F., Mathieu, P., Vincent, F., Piperno, M., Verret, C., Ionescu, M. & Vignon, E. 2008. Serum Concentrations Of Type Ii Collagen Biomarkers (C2c, C1, 2c And Cpii) Suggest Different Pathophysiologies In Patients With Hip Osteoarthritis. *Clinical And Experimental Rheumatology,* 26, 430–5.
26. D'andrea, M. R. 2004. Collagenase Predigestion On Paraffin Sections Enhances Collagen Immunohistochemical Detection Without Distorting Tissue Morphology. *Biotechnic And Histochemistry,* 79, 55–64.
27. Decaris, M. L., Gatmaitan, M., Florcruz, S., Luo, F., Li, K., Holmes, W. E., Hellerstein, M. K., Turner, S. M. & Emson, C. L. 2014. Proteomic Analysis Of Altered Extracellular Matrix Turnover In Bleomycin-Induced Pulmonary Fibrosis. *Mol Cell Proteomics,* 13, 1741–52.

28. Delgado, L. M., Shologu, N., Fuller, K. & Zeugolis, D. I. 2017. Acetic Acid And Pepsin Result In High Yield, High Purity And Low Macrophage Response Collagen For Biomedical Applications. *Biomedical Materials (Bristol, England),* 12, 065009.
29. Dolber, P. C. & Spach, M. S. 1993. Conventional And Confocal Fluorescence Microscopy Of Collagen Fibers In The Heart. *Journal Of Histochemistry And Cytochemistry,* 41, 465–9.
30. Endicott, J., Holden, P. & Fitzgerald, J. 2017. Authentication Of Collagen Vi Antibodies. *Bmc Research Notes,* 10, 358.
31. Eriksen, E. F., Charles, P., Melsen, F., Mosekilde, L., Risteli, L. & Risteli, J. 1993. Serum Markers Of Type I Collagen Formation And Degradation In Metabolic Bone Disease: Correlation With Bone Histomorphometry. *J Bone Miner Res,* 8, 127–32.
32. Eyre, D. 1992. New Biomarkers Of Bone Resorption. *J Clin Endocrinol Metab,* 74, 470a–470c.
33. Eyre, D. R. 1995. The Specificity Of Collagen Cross-Links As Markers Of Bone And Connective Tissue Degradation. *Acta Orthop Scand Suppl,* 266, 166–70.
34. Eyre, D. R., Weis, M. A. & Wu, J. J. 2008. Advances In Collagen Cross-Link Analysis. *Methods,* 45, 65–74.
35. Garnero, P. & Delmas, P. D. 1998. Biochemical Markers Of Bone Turnover. Applications For Osteoporosis. *Endocrinol Metab Clin North Am,* 27, 303–23.
36. Gopalakrishnan, B., Wang, W. M. & Greenspan, D. S. 2004. Biosynthetic Processing Of The Pro-Alpha1(V)Pro-Alpha2(V)Pro-Alpha3(V) Procollagen Heterotrimer. *Journal Of Biological Chemistry,* 279, 30904–12.
37. Haak, A. J., Tan, Q. & Tschumperlin, D. J. 2018. Matrix Biomechanics And Dynamics In Pulmonary Fibrosis. *Matrix Biology,* 73, 64–76.
38. Henke, E., Nandigama, R. & Ergun, S. 2019. Extracellular Matrix In The Tumor Microenvironment And Its Impact On Cancer Therapy. *Frontiers In Molecular Biosciences,* 6, 160.
39. Hill, R. C., Calle, E. A., Dzieciatkowska, M., Niklason, L. E. & Hansen, K. C. 2015. Quantification Of Extracellular Matrix Proteins From A Rat Lung Scaffold To Provide A Molecular Readout For Tissue Engineering. *Molecular And Cellular Proteomics,* 14, 961–73.
40. Imamura, Y., Steiglitz, B. M. & Greenspan, D. S. 1998. Bone Morphogenetic Protein-1 Processes The Nh2-Terminal Propeptide, And A Furin-Like Proprotein Convertase Processes The Cooh-Terminal Propeptide Of Pro-Alpha1(V) Collagen. *Journal Of Biological Chemistry,* 273, 27511–7.
41. Ishikawa, Y. & Bachinger, H. P. 2013. A Molecular Ensemble In The Rer For Procollagen Maturation. *Biochimica Et Biophysica Acta,* 1833, 2479–91.
42. Jaakkola, P., Mole, D. R., Tian, Y. M., Wilson, M. I., Gielbert, J., Gaskell, S. J., Von Kriegsheim, A., Hebestreit, H. F., Mukherji, M., Schofield, C. J., Maxwell, P. H., Pugh, C. W. & Ratcliffe, P. J. 2001. Targeting Of Hif-Alpha To The Von Hippel-Lindau Ubiquitylation Complex By O2-Regulated Prolyl Hydroxylation. *Science,* 292, 468–72.
43. Jenkins, R. G., Moore, B. B., Chambers, R. C., Eickelberg, O., Konigshoff, M., Kolb, M., Laurent, G. J., Nanthakumar, C. B., Olman, M. A., Pardo, A., Selman, M., Sheppard, D., Sime, P. J., Tager, A. M., Tatler, A. L., Thannickal, V. J., White, E. S., Cell, A. T. S. A. O. R. & Molecular, B. 2017. An Official American Thoracic Society Workshop Report: Use Of Animal Models For The Preclinical Assessment Of Potential Therapies For Pulmonary Fibrosis. *American Journal Of Respiratory Cell And Molecular Biology,* 56, 667–679.
44. Jenkins, R. G., Simpson, J. K., Saini, G., Bentley, J. H., Russell, A. M., Braybrooke, R., Molyneaux, P. L., Mckeever, T. M., Wells, A. U., Flynn, A., Hubbard, R. B., Leeming, D. J., Marshall, R. P., Karsdal, M. A., Lukey, P. T. & Maher, T. M. 2015. Longitudinal Change In Collagen Degradation Biomarkers In Idiopathic Pulmonary Fibrosis: An Analysis From The Prospective, Multicentre Profile Study. *Lancet Respiratory Medicine,* 3, 462–72.
45. Jones, M. G., Andriotis, O. G., Roberts, J. J., Lunn, K., Tear, V. J., Cao, L., Ask, K., Smart, D. E., Bonfanti, A., Johnson, P., Alzetani, A., Conforti, F., Doherty, R., Lai, C. Y., Johnson, B., Bourdakos, K. N., Fletcher, S. V., Marshall, B. G., Jogai, S., Brereton, C. J., Chee, S. J.,

Ottensmeier, C. H., Sime, P., Gauldie, J., Kolb, M., Mahajan, S., Fabre, A., Bhaskar, A., Jarolimek, W., Richeldi, L., O'reilly, K. M., Monk, P. D., Thurner, P. J. & Davies, D. E. 2018. Nanoscale Dysregulation Of Collagen Structure-Function Disrupts Mechano-Homeostasis And Mediates Pulmonary Fibrosis. *Elife,* 7.

46. Junqueira, L. C., Bignolas, G. & Brentani, R. R. 1979. Picrosirius Staining Plus Polarization Microscopy, A Specific Method For Collagen Detection In Tissue Sections. *Histochemical Journal,* 11, 447–55.
47. Jurgensen, H. J., Madsen, D. H., Ingvarsen, S., Melander, M. C., Gardsvoll, H., Patthy, L., Engelholm, L. H. & Behrendt, N. 2011. A Novel Functional Role Of Collagen Glycosylation: Interaction With The Endocytic Collagen Receptor Uparap/Endo180. *J Biol Chem,* 286, 32736–48.
48. Kadler, K. E., Holmes, D. F., Trotter, J. A. & Chapman, J. A. 1996. Collagen Fibril Formation. *Biochemical Journal,* 316 (Pt 1), 1–11.
49. Karsdal, M. A., Delvin, E. & Christiansen, C. 2011. Protein Fingerprints – Relying On And Understanding The Information Of Serological Protein Measurements. *Clinical Biochemistry,* 44, 1278–9.
50. Karsdal, M. A., Henriksen, K., Leeming, D. J., Woodworth, T., Vassiliadis, E. & Bay-Jensen, A. C. 2010. Novel Combinations Of Post-Translational Modification (Ptm) Neo-Epitopes Provide Tissue-Specific Biochemical Markers--Are They The Cause Or The Consequence Of The Disease? *Clinical Biochemistry,* 43, 793–804.
51. Karsdal, M. A., Nielsen, M. J., Sand, J. M., Henriksen, K., Genovese, F., Bay-Jensen, A. C., Smith, V., Adamkewicz, J. I., Christiansen, C. & Leeming, D. J. 2013. Extracellular Matrix Remodeling: The Common Denominator In Connective Tissue Diseases. Possibilities For Evaluation And Current Understanding Of The Matrix As More Than A Passive Architecture, But A Key Player In Tissue Failure. *Assay And Drug Development Technologies,* 11, 70–92.
52. Karsdal, M. A., Nielsen, S. H., Leeming, D. J., Langholm, L. L., Nielsen, M. J., Manon-Jensen, T., Siebuhr, A., Gudmann, N. S., Ronnow, S., Sand, J. M., Daniels, S. J., Mortensen, J. H. & Schuppan, D. 2017. The Good And The Bad Collagens Of Fibrosis – Their Role In Signaling And Organ Function. *Advanced Drug Delivery Reviews,* 121, 43–56.
53. Keikhosravi, A., Shribak, M., Conklin, M. W., Liu, Y., Li, B., Loeffler, A., Levenson, R. M. & Eliceiri, K. W. 2021. Real-Time Polarization Microscopy Of Fibrillar Collagen In Histopathology. *Scientific Reports,* 11, 19063.
54. Kliment, C. R., Englert, J. M., Crum, L. P. & Oury, T. D. 2011. A Novel Method For Accurate Collagen And Biochemical Assessment Of Pulmonary Tissue Utilizing One Animal. *International Journal Of Clinical And Experimental Pathology,* 4, 349–55.
55. Knüppel, L., Ishikawa, Y., Aichler, M., Heinzelmann, K., Hatz, R., Behr, J., Walch, A., Bachinger, H. P., Eickelberg, O. & Staab-Weijnitz, C. A. 2017. A Novel Antifibrotic Mechanism Of Nintedanib And Pirfenidone. Inhibition Of Collagen Fibril Assembly. *American Journal Of Respiratory Cell And Molecular Biology,* 57, 77–90.
56. Kruegel, J. & Miosge, N. 2010. Basement Membrane Components Are Key Players In Specialized Extracellular Matrices. *Cellular And Molecular Life Sciences,* 67, 2879–95.
57. Lammi, L., Ryhanen, L., Lakari, E., Risteli, J., Paakko, P., Kahlos, K., Lahde, S. & Kinnula, V. 1999. Type Iii And Type I Procollagen Markers In Fibrosing Alveolitis. *American Journal Of Respiratory And Critical Care Medicine,* 159, 818–23.
58. Lanfer, B., Seib, F. P., Freudenberg, U., Stamov, D., Bley, T., Bornhauser, M. & Werner, C. 2009. The Growth And Differentiation Of Mesenchymal Stem And Progenitor Cells Cultured On Aligned Collagen Matrices. *Biomaterials,* 30, 5950–8.
59. Lareu, R. R., Zeugolis, D. I., Abu-Rub, M., Pandit, A. & Raghunath, M. 2010. Essential Modification Of The Sircol Collagen Assay For The Accurate Quantification Of Collagen Content In Complex Protein Solutions. *Acta Biomaterialia,* 6, 3146–51.
60. Lattouf, R., Younes, R., Lutomski, D., Naaman, N., Godeau, G., Senni, K. & Changotade, S. 2014. Picrosirius Red Staining: A Useful Tool To Appraise Collagen Networks In Normal And Pathological Tissues. *Journal Of Histochemistry And Cytochemistry,* 62, 751–8.

61. Lauer-Fields, J. L., Malkar, N. B., Richet, G., Drauz, K. & Fields, G. B. 2003. Melanoma Cell Cd44 Interaction With The Alpha 1(Iv)1263–1277 Region From Basement Membrane Collagen Is Modulated By Ligand Glycosylation. *J Biol Chem,* 278, 14321–30.
62. Leeming, D., He, Y., Veidal, S., Nguyen, Q., Larsen, D., Koizumi, M., Segovia-Silvestre, T., Zhang, C., Zheng, Q., Sun, S., Cao, Y., Barkholt, V., Hägglund, P., Bay-Jensen, A., Qvist, P. & Karsdal, M. 2011. A Novel Marker For Assessment Of Liver Matrix Remodeling: An Enzyme-Linked Immunosorbent Assay (Elisa) Detecting A Mmp Generated Type I Collagen Neo-Epitope (C1m). *Biomarkers,* 16, 616–28.
63. Leeming, D. J., Larsen, D. V., Zhang, C., Hi, Y., Veidal, S. S., Nielsen, R. H., Henriksen, K., Zheng, Q., Barkholt, V., Riis, B. J., Byrjalsen, I., Qvist, P. & Karsdal, M. A. 2010. Enzyme-Linked Immunosorbent Serum Assays (Elisas) For Rat And Human N-Terminal Pro-Peptide Of Collagen Type I (Pinp)--Assessment Of Corresponding Epitopes. *Clinical Biochemistry,* 43, 1249–56.
64. Leeming, D. J., Nielsen, M. J., Dai, Y., Veidal, S. S., Vassiliadis, E., Zhang, C., He, Y., Vainer, B., Zheng, Q. & Karsdal, M. A. 2012a. Enzyme-Linked Immunosorbent Serum Assay Specific For The 7s Domain Of Collagen Type Iv (P4np 7s): A Marker Related To The Extracellular Matrix Remodeling During Liver Fibrogenesis. *Hepatology Research,* 42, 482–93.
65. Leeming, D. J., Sand, J. M., Nielsen, M. J., Genovese, F., Martinez, F. J., Hogaboam, C. M., Han, M. K., Klickstein, L. B. & Karsdal, M. A. 2012b. Serological Investigation Of The Collagen Degradation Profile Of Patients With Chronic Obstructive Pulmonary Disease Or Idiopathic Pulmonary Fibrosis. *Biomarker Insights,* 7, 119–26.
66. Light, N. D. & Bailey, A. J. 1982. Covalent Cross-Links In Collagen. *Methods Enzymol,* 82 Pt A, 360–72.
67. Liu, J., Xu, M. Y., Wu, J., Zhang, H., Yang, L., Lun, D. X., Hu, Y. C. & Liu, B. 2021. Picrosirius-Polarization Method For Collagen Fiber Detection In Tendons: A Mini-Review. *Orthopaedic Surgery,* 13, 701–707.
68. Liu, Y., Keikhosravi, A., Mehta, G. S., Drifka, C. R. & Eliceiri, K. W. 2017. Methods For Quantifying Fibrillar Collagen Alignment. *Methods In Molecular Biology,* 1627, 429–451.
69. Lopez De Padilla, C. M., Coenen, M. J., Tovar, A., De La Vega, R. E., Evans, C. H. & Muller, S. A. 2021. Picrosirius Red Staining: Revisiting Its Application To The Qualitative And Quantitative Assessment Of Collagen Type I And Type Iii In Tendon. *Journal Of Histochemistry And Cytochemistry,* 69, 633–643.
70. Ma, Z. Q., Dasari, S., Chambers, M. C., Litton, M. D., Sobecki, S. M., Zimmerman, L. J., Halvey, P. J., Schilling, B., Drake, P. M., Gibson, B. W. & Tabb, D. L. 2009. Idpicker 2.0: Improved Protein Assembly With High Discrimination Peptide Identification Filtering. *J Proteome Res,* 8, 3872–81.
71. Marini, J. C., Forlino, A., Bächinger, H. P., Bishop, N. J., Byers, P. H., Paepe, A. D., Fassier, F., Fratzl-Zelman, N., Kozloff, K. M., Krakow, D., Montpetit, K. & Semler, O. 2017. Osteogenesis Imperfecta. *Nature Reviews Disease Primers,* 3, 17052.
72. Matsuse, T., Ohga, E., Teramoto, S., Fukayama, M., Nagai, R., Horiuchi, S. & Ouchi, Y. 1998. Immunohistochemical Localisation Of Advanced Glycation End Products In Pulmonary Fibrosis. *J Clin Pathol,* 51, 515–9.
73. Mercer, P. F., Woodcock, H. V., Eley, J. D., Plate, M., Sulikowski, M. G., Durrenberger, P. F., Franklin, L., Nanthakumar, C. B., Man, Y., Genovese, F., Mcanulty, R. J., Yang, S., Maher, T. M., Nicholson, A. G., Blanchard, A. D., Marshall, R. P., Lukey, P. T. & Chambers, R. C. 2016. Exploration Of A Potent Pi3 Kinase/Mtor Inhibitor As A Novel Anti-Fibrotic Agent In Ipf. *Thorax,* 71, 701–11.
74. Merl-Pham, J., Basak, T., Knüppel, L., Ramanujam, D., Athanason, M., Behr, J., Engelhardt, S., Eickelberg, O., Hauck, S. M., Vanacore, R. & Staab-Weijnitz, C. A. 2019. Quantitative Proteomic Profiling Of Extracellular Matrix And Site-Specific Collagen Post-Translational Modifications In An In Vitro Model Of Lung Fibrosis. *Matrix Biology Plus,* 1, 100005.
75. Miller, E. J. & Rhodes, R. K. 1982. Preparation And Characterization Of The Different Types Of Collagen. *Methods In Enzymology,* 82 Pt A, 33–64.

76. Miranda, K., Girard-Dias, W., Attias, M., De Souza, W. & Ramos, I. 2015. Three Dimensional Reconstruction By Electron Microscopy In The Life Sciences: An Introduction For Cell And Tissue Biologists. *Molecular Reproduction And Development,* 82, 530–47.
77. Naba, A., Clauser, K. R., Hoersch, S., Liu, H., Carr, S. A. & Hynes, R. O. 2012. The Matrisome: In Silico Definition And In Vivo Characterization By Proteomics Of Normal And Tumor Extracellular Matrices. *Molecular And Cellular Proteomics,* 11, M111 014647.
78. Naffa, R. & Pesek, J. 2019. Separation Of Natural Collagen Crosslinks Using Buffer And Ion-Pairing Agent Free Solvents On Silica Hydride Column For Mass Spectrometry Detection. *Bio Protoc,* 9, E3224.
79. Naffa, R., Watanabe, S., Zhang, W., Maidment, C., Singh, P., Chamber, P., Matyska, M. T. & Pesek, J. J. 2019. Rapid Analysis Of Pyridinoline And Deoxypyridinoline In Biological Samples By Liquid Chromatography With Mass Spectrometry And A Silica Hydride Column. *J Sep Sci,* 42, 1482–1488.
80. Nielsen, M. J., Nedergaard, A. F., Sun, S., Veidal, S. S., Larsen, L., Zheng, Q., Suetta, C., Henriksen, K., Christiansen, C., Karsdal, M. A. & Leeming, D. J. 2013. The Neo-Epitope Specific Pro-C3 Elisa Measures True Formation Of Type Iii Collagen Associated With Liver And Muscle Parameters. *American Journal Of Translational Research,* 5, 303–15.
81. Nimptsch, A., Schibur, S., Ihling, C., Sinz, A., Riemer, T., Huster, D. & Schiller, J. 2011. Quantitative Analysis Of Denatured Collagen By Collagenase Digestion And Subsequent Maldi-Tof Mass Spectrometry. *Cell And Tissue Research,* 343, 605–17.
82. Norman, M. D. A., Ferreira, S. A., Jowett, G. M., Bozec, L. & Gentleman, E. 2021. Measuring The Elastic Modulus Of Soft Culture Surfaces And Three-Dimensional Hydrogels Using Atomic Force Microscopy. *Nature Protocols,* 16, 2418–2449.
83. Onursal, C., Dick, E., Angelidis, I., Schiller, H. B. & Staab-Weijnitz, C. A. 2021. Collagen Biosynthesis, Processing, And Maturation In Lung Ageing. *Frontiers In Medicine,* 8, 593874.
84. Organ, L. A., Duggan, A. R., Oballa, E., Taggart, S. C., Simpson, J. K., Kang'ombe, A. R., Braybrooke, R., Molyneaux, P. L., North, B., Karkera, Y., Leeming, D. J., Karsdal, M. A., Nanthakumar, C. B., Fahy, W. A., Marshall, R. P., Jenkins, R. G. & Maher, T. M. 2019. Biomarkers Of Collagen Synthesis Predict Progression In The Profile Idiopathic Pulmonary Fibrosis Cohort. *Respiratory Research,* 20, 148.
85. Perdivara, I., Perera, L., Sricholpech, M., Terajima, M., Pleshko, N., Yamauchi, M. & Tomer, K. B. 2013. Unusual Fragmentation Pathways In Collagen Glycopeptides. *J Am Soc Mass Spectrom,* 24, 1072–81.
86. Pizzo, A. M., Kokini, K., Vaughn, L. C., Waisner, B. Z. & Voytik-Harbin, S. L. 2005. Extracellular Matrix (Ecm) Microstructural Composition Regulates Local Cell-Ecm Biomechanics And Fundamental Fibroblast Behavior: A Multidimensional Perspective. *Journal Of Applied Physiology,* 98, 1909–21.
87. Pompili, S., Latella, G., Gaudio, E., Sferra, R. & Vetuschi, A. 2021. The Charming World Of The Extracellular Matrix: A Dynamic And Protective Network Of The Intestinal Wall. *Frontiers In Medicine (Lausanne),* 8, 610189.
88. Qiu, B., Wei, F., Sun, X., Wang, X., Duan, B., Shi, C., Zhang, J., Zhang, J., Qiu, W. & Mu, W. 2014. Measurement Of Hydroxyproline In Collagen With Three Different Methods. *Molecular Medicine Reports,* 10, 1157–63.
89. Rajan, N., Habermehl, J., Cote, M. F., Doillon, C. J. & Mantovani, D. 2006. Preparation Of Ready-To-Use, Storable And Reconstituted Type I Collagen From Rat Tail Tendon For Tissue Engineering Applications. *Nature Protocols,* 1, 2753–8.
90. Rappu, P., Salo, A. M., Myllyharju, J. & Heino, J. 2019. Role Of Prolyl Hydroxylation In The Molecular Interactions Of Collagens. *Essays In Biochemistry,* 63, 325–335.
91. Reimann, C., Brangsch, J., Colletini, F., Walter, T., Hamm, B., Botnar, R. M. & Makowski, M. R. 2017. Molecular Imaging Of The Extracellular Matrix In The Context Of Atherosclerosis. *Advanced Drug Delivery Reviews,* 113, 49–60.

92. Reiser, K. M., Tryka, A. F., Lindenschmidt, R. C., Last, J. A. & Witschi, H. R. 1986. Changes In Collagen Cross-Linking In Bleomycin-Induced Pulmonary Fibrosis. *J Biochem Toxicol,* 1, 83–91.
93. Rickelt, S. & Hynes, R. O. 2018. Antibodies And Methods For Immunohistochemistry Of Extracellular Matrix Proteins. *Matrix Biol,* 71–72, 10–27.
94. Rosell-Garcia, T. & Rodriguez-Pascual, F. 2018. Enhancement Of Collagen Deposition And Cross-Linking By Coupling Lysyl Oxidase With Bone Morphogenetic Protein-1 And Its Application In Tissue Engineering. *Sci Rep,* 8, 10780.
95. Salarian, M., Ibhagui, O. Y. & Yang, J. J. 2020. Molecular Imaging Of Extracellular Matrix Proteins With Targeted Probes Using Magnetic Resonance Imaging. *Wiley Interdisciplinary Reviews: Nanomedicine And Nanobiotechnology,* 12, E1622.
96. Sandberg, L. B., Weissman, N. & Smith, D. W. 1969. The Purification And Partial Characterization Of A Soluble Elastin-Like Protein From Copper-Deficient Porcine Aorta. *Biochemistry,* 8, 2940–5.
97. Sato, K., Ebihara, T., Adachi, E., Kawashima, S., Hattori, S. & Irie, S. 2000. Possible Involvement Of Aminotelopeptide In Self-Assembly And Thermal Stability Of Collagen I As Revealed By Its Removal With Proteases. *Journal Of Biological Chemistry,* 275, 25870–5.
98. Schiller, H. B., Fernandez, I. E., Burgstaller, G., Schaab, C., Scheltema, R. A., Schwarzmayr, T., Strom, T. M., Eickelberg, O. & Mann, M. 2015. Time- And Compartment-Resolved Proteome Profiling Of The Extracellular Niche In Lung Injury And Repair. *Molecular Systems Biology,* 11, 819.
99. Schiller, H. B., Mayr, C. H., Leuschner, G., Strunz, M., Staab-Weijnitz, C., Preisendorfer, S., Eckes, B., Moinzadeh, P., Krieg, T., Schwartz, D. A., Hatz, R. A., Behr, J., Mann, M. & Eickelberg, O. 2017. Deep Proteome Profiling Reveals Common Prevalence Of Mzb1-Positive Plasma B Cells In Human Lung And Skin Fibrosis. *Am J Respir Crit Care Med,* 196, 1298–1310.
100. Seo, B. R., Chen, X., Ling, L., Song, Y. H., Shimpi, A. A., Choi, S., Gonzalez, J., Sapudom, J., Wang, K., Andresen Eguiluz, R. C., Gourdon, D., Shenoy, V. B. & Fischbach, C. 2020. Collagen Microarchitecture Mechanically Controls Myofibroblast Differentiation. *Proceedings Of The National Academy Of Sciences Of The United States Of America,* 117, 11387–11398.
101. Smolenski, K. A., Avery, N. C. & Light, N. D. 1983. A New Rapid Method For The Identification Of Reducible Collagen Cross-Links In Small Tissue Samples. *Biochem J,* 213, 525–32.
102. Staab-Weijnitz, C. A., Fernandez, I. E., Knuppel, L., Maul, J., Heinzelmann, K., Juan-Guardela, B. M., Hennen, E., Preissler, G., Winter, H., Neurohr, C., Hatz, R., Lindner, M., Behr, J., Kaminski, N. & Eickelberg, O. 2015. Fk506-Binding Protein 10, A Potential Novel Drug Target For Idiopathic Pulmonary Fibrosis. *American Journal Of Respiratory And Critical Care Medicine,* 192, 455–67.
103. Starborg, T., Kalson, N. S., Lu, Y., Mironov, A., Cootes, T. F., Holmes, D. F. & Kadler, K. E. 2013. Using Transmission Electron Microscopy And 3view To Determine Collagen Fibril Size And Three-Dimensional Organization. *Nature Protocols,* 8, 1433–48.
104. Starborg, T., Lu, Y., Kadler, K. E. & Holmes, D. F. 2008. Electron Microscopy Of Collagen Fibril Structure In Vitro And In Vivo Including Three-Dimensional Reconstruction. *Methods In Cell Biology,* 88, 319–45.
105. Strupler, M., Pena, A. M., Hernest, M., Tharaux, P. L., Martin, J. L., Beaurepaire, E. & Schanne-Klein, M. C. 2007. Second Harmonic Imaging And Scoring Of Collagen In Fibrotic Tissues. *Optics Express,* 15, 4054–65.
106. Tabb, D. L., Fernando, C. G. & Chambers, M. C. 2007. Myrimatch: Highly Accurate Tandem Mass Spectral Peptide Identification By Multivariate Hypergeometric Analysis. *J Proteome Res,* 6, 654–61.

107. Taga, Y., Kusubata, M., Ogawa-Goto, K. & Hattori, S. 2014. Stable Isotope-Labeled Collagen: A Novel And Versatile Tool For Quantitative Collagen Analyses Using Mass Spectrometry. *J Proteome Res,* 13, 3671–8.
108. Taga, Y., Tanaka, K., Hamada, C., Kusubata, M., Ogawa-Goto, K. & Hattori, S. 2017. Hydroxyhomocitrulline Is A Collagen-Specific Carbamylation Mark That Affects Cross-Link Formation. *Cell Chem Biol,* 24, 1276–1284 E3.
109. Tatler, A. L. 2019. Recent Advances In The Non-Invasive Assessment Of Fibrosis Using Biomarkers. *Current Opinion In Pharmacology,* 49, 110–115.
110. Van De Vlekkert, D., Machado, E. & D'azzo, A. 2020. Analysis Of Generalized Fibrosis In Mouse Tissue Sections With Masson's Trichrome Staining. *Bio-Protocol,* 10, E3629.
111. Veidal, S. S., Karsdal, M. A., Nawrocki, A., Larsen, M. R., Dai, Y., Zheng, Q., Hägglund, P., Vainer, B., Skjøt-Arkil, H. & Leeming, D. J. 2011. Assessment Of Proteolytic Degradation Of The Basement Membrane: A Fragment Of Type Iv Collagen As A Biochemical Marker For Liver Fibrosis. *Fibrogenesis Tissue Repair,* 4, 22.
112. Vogel, B., Siebert, H., Hofmann, U. & Frantz, S. 2015. Determination Of Collagen Content Within Picrosirius Red Stained Paraffin-Embedded Tissue Sections Using Fluorescence Microscopy. *Methodsx,* 2, 124–34.
113. Waghorn, P. A., Jones, C. M., Rotile, N. J., Koerner, S. K., Ferreira, D. S., Chen, H. H., Probst, C. K., Tager, A. M. & Caravan, P. 2017. Molecular Magnetic Resonance Imaging Of Lung Fibrogenesis With An Oxyamine-Based Probe. *Angewandte Chemie, International Edition In English,* 56, 9825–9828.
114. Watt, F. M. & Huck, W. T. 2013. Role Of The Extracellular Matrix In Regulating Stem Cell Fate. *Nature Reviews: Molecular Cell Biology,* 14, 467–73.
115. Wegner, K. A., Keikhosravi, A., Eliceiri, K. W. & Vezina, C. M. 2017. Fluorescence Of Picrosirius Red Multiplexed With Immunohistochemistry For The Quantitative Assessment Of Collagen In Tissue Sections. *Journal Of Histochemistry And Cytochemistry,* 65, 479–490.
116. Wei, Y., Kim, T. J., Peng, D. H., Duan, D., Gibbons, D. L., Yamauchi, M., Jackson, J. R., Le Saux, C. J., Calhoun, C., Peters, J., Derynck, R., Backes, B. J. & Chapman, H. A. 2017. Fibroblast-Specific Inhibition Of Tgf-Beta1 Signaling Attenuates Lung And Tumor Fibrosis. *J Clin Invest,* 127, 3675–3688.
117. Weis, M. A., Hudson, D. M., Kim, L., Scott, M., Wu, J. J. & Eyre, D. R. 2010. Location Of 3-Hydroxyproline Residues In Collagen Types I, Ii, Iii, And V/Xi Implies A Role In Fibril Supramolecular Assembly. *J Biol Chem,* 285, 2580–90.
118. Williams, B. R., Gelman, R. A., Poppke, D. C. & Piez, K. A. 1978. Collagen Fibril Formation. Optimal In Vitro Conditions And Preliminary Kinetic Results. *Journal Of Biological Chemistry,* 253, 6578–85.
119. Wolman, M. 1975. Polarized Light Microscopy As A Tool Of Diagnostic Pathology. *Journal Of Histochemistry And Cytochemistry,* 23, 21–50.
120. Xie, J., Bao, M., Bruekers, S. M. C. & Huck, W. T. S. 2017. Collagen Gels With Different Fibrillar Microarchitectures Elicit Different Cellular Responses. *Acs Applied Material & Interfaces,* 9, 19630–19637.
121. Xydias, D., Ziakas, G., Psilodimitrakopoulos, S., Lemonis, A., Bagli, E., Fotsis, T., Gravanis, A., Tzeranis, D. S. & Stratakis, E. 2021. Three-Dimensional Characterization Of Collagen Remodeling In Cell-Seeded Collagen Scaffolds Via Polarization Second Harmonic Generation. *Biomedical Optics Express,* 12, 1136–1153.
122. Yamauchi, M. & Sricholpech, M. 2012. Lysine Post-Translational Modifications Of Collagen. *Essays Biochem,* 52, 113–33.
123. Zhou, Y., Horowitz, J. C., Naba, A., Ambalavanan, N., Atabai, K., Balestrini, J., Bitterman, P. B., Corley, R. A., Ding, B. S., Engler, A. J., Hansen, K. C., Hagood, J. S., Kheradmand, F., Lin, Q. S., Neptune, E., Niklason, L., Ortiz, L. A., Parks, W. C., Tschumperlin, D. J., White, E. S., Chapman, H. A. & Thannickal, V. J. 2018. Extracellular Matrix In Lung Development, Homeostasis And Disease. *Matrix Biology,* 73, 77–104.

124. Harris, W.T., et al., *Myofibroblast differentiation and enhanced TGF-B signaling in cystic fibrosis lung disease.* PLoS One, 2013. 8(**8**): p. e70196.
125. Egger, C., et al., *Administration of bleomycin via the oropharyngeal aspiration route leads to sustained lung fibrosis in mice and rats as quantified by UTE-MRI and histology.* PLoS One, 2013. **8**(5): p. e63432.
126. Booth, A.J., et al., *Acellular normal and fibrotic human lung matrices as a culture system for in vitro investigation.* American Journal of Respiratory and Critical Care Medicine, 2012. **186**(9): p. 866–76.
127. Tjin, G., et al., *Lysyl oxidases regulate fibrillar collagen remodelling in idiopathic pulmonary fibrosis.* Disease Models & Mechanisms, 2017. **10**(11): p. 1301–1312.

# Part IV
# Engineering and Modeling the Pulmonary Vasculature

# Chapter 12
# Understanding and Engineering the Pulmonary Vasculature

**Wai Hoe Ng, Barbie Varghese, and Xi Ren**

## 12.1 Pulmonary Vasculature in Development and Diseases

Blood circulation in the body is essential for transporting oxygen and nutrients to nearly all tissues. It also plays crucial roles in regulating organ and tissue homeostasis [1]. The mammalian circulation system can be divided into systemic and pulmonary circulation. Here, we will be focusing on the pulmonary circulation, which involves transporting blood in between the heart and lung. Deoxygenated blood is transported from the right ventricle of the heart to the lung via pulmonary arteries that further branch into smaller arterioles, and finally, capillaries, where the blood vessels contact the air-filled alveoli to facilitate gas exchange. Blood that is being oxygenated as it flows through the pulmonary capillaries merges into the pulmonary veins and eventually reaches the left atrium of the heart prior to systemic circulation.

Gas exchange involves the diffusion of $CO_2$ and $O_2$ between the alveoli and capillaries in the distal lung. The human lung is made up of an estimated 480 million alveoli that are paired up with 280 billion capillary segments at the terminal ends of the respiratory tree, covering a surface area of 40–80 square meters to allow effective gas diffusion between the two compartments [2, 3]. During mouse embryogenesis, E9.5 marks the earliest sighting of the mesoderm-derived endothelial cells (ECs) in the areas ventral and lateral to the foregut endoderm [4], followed by their expansion and integration with the surrounding epithelium throughout all stages of lung organogenesis [5]. During human lung development, the first air-blood interface formation is observed during the canalicular stage (gestation week 16–26), which continues to mature during the saccular stage (gestational week 24–38) [6].

W. H. Ng · B. Varghese · X. Ren (✉)
Department of Biomedical Engineering, Carnegie Mellon University, Pittsburgh, PA, USA
e-mail: xiren@cmu.edu

 
C. M. Magin (ed.), *Engineering Translational Models of Lung Homeostasis and Disease*, Advances in Experimental Medicine and Biology 1413,
https://doi.org/10.1007/978-3-031-26625-6_12

Disruption of alveolar development impairs vascular development and vice versa [6]. Circulation between the heart and distal lung vascular plexus is initiated during the pseudoglandular stage (38 days of gestation) [7].

During embryonic vascular development, the Vascular Endothelial Growth Factor (VEGF) receptors play crucial roles in EC specification and are also commonly used as early markers to label ECs [8, 9]. At E11.5 in mice, soon after the initiation of lung organogenesis, VEGF is uniformly distributed in the airway epithelium and the subepithelial matrix. Later, at E13.5 and E15.5, its expression becomes restricted to the branching tips of the airways in the distal lung [10]. These emphasize the role of matrix-associated VEGF in coordinating neovascularization and lung epithelial morphogenesis [10]. Reporter targeting flk-1 (VEGF receptor 2) in animal models has been used to identify the signals regulating pulmonary vascular development: (1) WNT signaling is required for proper lung mesenchymal growth and vascular smooth muscle development [11]; (2) Blocking Sonic Hedgehog (SHH) disrupts pulmonary vascular plexus formation [12]; (3) VEGF-A stimulates ECs to undergo vasculogenesis and angiogenesis [13]; (4) Transforming Growth Factor (TGF)-β also directs the differentiation and maturation of new vessels through vasculogenesis and angiogenesis [14]; and (5) Lack of Bone Morphogenetic Protein (BMP) stimulator reduces the number of smooth muscle cells and enlarges vessels [15]. Further, the vasculature grows along with foregut endoderm and continues to branch alongside the lung bud [13]. In the absence of lung epithelium, pulmonary vascular ECs fail to proliferate [13]. Multiple factors, including TGF-β, retinoic acid, BMPs, Fibroblast Growth Factor 10 (FGF10), and WNTs, are primarily supplied by the pulmonary mesenchyme and have been implicated to mediate the specification and expansion of both lung epithelium and endothelium during development [16–19].

Defective vascular development is associated with pulmonary dysfunction and an array of respiratory diseases. Bronchopulmonary dysplasia is a condition marked by the disruption of pulmonary vascular development as well as a reduction in alveolar number [20]. In experimental models, blockage of VEGF or its receptor (VEGFR2) during fetal lung development results in defective angiogenesis and alveolar hypoplasia, producing pathological phenotypes resembling bronchopulmonary dysplasia [21, 22]. Alveolar capillary dysplasia is a disease caused by defective formation of the air-blood barrier in the lung due to misarrangement of veins in the center of lung lobules [23, 24]. Congenital cystic pulmonary disorders are caused by maldevelopment of the pre-acinar airways, resulting in immature and abnormal vascularization [25]. In another scenario, congenital diaphragmatic hernia, where the lungs are mechanically forced to one side within the thoracic cavity, causes hypoplastic lungs with an abnormally thickened pulmonary arterial wall [26]. These conditions hint that pulmonary vasculature is physiologically or pathologically linked with pulmonary epithelial specification. Thus, understanding and recapitulating the inter-lineage endothelial-epithelial crosstalk is of critical importance for lung bioengineering.

## 12.2 Pulmonary ECs and Their Angiocrine Functions

Although the vasculature permeates almost every organ, the microvascular (capillary) ECs of different organs exhibit distinct organotypic phenotypes and functions (Fig. 12.1) [27]. For example, capillary ECs in the nervous system form a stringent blood brain barrier to restrict molecule exchange between the brain tissue and circulating blood, while capillary ECs of the kidney glomeruli are fenestrated enabling effective solute filtration [28]. Bone marrow and liver vasculature are composed of fenestrated sinusoidal ECs [29], with the appearance of open pores lacking basal lamina underneath the endothelium, to enable filtration of large molecules and circulating cells through these fenestrations [30, 31]. Besides acting as a key component to support the physiological functions of each organ, tissue-specific ECs are also capable of secreting angiocrine factors to influence the surrounding parenchymal and non-parenchymal cells to regulate tissue homeostasis, injury response, and

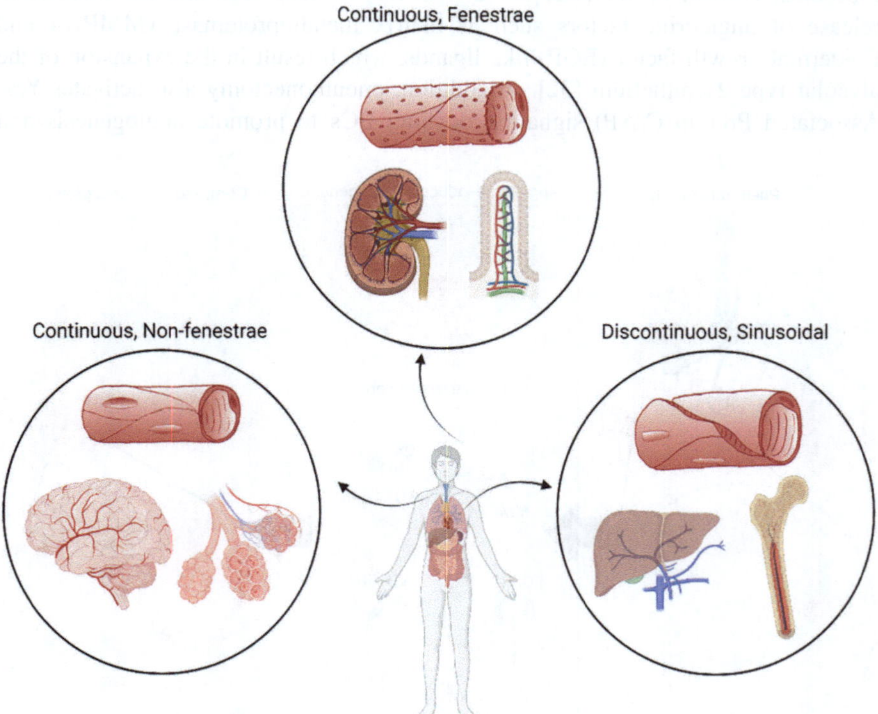


**Fig. 12.1** Tissue-specific endothelial heterogeneity in human. ECs across different vascular beds exhibit distinct phenotypes and functions. Brain and lung microvascular ECs are continuous and without fenestration, and can form a stringent barrier to limit biomolecular transfer. Kidney and small intestinal microvascular ECs are fenestrated, and function to aid in solute filtration and nutrient absorption, respectively. The sinusoidal ECs in the liver are discontinuous, allowing passage of macromolecules and transmigratory cells. Schematics created using Biorender

regeneration [27, 28]. Generally, capillary ECs are identified by their expression of CD31, CD34, FGFR-1, VEGFR-1, and VEGFR2 [32]. In addition to these, the expression of VE-cadherin, CD36, and TBX3 marks the unique identity of microvascular ECs in the lung [27]. Furthermore, two distinct microvascular endothelial phenotypes have recently been identified in the lung: aerocyte (aCAP) that is unique to the lung and is specialized for gas exchange and leukocyte trafficking, and general capillary cell (gCAP) that serves as the capillary stem cell [33].

Pulmonary capillary ECs play crucial roles in modulating lung regeneration following injury. New lung tissue formation is usually restricted to embryonic and neonatal lung development. In an elegant model of adult lung regeneration, surgical lung resection (usually removal of the entire left lung in rodent models) triggers regeneration of the remaining lung tissue and allows for the investigation of key signaling pathways involved in the re-activation of this developmentally relevant program [34]. This process, termed pneumonectomy induced compensatory lung growth, involves the formation of new alveolar units (Fig. 12.2) [35]. A key driving mechanism underlying this regenerative response is the activation of VEGFR2 and Fibroblast Growth Factor Receptor 2 (FGFR2) on lung ECs and the subsequent release of angiocrine factors such as matrix metalloproteinase (MMP)14 and Epidermal growth factor (EGF)-like ligands, which result in the expansion of the alveolar type 2 epithelium [32]. In addition, pneumonectomy also activates Yes-Associated Protein (YAP) signaling on lung ECs to promote angiogenesis and

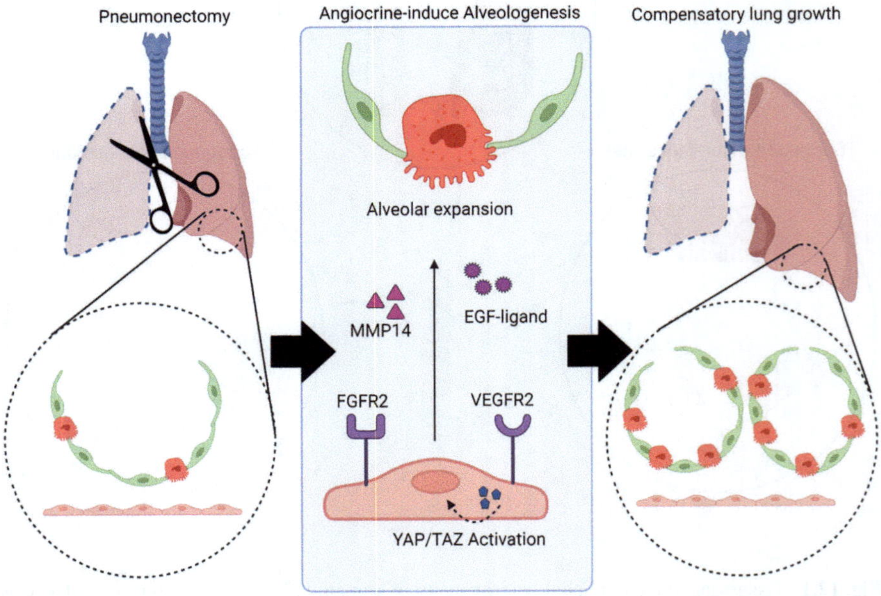


**Fig. 12.2** Pneumonectomy induces compensatory lung growth. Following surgical removal of the left lung, activated lung microvascular ECs respond to injury signals and secrete tissue-specific angiocrine factors such as MMP14 and EGF to induce alveologenesis. Expansion of alveolar epithelial cells result in enlargement of the remaining right lungs. Schematics created using Biorender

alveolarization [36]. In an elastase-induced emphysema model in mice, damaged aCAP can be replaced through differentiation from capillary stem cells (gCap) that underwent extensive proliferation following injury [33]. Bleomycin-induced alveolar injury resulted in attenuation of C-X-C motif chemokine receptor (CXCR) 7 expression on lung ECs, hence, stimulated recruitment of macrophages into injured area, and boosted the progression of fibrosis [37]. This process was reversible with the addition of CXCR7 agonist, suggesting the importance of CXCR7 in maintaining alveolar integrity [37]. In summary, given the unique functions of pulmonary ECs in mediating lung injury response, repair, and regeneration, the use of lung-specific ECs is an important step in the bioengineering of vascularized lung grafts.

Maladapted ECs contribute to multiple pathological conditions such as fibrosis and cancer. Physiologically, capillary EC-derived Hepatocyte Growth Factor (HGF) prevents acute injury-induced lung fibrosis in mice. In contrast, ECs suffering from chronic lung injury promote fibroblast activation and induce lung fibrosis [37], mainly through TGF-β signaling [38]. In addition, the structure of alveolar capillary blood vessels exhibits severe disorganization with aging in mice, and the pulmonary ECs from aged mice show dysregulated angiogenic ability and are unable to maintain physiological interaction with the alveolar epithelium [36]. Further, compensatory lung growth following pneumonectomy is also compromised in aged animals [36]. These findings suggest that aging-associated alterations of the vascular niche may contribute to the vulnerability of aged alveoli to a wide range of respiratory injuries.

## 12.3 Engineering the Pulmonary Vasculature

### *12.3.1 Generation of Vascularized Organoids*

With advances over the past two decades in the reprograming and directed differentiation of induced pluripotent stem cell (iPSC), an extensive list of organotypic cell types, including ECs and lung epithelial cells, can be induced from patient-derived iPSCs [39–42]. These in vitro cellular specification processes at least partially recapitulate native organogenesis and provide valuable models for studying developmental vascular-parenchymal interactions, and open the opportunity to incorporate and investigate the vascular niche in current regenerative and engineering strategy towards repairing injured lungs.

In hydrogel-supported culture, alveolar epithelial cells can self-organize into 3D organoid structures and undergo further lineage and morphological maturation [39, 40, 43]. Most of these alveolar organoids exhibit enclosed spherical morphology formed by polarized epithelium. However, current strategies to promoting alveolar maturation generally requires the cells to be embedded in the undefined Matrigel [39, 40, 43], rendering difficulty in scaling up the production of cells to meet the need for engineering lung grafts for regenerative therapy. Furthermore, there remain limited studies on how alveolar type 2 cells can be differentiated into alveolar type 1 cells, which is the major cell type covering the gas-exchange interface in the distal lung, limiting the development of functional bioengineered lungs.

The role of ECs in engineered alveologenesis remains to be fully understood. In epithelial-endothelial coculture models, ECs stimulated lung epithelial morphogenesis in vitro [36, 44, 45]. During embryogenesis, cells constituting the pulmonary vasculature are initiated by a group of progenitor cells derived from the secondary heart field, referred to as cardiopulmonary progenitor cells [12], suggesting crucial signaling crosstalk among the heart, lung, and vasculature. However, there has long been a lack of proper models for investigating such complex inter-lineage crosstalk, especially when the participating lineages are derived from different germ layers, such as the endoderm (giving rise to the lung epithelium) and mesoderm (giving rise to the heart and vasculature). In an effort to address this challenge, we recently demonstrated that human iPSCs can be co-differentiated into endoderm-derived lung and mesoderm-derived heart progenitors simultaneously, and observed expedited alveolar induction in the presence of accompanying cardiac lineage in 3D hydrogel-free suspension culture [46]. Likewise, co-emergence of foregut endoderm has recently been demonstrated in engineered cardiac organoids [47]. Given the shared mesodermal origin and close lineage relationship between the cardiac and vascular lineages, these studies suggest the possibility to recapitulate the co-development of lung epithelium and endothelium within in vitro iPSC differentiation.

Emerging evidence suggests the feasibility and benefits of simultaneous induction of ECs with organotypic parenchymal lineages. Human blood-brain-barrier ECs can be derived via co-differentiation with neural cells from iPSCs [48], which demonstrated barrier property and expression of functional transport apparatus similar to what is observed in native brain microvascular ECs [48]. Co-differentiation of cardiomyocytes and ECs can also be achieved by introducing VEGF during cardiac differentiation [49, 50]. Co-culture of iPSC-ECs with iPSC-derived cardiomyocytes further led to the acquisition of cardiac endothelial phenotypes [51]. Vascular ECs acquired intestinal specific EC transcriptomic signature in vitro when being co-induced together with human intestinal organoids from iPSCs [52]. These studies again imply the importance of an organotypic cellular niche in driving tissue-specific EC phenotypes.

Effective vascularization and timely establishment of host blood perfusion is usually a prerequisite for successful implantation of tissue engineered constructs. Accordingly, there has been a lasting interest in engineering pre-vascularized tissues prior to implantation. By co-culturing generic ECs, such as human umbilical vein endothelial cells (HUVECs), together with pluripotent stem cells (PSCs) undergoing neural induction, vascularized cerebral organoids can be obtained that demonstrated functional human blood vessel formation post-implantation that further promoted organoid graft survival in vivo [53]. Human iPSC-derived hepatocytes self-organized into liver buds when combined with ECs and mesenchymal stem cells (MSCs), which formed perfusable vascular networks within 48 hours post-implantation [54, 55]. Using a cell condensation approach, lung epithelial cells isolated from mice generated 3D vascularized organ buds in vitro following coculture with ECs and MSCs [56]. Transplantation of pre-vascularized islets in diabetic mice improved islet survival, promoted microvascular perfusion, and restored normoglycemia in chemically induced diabetic mice [57–59]. Given that the physiological alveoli are covered not only by alveolar epithelium but also by capillary endothelium to form the air-blood interface, further engineering of conventional lung organoid systems to improve the recapitulation of physiologically relevant epithelial-endothelial interface will be necessary to better characterize the alveolar vascular niche and its function in regulating alveolar tissue morphogenesis and homeostasis (Fig. 12.3).

Although the above studies have demonstrated vascular incorporation into the engineered parenchymal tissues, these vascularized organoids exhibit distinct morphology and structure when compared with their native counterparts. It remains a hurdle to simulate complex multi-lineage tissue morphogenesis resembling native organogenesis and thus, to generate fully functional tissue replacement or to

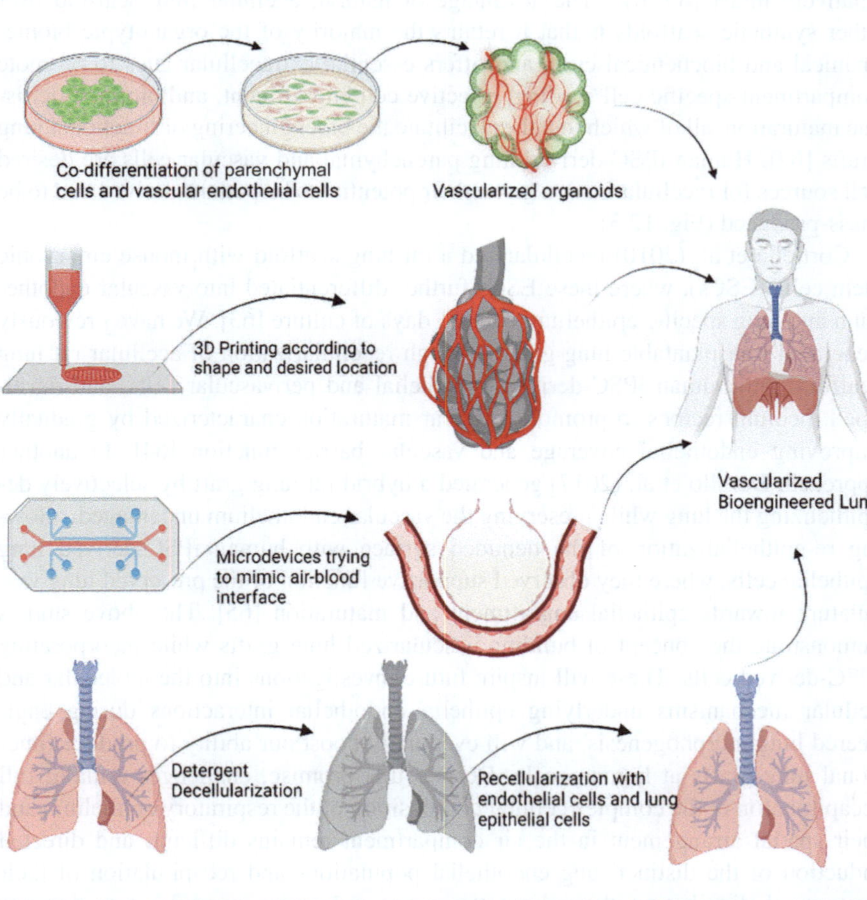


**Fig. 12.3** Current strategies to bioengineer vascularized lung tissues. Schematics created using Biorender

accurately model disease phenotypes and predict treatment outcome. Nonetheless, the iPSC platform provides an unprecedented opportunity of using patients' own cells for vascularized tissue engineering, which offers enormous advantage in regenerative therapy, alleviating the uncertainties in results obtained from animal studies.

### 12.3.2 Bioengineered Lung and Vasculature Using Acellular Native Lung Scaffold

The organotypic arrangement of the vascular and parenchymal tissues is documented in the organ-specific extracellular matrix (ECM) framework. One popular strategy for deriving such a framework is to treat a harvested organ with a series of detergents to remove all the cellular content while leaving the ECM ultrastructure relatively intact [60, 61]. The advantage of natural, acellular lung scaffold over other synthetic scaffolds is that it retains the majority of the organotypic biomechanical and biochemical cues, and offers essential extracellular cues to promote compartment-specific cell seeding, effective cell engraftment, and biomimetic tissue maturation, all of which together facilitate the bioengineering of functional lung grafts [62]. Human iPSC-derived lung parenchymal and vascular cells are desired cell sources for recellularization given their potential to be patient-derived and to be mass-produced (Fig. 12.3).

Cortiella et al. (2010) recellularized a rat lung scaffold with mouse embryonic stem cells (ESCs), where these ESCs further differentiated into vascular endothelium and lung specific epithelium over 21 days of culture [63]. We have previously generated transplantable lung grafts through recellularization of acellular rat lung scaffold with human iPSC-derived endothelial and perivascular cells, and developed medium recipes to promote vascular maturation characterized by gradually improving endothelial coverage and vascular barrier function [64]. In another approach, Dorello et al. (2017) generated a hybrid rat lung graft by selectively de-epithelizing the lung while preserving the vascular endothelium undamaged, allowing re-epithelialization of the denuded surface with human iPSC-derived lung epithelial cells, where they observed supportive function of the preserved lung vasculature towards epithelial engraftment and maturation [65]. The above studies demonstrate the concept of building vascularized lung grafts while incorporating iPSC-derived cells. These will inspire future investigations into the molecular and cellular mechanisms underlying epithelial-endothelial interactions during engineered lung morphogenesis, and will eventually boost our ability to produce functional lung grafts at human scale. Despite this promise, challenges remain: full recapitulation of the complex cellular composition of the respiratory epithelium and their spatial arrangement in the air compartment remains difficult; and directed induction of the distinct lung endothelial populations and recapitulation of their anatomical distribution throughout the entire pulmonary vascular tree remains unachieved [33].

### *12.3.3 Vascularized Lung-on-a-Chip*

Microfluidic devices such as lung-on-a-chip can mimic the paralleled lining of the lung epithelium and blood vessel (Fig. 12.3). Human lung-alveolus-on-a-chip has been generated by combining two superimposed microfluidic channels, harboring human alveolar epithelial cells and pulmonary microvascular ECs respectively to mimic lung microenvironment [66]. These two participating cell lineages were separated by a 10-μm-thick flexible microporous membrane, with alveolar epithelial cells exposed to the air and pulmonary ECs exposed to constant fluid flow, thereby mimicking the alveolar-capillary barrier seen in the native lung alveoli [66]. Such lung-alveolus-on-a-chip system allows effective modeling and investigation of respiratory diseases [67] as well as the effect of mechanical stress on respiratory barrier permeability [68]. Coculture of patient-derived alveolar epithelial type 1 and type 2 cells with pulmonary ECs on lung-on-a-chip were reported to mimic native respiratory microenvironment, leading to preserved cellular phenotypes, functions, and barrier integrity, which can be maintained for several days [69].

The establishment of microfluidic lung-alveolus-on-a-chip device has led to the ability to unravel complex responses of human primary alveolar epithelium and ECs on vascular hemodynamics and endotoxin-induced intravascular thrombosis, making it suitable for developing antithrombotic therapy [70]. The microdevice can be further improved through the introduction of a stretchable biological membrane made of collagen type 1 and elastin, sandwiched between alveolar epithelium and ECs, leading to mechanically active air-blood barrier [71]. Such a membrane can also be replaced with a microporous polycarbonate film, with an added advantage that it allows exchange of soluble biomolecules, thus, creating a convenient system to study the impact of airborne pollutants on the respiratory system [72].

Continued efforts on lung-on-a-chip device engineering should focus on the introduction of microvasculature and recapitulating vascular hierarchy. AngioChip was built using a biodegradable material, poly(octamethylenemaleate(anhydride) citrate) (POMaC) with internal channels for vasculogenesis of the incorporated ECs and outer space for hydrogel containing parenchymal cells to encapsulate the internal channels, leading to biomimetic tissue systems such as vascularized liver and cardiac tissues [73]. By tuning collagen and EC density, Morgan et al. (2018) were able to generate various vessel sizes ranging from 3 to 10 μm for capillary to 0.5 mm for small arteries and veins [74].

Vascularized lung-on-a-chip engineering remains a rapidly expanding field. Future excitement will likely come from further improvement in recapitulating the morphological and functional characteristics of native air-blood interface. Such an improvement is not limited to the type of materials being used, and also includes incorporation of other environmental cues, biological, mechanical, chemical, or structural, to promote seamless vascular integration into the engineered lung parenchyma. In parallel, further augmentation in device complexity to match that of native alveoli can be derived from guiding vascular structure formation to follow a defined geometry and spatial arrangement relative to the lung epithelium.

### 12.3.4 Guided Vascularization Through 3D Bioprinting

The uniqueness of 3D printing technology is that it enables the placement of desired materials and cells in a spatially specific manner in 3D space to facilitate tissue fabrication with high reproducibility. 3D printing of alveolar structures at a native-like scale remains an elusive goal due to its intrinsic geometrical complexity and the need for micro- and nano-scale features. However, current and continued progresses in additive manufacturing techniques are poised to bring us closer to this ultimate goal.

Lewis et al. (2015) used photodegradable microspheres to create hollow epithelial cysts on the scale of alveolar structures [75]. These microspheres were made from polyethylene glycol (PEG) that was crosslinked using photolabile, nitrobenzyl ether crosslinker to allow linkage to be cleaved upon light exposure. The surface of these microspheres was seeded with alveolar epithelial cells and embedded in a second hydrogel. Light-triggered degradation of PEG hydrogel resulted in 120-μm cyst structures lined by alveolar epithelium [75]. This success in alveolar patterning can be improved further by the incorporation of a vascular compartment.

To engineer the vasculature using additive manufacturing techniques, Horvath et al. (2015) printed alternating layers of Matrigel with ECs and epithelial cells over a porous membrane [76]. Kolesky et al. (2016) biofabricated 3D perfusion chips using multiple cell-laden vascular inks and extracellular matrix inks. The vascular ink was composed of polyethylene oxide (PEO)-polypropylene oxide (PPO-PEO), which was thermo-responsive and could be removed upon cooling to 4 °C to yield perfusable network of interconnected channels [77]. In addition, Gao et al. (2018) established an in vitro vascular model using coaxial printing and blood vessel-derived ECM bioink, resulting in complex vascular patterns with the presence of hollow channels to allow perfusion. This model allows investigation of 3D vascular response to proangiogenic and inflammatory cytokines, mimicking native blood vessel pathophysiology [78].

Grigoryan et al. (2019) 3D printed distal lung models with complex vascular and airway spaces using poly(ethylene glycol) diacrylate (PEGDA), which is a photopolymerizable hydrogel, together with biocompaticle food dye additives that serve as photoabsorbers during projection stereolithography printing [79]. In combination with lung-mimetic architectures, they created a breathing lung model with tidal air ventilation and blood flow that demonstrated effective oxygen delivery when perfused with deoxygenated red blood cells, recapitulating the native alveoli-capillary system [79].

Feinberg and co-workers have established a 3D biofabrication technique referred to as freeform reversible embedding of suspended hydrogels (FRESH), which allows extrusion printing of large and complex tissue structures using soft biomaterials such as collagen and fibrin using a thermoreversible support bath that holds the extruded bioink in place until it cures [80]. FRESH bioprinting has been able to generate a full-size human heart model with intricate internal architecture, a life-size human heart valve with cyclical opening and closing function under pulsatile flow, and perfusable human coronary vessels [80, 81]. The ability of FRESH to recapitulate the biomechanical properties of native tissues makes it an ideal technology for future exploration of additive biofabrication of lung tissues where tissue elasticity is central to the cyclic respiratory motion.

3D printing is a promising strategy to mimic multi-layered blood vessel models with precise vascular patterning and high tissue complexity. While it is possible to bioprint and endothelialize perfusable vascular networks suitable for transplantation, it remains a challenge to print at the capillary level with a diameter of 5 to 10 μm. One possible strategy that is currently under extensive investigation is to combine pre-vascularized 3D printed constructs in the presence of guided angiogenic cues to allow sprouting of pre-existing vessels to form capillaries via angiogenesis. Furthermore, efforts are needed to better understand the intricate lung vascular architecture as well as the spatial distribution of native ECM throughout the entire lung vascular bed (Fig. 12.3). The resulting knowledge will then guide future development of biofabricated lung vasculature to the next level, incorporating biomimetic vascular density, complexity and spatially regulated biomechanics.

## 12.4 Pulmonary Vascular Diseases

Pulmonary vascular diseases (PVD) refer to conditions affecting the blood vessels in the lung. A pulmonary vascular disorder does not necessarily affect the lungs alone and may also lead to cardiovascular abnormalities in other organ systems. Preterm babies are often associated with simplified and immature alveolar microvascular development. In a cohort study, young adults who were born premature have a higher tendency to develop early pulmonary vascular diseases and increased risk of developing pulmonary hypertension (PH) [82, 83]. Right heart catheterization (RHC) at rest is the gold standard to assess pulmonary hemodynamics in patients with chronic obstructive pulmonary disease (COPD) and PH [84]. Other diagnostic methods include echocardiography [84], chest computed tomography [85, 86], and Magnetic resonance imaging of the blood flow in the main pulmonary artery [87, 88].

PH is one of the main diseases affecting pulmonary vascular health, with the key pathological hallmark being increased resistance and blood pressure in pulmonary arteries stemming from aberrant chronic vascular remodeling [89, 90]. PH is defined as an elevated mean pulmonary arterial pressure of >25 mm Hg at rest or 30 mm Hg on exercise [91]. To model PH in animal models, monocrotaline (MCT) injected peritoneally is commonly used, which upon being metabolized by the liver to its active form has well established effects to cause pulmonary vascular hypertrophy that closely resembles the clinical presentation of PH in human [92, 93]. MCT has been shown to selectively damage pulmonary artery ECs [94, 95], causing dysregulation of endothelial nitric oxide signaling, disruption of cell membrane, and ultimately vascular remodeling [96, 97]. Consistent with the central role of VEGF signaling in maintaining endothelial homeostasis, severe angioproliferative rat model of PH can be induced via the combination of a single injection of VEGFR-2 inhibitor (SU5416) and 3–4 weeks of hypoxia followed by 2 weeks of normoxia exposure [98–100]. These SU5416-induced PH rats exhibit increased pulmonary arterial occlusion, elevation in muscularization, and extensive vascular remodeling [101, 102]. The stiffness of pulmonary arteries is often elevated in PH [103, 104], which together with dysregulated vascular ECM deposition plays crucial roles in the pathological transformation of endothelial phenotypes during PH progression [105].

Pulmonary embolism is associated with an increased risk of right ventricular dilation and a 50% reduction in small venous volume [106] and diagnosis of the disease requires careful inspection of the pulmonary vascular tree for the presence of emboli [107]. Pulmonary embolism is characterized by vascular occlusion that causes an increase in pulmonary arterial pressure (>50 mm Hg), resulting in irregular blood circulation and ventilation [108, 109]. Pulmonary fibrosis is also a lethal condition that is closely associated with functional changes in lung ECs. In a classic model of pulmonary fibrosis in mice, bleomycin-induced lung injury leads to massive lung inflammation, further leading to endothelial-to-mesenchymal transition, myofibroblast activation, and excessive deposition of ECM such as fibrillar collagen, which ultimately results in progressive fibrogenic transformation that replaces functional lung tissues with scar tissues [110]. Pulmonary fibrosis is associated with aberrant vascular remodeling, with significant decrease in vessel density and development of intimal fibrosis [111, 112]. Furthermore, many studies have identified perivascular cells as the main cell source of myofibroblasts in lung fibrogenesis [113–115].

Similar to pulmonary vascular involvement in lung development and regeneration, pulmonary vascular diseases are intricately linked to injury and dysregulation of the lung epithelium and stroma. Aberrant remodeling of the pulmonary parenchyma, vasculature, stroma and ECM usually forms a pathogenic feedback loop that deteriorate lung health in a progressive manner.

## 12.5 Conclusion

The lung is a highly vascularized organ, which specializes in gas exchange. Moreover, the lung-specific ECs involve in reparative mechanisms of the lung upon injury and ensure optimal functional recovery. Therefore, the attempts to establish a vascular network within bioengineered lung tissues open the opportunity to understand how these lung-specific ECs are being specified and unravel the epithelial-endothelial crosstalk mechanism during lung development and injury repair. From the regenerative medicine perspective, co-transplantation of ECs together with the lung epithelium is essential for connecting transplanted cells with the host lung tissue, thus improving survival and engraftment and driving maturation of the transplanted lung cells. From the tissue engineering perspective, biofabricated lung tissues with pre-established vasculature are more likely to mimic the native pathophysiological conditions compared to lung epithelial tissue culture alone. Thus systems, such as the lung-on-a-chip, represent more effective platforms for respiratory disease modeling and therapeutic development. Overall, engineering pulmonary vasculature is an essential pillar in the efforts to solve current limitations in lung tissue repair and regeneration, and will pave new ways towards alleviating respiratory diseases and overcome shortage of donor lung grafts.

## References

1. Carmeliet, P., *Angiogenesis in health and disease.* Nature Medicine, 2003. **9**(6): p. 653–660.
2. Weibel, E.R. and D.M. Gomez, *Architecture of the human lung. Use of quantitative methods establishes fundamental relations between size and number of lung structures.* Science, 1962. **137**(3530): p. 577–85.
3. Ochs, M., et al., *The number of alveoli in the human lung.* Am J Respir Crit Care Med, 2004. **169**(1): p. 120–4.
4. Parera, M.C., et al., *Distal angiogenesis: a new concept for lung vascular morphogenesis.* Am J Physiol Lung Cell Mol Physiol, 2005. **288**(1): p. L141–9.
5. Schachtner, S.K., Y. Wang, and H. Scott Baldwin, *Qualitative and quantitative analysis of embryonic pulmonary vessel formation.* Am J Respir Cell Mol Biol, 2000. **22**(2): p. 157–65.
6. Thébaud, B. and S.H. Abman, *Bronchopulmonary dysplasia: where have all the vessels gone? Roles of angiogenic growth factors in chronic lung disease.* Am J Respir Crit Care Med, 2007. **175**(10): p. 978–85.
7. Hall, S.M., et al., *Prenatal origins of human intrapulmonary arteries: formation and smooth muscle maturation.* Am J Respir Cell Mol Biol, 2000. **23**(2): p. 194–203.
8. Shalaby, F., et al., *Failure of blood-Island formation and vasculogenesis in Flk-1-deficient mice.* Nature, 1995. **376**(6535): p. 62–6.
9. Yamaguchi, T.P., et al., *flk-1, an flt-related receptor tyrosine kinase is an early marker for endothelial cell precursors.* Development, 1993. **118**(2): p. 489–98.
10. Healy, A.M., et al., *VEGF is deposited in the subepithelial matrix at the leading edge of branching airways and stimulates neovascularization in the murine embryonic lung.* Dev Dyn, 2000. **219**(3): p. 341–52.
11. Shu, W., et al., *Wnt7b regulates mesenchymal proliferation and vascular development in the lung.* Development, 2002. **129**(20): p. 4831–42.
12. Peng, T., et al., *Coordination of heart and lung co-development by a multipotent cardiopulmonary progenitor.* Nature, 2013. **500**(7464): p. 589–92.
13. Gebb, S.A. and J.M. Shannon, *Tissue interactions mediate early events in pulmonary vasculogenesis.* Dev Dyn, 2000. **217**(2): p. 159–69.
14. Roberts, A.B. and M.B. Sporn, *Regulation of endothelial cell growth, architecture, and matrix synthesis by TGF-beta.* Am Rev Respir Dis, 1989. **140**(4): p. 1126–8.
15. Cai, J., et al., *BMP signaling in vascular diseases.* FEBS Lett, 2012. **586**(14): p. 1993–2002.
16. Bellusci, S., et al., *Fibroblast growth factor 10 (FGF10) and branching morphogenesis in the embryonic mouse lung.* Development, 1997. **124**(23): p. 4867–78.
17. Weaver, M., N.R. Dunn, and B.L. Hogan, *Bmp4 and Fgf10 play opposing roles during lung bud morphogenesis.* Development, 2000. **127**(12): p. 2695–704.
18. Chen, F., et al., *A retinoic acid-dependent network in the foregut controls formation of the mouse lung primordium.* J Clin Invest, 2010. **120**(6): p. 2040–8.
19. Goss, A.M., et al., *Wnt2 signaling is necessary and sufficient to activate the airway smooth muscle program in the lung by regulating myocardin/Mrtf-B and Fgf10 expression.* Dev Biol, 2011. **356**(2): p. 541–52.

20. Husain, A.N., N.H. Siddiqui, and J.T. Stocker, *Pathology of arrested acinar development in postsurfactant bronchopulmonary dysplasia.* Hum Pathol, 1998. **29**(7): p. 710–7.
21. McGrath-Morrow, S.A., et al., *Vascular endothelial growth factor receptor 2 blockade disrupts postnatal lung development.* Am J Respir Cell Mol Biol, 2005. **32**(5): p. 420–7.
22. Thébaud, B., et al., *Vascular endothelial growth factor gene therapy increases survival, promotes lung angiogenesis, and prevents alveolar damage in hyperoxia-induced lung injury: evidence that angiogenesis participates in alveolarization.* Circulation, 2005. **112**(16): p. 2477–86.
23. Miranda, J., et al., *Alveolar Capillary Dysplasia with Misalignment of Pulmonary Veins (ACD/MPV): A Case Series.* Case Reports in Critical Care, 2013. **2013**: p. 327250.
24. Sirkin, W., et al., *Alveolar capillary dysplasia: lung biopsy diagnosis, nitric oxide responsiveness, and bronchial generation count.* Pediatr Pathol Lab Med, 1997. **17**(1): p. 125–32.
25. Cangiarella, J., et al., *Congenital cystic adenomatoid malformation of the lung: insights into the pathogenesis utilizing quantitative analysis of vascular marker CD34 (QBEND-10) and cell proliferation marker MIB-1.* Mod Pathol, 1995. **8**(9): p. 913–8.
26. Levin, D.L., *Morphologic analysis of the pulmonary vascular bed in congenital left-sided diaphragmatic hernia.* J Pediatr, 1978. **92**(5): p. 805–9.
27. Nolan, D.J., et al., *Molecular signatures of tissue-specific microvascular endothelial cell heterogeneity in organ maintenance and regeneration.* Dev Cell, 2013. **26**(2): p. 204–19.
28. Rafii, S., J.M. Butler, and B.S. Ding, *Angiocrine functions of organ-specific endothelial cells.* Nature, 2016. **529**(7586): p. 316–25.
29. Schildberg, F.A., et al., *Liver sinusoidal endothelial cells veto CD8 T cell activation by antigen-presenting dendritic cells.* Eur J Immunol, 2008. **38**(4): p. 957–67.
30. Naito, M. and E. Wisse, *Filtration effect of endothelial fenestrations on chylomicron transport in neonatal rat liver sinusoids.* Cell Tissue Res, 1978. **190**(3): p. 371–82.
31. Hennigs, J.K., et al., *Vascular Endothelial Cells: Heterogeneity and Targeting Approaches.* Cells, 2021. **10**(10): p. 2712.
32. Ding, B.S., et al., *Endothelial-derived angiocrine signals induce and sustain regenerative lung alveolarization.* Cell, 2011. **147**(3): p. 539–53.
33. Gillich, A., et al., *Capillary cell-type specialization in the alveolus.* Nature, 2020. **586**(7831): p. 785–789.
34. Liu, S., J. Cimprich, and B.M. Varisco, *Mouse pneumonectomy model of compensatory lung growth.* Journal of visualized experiments: JoVE, 2014(94): p. 52294.
35. Hogan, B.L., et al., *Repair and regeneration of the respiratory system: complexity, plasticity, and mechanisms of lung stem cell function.* Cell Stem Cell, 2014. **15**(2): p. 123–38.
36. Mammoto, T., M. Muyleart, and A. Mammoto, *Endothelial YAP1 in Regenerative Lung Growth through the Angiopoietin-Tie2 Pathway.* Am J Respir Cell Mol Biol, 2019. **60**(1): p. 117–127.
37. Cao, Z., et al., *Targeting of the pulmonary capillary vascular niche promotes lung alveolar repair and ameliorates fibrosis.* Nat Med, 2016. **22**(2): p. 154–62.
38. Pardali, E., et al., *TGF-β-Induced Endothelial-Mesenchymal Transition in Fibrotic Diseases.* Int J Mol Sci, 2017. **18**(10).
39. Jacob, A., et al., *Differentiation of Human Pluripotent Stem Cells into Functional Lung Alveolar Epithelial Cells.* Cell Stem Cell, 2017. **21**(4): p. 472–488.
40. Huang, S.X.L., et al., *Efficient generation of lung and airway epithelial cells from human pluripotent stem cells.* Nat Biotech, 2014. **32**(1): p. 84–91.
41. Ikuno, T., et al., *Efficient and robust differentiation of endothelial cells from human induced pluripotent stem cells via lineage control with VEGF and cyclic AMP.* PLoS One, 2017. **12**(3): p. e0173271.
42. Williams, I.M. and J.C. Wu, *Generation of Endothelial Cells From Human Pluripotent Stem Cells.* Arterioscler Thromb Vasc Biol, 2019. **39**(7): p. 1317–1329.
43. Hawkins, F., et al., *Prospective isolation of NKX2-1-expressing human lung progenitors derived from pluripotent stem cells.* J Clin Invest, 2017. **127**(6): p. 2277–2294.

44. Mammoto, T., et al., *Acceleration of Lung Regeneration by Platelet-Rich Plasma Extract through the Low-Density Lipoprotein Receptor-Related Protein 5-Tie2 Pathway.* American journal of respiratory cell and molecular biology, 2016. **54**(1): p. 103–113.
45. Lee, J.H., et al., *Lung stem cell differentiation in mice directed by endothelial cells via a BMP4-NFATc1-thrombospondin-1 axis.* Cell, 2014. **156**(3): p. 440–55.
46. Ng, W.H., et al., *Recapitulate Human Cardio-pulmonary Co-development Using Simultaneous Multilineage Differentiation of Pluripotent Stem Cells.* bioRxiv, 2021: p. 2021.03.03.433714.
47. Drakhlis, L., et al., *Human heart-forming organoids recapitulate early heart and foregut development.* Nat Biotechnol, 2021. **39**(6): p. 737–746.
48. Lippmann, E.S., et al., *Derivation of blood-brain barrier endothelial cells from human pluripotent stem cells.* Nature biotechnology, 2012. **30**(8): p. 783–791.
49. Giacomelli, E., et al., *Three-dimensional cardiac microtissues composed of cardiomyocytes and endothelial cells co-differentiated from human pluripotent stem cells.* Development (Cambridge, England), 2017. **144**(6): p. 1008–1017.
50. Giacomelli, E., et al., *Co-Differentiation of Human Pluripotent Stem Cells-Derived Cardiomyocytes and Endothelial Cells from Cardiac Mesoderm Provides a Three-Dimensional Model of Cardiac Microtissue.* Curr Protoc Hum Genet, 2017. **95**: p. 21.9.1–21.9.22.
51. Helle, E., et al., *HiPS-Endothelial Cells Acquire Cardiac Endothelial Phenotype in Co-culture With hiPS-Cardiomyocytes.* Front Cell Dev Biol, 2021. **9**: p. 715093.
52. Holloway, E.M., et al., *Differentiation of Human Intestinal Organoids with Endogenous Vascular Endothelial Cells.* Dev Cell, 2020. **54**(4): p. 516–528.e7.
53. Shi, Y., et al., *Vascularized human cortical organoids (vOrganoids) model cortical development in vivo.* PLoS biology, 2020. **18**(5): p. e3000705–e3000705.
54. Takebe, T., et al., *Vascularized and functional human liver from an iPSC-derived organ bud transplant.* Nature, 2013. **499**(7459): p. 481–484.
55. Takebe, T., et al., *Generation of a vascularized and functional human liver from an iPSC-derived organ bud transplant.* Nature Protocols, 2014. **9**(2): p. 396–409.
56. Takebe, T., et al., *Vascularized and Complex Organ Buds from Diverse Tissues via Mesenchymal Cell-Driven Condensation.* Cell Stem Cell, 2015. **16**(5): p. 556–65.
57. Liu, Y., et al., *A novel prevascularized tissue-engineered chamber as a site for allogeneic and xenogeneic islet transplantation to establish a bioartificial pancreas.* PLoS One, 2020. **15**(12): p. e0234670.
58. Song, W., et al., *Engineering transferrable microvascular meshes for subcutaneous islet transplantation.* Nature Communications, 2019. **10**(1): p. 4602.
59. Nalbach, L., et al., *Improvement of islet transplantation by the fusion of islet cells with functional blood vessels.* EMBO Mol Med, 2021. **13**(1): p. e12616.
60. Ott, H.C., et al., *Regeneration and orthotopic transplantation of a bioartificial lung.* Nat Med, 2010. **16**(8): p. 927–933.
61. Ott, H.C., et al., *Perfusion-decellularized matrix: using nature's platform to engineer a bioartificial heart.* Nat Med, 2008. **14**(2): p. 213–221.
62. Petersen, T.H., et al., *Matrix composition and mechanics of decellularized lung scaffolds.* Cells Tissues Organs, 2012. **195**(3): p. 222–231.
63. Cortiella, J., et al., *Influence of acellular natural lung matrix on murine embryonic stem cell differentiation and tissue formation.* Tissue Eng Part A, 2010. **16**(8): p. 2565–80.
64. Ren, X., et al., *Engineering pulmonary vasculature in decellularized rat and human lungs.* Nature Biotechnology, 2015. **33**(10): p. 1097–102.
65. Dorrello, N.V., et al., *Functional vascularized lung grafts for lung bioengineering.* Science Advances, 2017. **3**(8): p. e1700521.
66. Huh, D., et al., *Reconstituting organ-level lung functions on a chip.* Science, 2010. **328**(5986): p. 1662–1668.
67. Huh, D., et al., *A human disease model of drug toxicity-induced pulmonary edema in a lung-on-a-chip microdevice.* Sci Transl Med, 2012. **4**(159): p. 159ra147.

68. Stucki, A.O., et al., *A lung-on-a-chip array with an integrated bio-inspired respiration mechanism.* Lab Chip, 2015. **15**(5): p. 1302–10.
69. Stucki, J.D., et al., *Medium throughput breathing human primary cell alveolus-on-chip model.* Scientific Reports, 2018. **8**(1): p. 14359.
70. Jain, A., et al., *Primary Human Lung Alveolus-on-a-chip Model of Intravascular Thrombosis for Assessment of Therapeutics.* Clin Pharmacol Ther, 2018. **103**(2): p. 332–340.
71. Zamprogno, P., et al., *Second-generation lung-on-a-chip with an array of stretchable alveoli made with a biological membrane.* Communications Biology, 2021. **4**(1): p. 168.
72. Guan, M., et al., *Development of alveolar-capillary-exchange (ACE) chip and its application for assessment of PM(2.5)-induced toxicity.* Ecotoxicol Environ Saf, 2021. **223**: p. 112601.
73. Zhang, B., et al., *Biodegradable scaffold with built-in vasculature for organ-on-a-chip engineering and direct surgical anastomosis.* Nature materials, 2016. **15**(6): p. 669–678.
74. Morgan, J.T., et al., *Fabrication of centimeter-scale and geometrically arbitrary vascular networks using in vitro self-assembly.* Biomaterials, 2019. **189**: p. 37–47.
75. Lewis, K.J., et al., *In vitro model alveoli from photodegradable microsphere templates.* Biomater Sci, 2015. **3**(6): p. 821–32.
76. Horváth, L., et al., *Engineering an in vitro air-blood barrier by 3D bioprinting.* Scientific Reports, 2015. **5**(1): p. 7974.
77. Kolesky, D.B., et al., *Three-dimensional bioprinting of thick vascularized tissues.* Proc Natl Acad Sci U S A, 2016. **113**(12): p. 3179–84.
78. Gao, G., et al., *Coaxial Cell Printing of Freestanding, Perfusable, and Functional In Vitro Vascular Models for Recapitulation of Native Vascular Endothelium Pathophysiology.* Adv Healthc Mater, 2018. **7**(23): p. e1801102.
79. Grigoryan, B., et al., *Multivascular networks and functional intravascular topologies within biocompatible hydrogels.* Science (New York, N.Y.), 2019. **364**(6439): p. 458–464.
80. Lee, A., et al., *3D bioprinting of collagen to rebuild components of the human heart.* Science, 2019. **365**(6452): p. 482–487.
81. Mirdamadi, E., et al., *FRESH 3D Bioprinting a Full-Size Model of the Human Heart.* ACS Biomaterials Science & Engineering, 2020. **6**(11): p. 6453–6459.
82. Goss, K.N., et al., *Early Pulmonary Vascular Disease in Young Adults Born Preterm.* Am J Respir Crit Care Med, 2018. **198**(12): p. 1549–1558.
83. Naumburg, E., et al., *Risk factors for pulmonary arterial hypertension in children and young adults.* Pediatr Pulmonol, 2017. **52**(5): p. 636–641.
84. Galiè, N., et al., *2015 ESC/ERS Guidelines for the diagnosis and treatment of pulmonary hypertension: The Joint Task Force for the Diagnosis and Treatment of Pulmonary Hypertension of the European Society of Cardiology (ESC) and the European Respiratory Society (ERS): Endorsed by: Association for European Paediatric and Congenital Cardiology (AEPC), International Society for Heart and Lung Transplantation (ISHLT).* Eur Heart J, 2016. **37**(1): p. 67–119.
85. Dornia, C., et al., *Multidetector computed tomography for detection and characterization of pulmonary hypertension in consideration of WHO classification.* J Comput Assist Tomogr, 2012. **36**(2): p. 175–80.
86. Coste, F., et al., *CT evaluation of small pulmonary vessels area in patients with COPD with severe pulmonary hypertension.* Thorax, 2016. **71**(9): p. 830–7.
87. Reiter, G., et al., *Magnetic resonance-derived 3-dimensional blood flow patterns in the main pulmonary artery as a marker of pulmonary hypertension and a measure of elevated mean pulmonary arterial pressure.* Circ Cardiovasc Imaging, 2008. **1**(1): p. 23–30.
88. Hueper, K., et al., *Pulmonary Microvascular Blood Flow in Mild Chronic Obstructive Pulmonary Disease and Emphysema. The MESA COPD Study.* Am J Respir Crit Care Med, 2015. **192**(5): p. 570–80.
89. McLaughlin, V.V., et al., *ACCF/AHA 2009 expert consensus document on pulmonary hypertension a report of the American College of Cardiology Foundation Task Force on Expert Consensus Documents and the American Heart Association developed in collaboration*

*with the American College of Chest Physicians; American Thoracic Society, Inc.; and the Pulmonary Hypertension Association.* J Am Coll Cardiol, 2009. **53**(17): p. 1573–619.
90. Montani, D., et al., *Pulmonary arterial hypertension.* Orphanet Journal of Rare Diseases, 2013. **8**(1): p. 97.
91. Augustine, D.X., et al., *Echocardiographic assessment of pulmonary hypertension: a guideline protocol from the British Society of Echocardiography.* Echo research and practice, 2018. **5**(3): p. G11–G24.
92. Wilson, D.W., et al., *Mechanisms and pathology of monocrotaline pulmonary toxicity.* Crit Rev Toxicol, 1992. **22**(5–6): p. 307–25.
93. Gomez-Arroyo, J.G., et al., *The monocrotaline model of pulmonary hypertension in perspective.* Am J Physiol Lung Cell Mol Physiol, 2012. **302**(4): p. L363–9.
94. Kay, J.M., P. Harris, and D. Heath, *Pulmonary hypertension produced in rats by ingestion of Crotalaria spectabilis seeds.* Thorax, 1967. **22**(2): p. 176–179.
95. Rosenberg, H.C. and M. Rabinovitch, *Endothelial injury and vascular reactivity in monocrotaline pulmonary hypertension.* Am J Physiol, 1988. **255**(6 Pt 2): p. H1484–91.
96. Sehgal, P.B. and S. Mukhopadhyay, *Dysfunctional intracellular trafficking in the pathobiology of pulmonary arterial hypertension.* Am J Respir Cell Mol Biol, 2007. **37**(1): p. 31–7.
97. Huang, J., et al., *Progressive endothelial cell damage in an inflammatory model of pulmonary hypertension.* Exp Lung Res, 2010. **36**(1): p. 57–66.
98. Taraseviciene-Stewart, L., et al., *Inhibition of the VEGF receptor 2 combined with chronic hypoxia causes cell death-dependent pulmonary endothelial cell proliferation and severe pulmonary hypertension.* Faseb j, 2001. **15**(2): p. 427–38.
99. Toba, M., et al., *Temporal hemodynamic and histological progression in Sugen5416/hypoxia/normoxia-exposed pulmonary arterial hypertensive rats.* Am J Physiol Heart Circ Physiol, 2014. **306**(2): p. H243–50.
100. Nicolls, M.R., et al., *New models of pulmonary hypertension based on VEGF receptor blockade-induced endothelial cell apoptosis.* Pulm Circ, 2012. **2**(4): p. 434–42.
101. Chen, Y., et al., *A novel rat model of pulmonary hypertension induced by mono treatment with SU5416.* Hypertension Research, 2020. **43**(8): p. 754–764.
102. Mizuno, S., et al., *Severe pulmonary arterial hypertension induced by SU5416 and ovalbumin immunization.* American journal of respiratory cell and molecular biology, 2012. **47**(5): p. 679–687.
103. Sanz, J., et al., *Evaluation of pulmonary artery stiffness in pulmonary hypertension with cardiac magnetic resonance.* JACC Cardiovasc Imaging, 2009. **2**(3): p. 286–95.
104. Hunter, K.S., S.R. Lammers, and R. Shandas, *Pulmonary vascular stiffness: measurement, modeling, and implications in normal and hypertensive pulmonary circulations.* Comprehensive Physiology, 2011. **1**(3): p. 1413–1435.
105. Scarritt, M.E., et al., *Hypertensive rat lungs retain hallmarks of vascular disease upon decellularization but support the growth of mesenchymal stem cells.* Tissue engineering. Part A, 2014. **20**(9–10): p. 1426–1443.
106. Minhas, J., et al., *Loss of Pulmonary Vascular Volume as a Predictor of Right Ventricular Dysfunction and Mortality in Acute Pulmonary Embolism.* Circ Cardiovasc Imaging, 2021. **14**(9): p. e012347.
107. Morrone, D. and V. Morrone, *Acute Pulmonary Embolism: Focus on the Clinical Picture.* kcj, 2018. **48**(5): p. 365–381.
108. Dantzker, D.R. and J.S. Bower, *Alterations in gas exchange following pulmonary thromboembolism.* Chest, 1982. **81**(4): p. 495–501.
109. Fernandes, C.J., et al., *Pulmonary Embolism and Gas Exchange.* Respiration, 2019. **98**(3): p. 253–262.
110. Kato, S., et al., *Changes in pulmonary endothelial cell properties during bleomycin-induced pulmonary fibrosis.* Respiratory Research, 2018. **19**(1): p. 127.
111. Ebina, M., et al., *Heterogeneous increase in CD34-positive alveolar capillaries in idiopathic pulmonary fibrosis.* Am J Respir Crit Care Med, 2004. **169**(11): p. 1203–8.

112. Colombat, M., et al., *Pulmonary vascular lesions in end-stage idiopathic pulmonary fibrosis: Histopathologic study on lung explant specimens and correlations with pulmonary hemodynamics.* Hum Pathol, 2007. **38**(1): p. 60–5.
113. Greenhalgh, S.N., J.P. Iredale, and N.C. Henderson, *Origins of fibrosis: pericytes take centre stage.* F1000prime reports, 2013. **5**: p. 37–37.
114. Hung, C., et al., *Role of lung pericytes and resident fibroblasts in the pathogenesis of pulmonary fibrosis.* Am J Respir Crit Care Med, 2013. **188**(7): p. 820–30.
115. Cao, Z., et al., *Targeting the vascular and perivascular niches as a regenerative therapy for lung and liver fibrosis.* Sci Transl Med, 2017. **9**(405).

# Chapter 13
# An Overview of Organ-on-a-Chip Models for Recapitulating Human Pulmonary Vascular Diseases

**Trieu Nguyen and Fakhrul Ahsan**

## 13.1 Introduction

Traditionally, animal models have been used for recapitulating human physiology and for studying the pathological basis of many diseases affecting humankind. Indeed, over the centuries, animal models helped advance our understanding of the biology and pathology of drug therapy for humans. However, with the advent of genomics and pharmacogenomics, we now know that conventional models cannot accurately capture the pathological conditions and biological processes in humans, although humans share many physiological and anatomical features with many animals [1–3]. Species to species variation have raised concerns about the validity and suitability of animal models for studying human conditions. Over the past decade, the development and advances in microfabrication and biomaterials have spurred the growth in micro-engineered tissue and organ models (organs-on-a-chip, OoC) as alternatives to animal and cellular models [4]. This state-of-the-art technology has been used to emulate human physiology for investigating multitudes of cellular and biomolecular processes implicated in the pathological basis of disease (Fig. 13.1) [4]. Because of their tremendous potential, OoC-based models have been listed as one of the top 10 emerging technologies in the 2016 World Economic Forum [2].

T. Nguyen · F. Ahsan (✉)
Department of Pharmaceutical and Biomedical Sciences, California Northstate University, Elk Grove, CA, USA

East Bay Institute for Research and Education, Mather, CA, USA
e-mail: Fakhrul.Ahsan@cnsu.edu

 
C. M. Magin (ed.), *Engineering Translational Models of Lung Homeostasis and Disease*, Advances in Experimental Medicine and Biology 1413,
https://doi.org/10.1007/978-3-031-26625-6_13

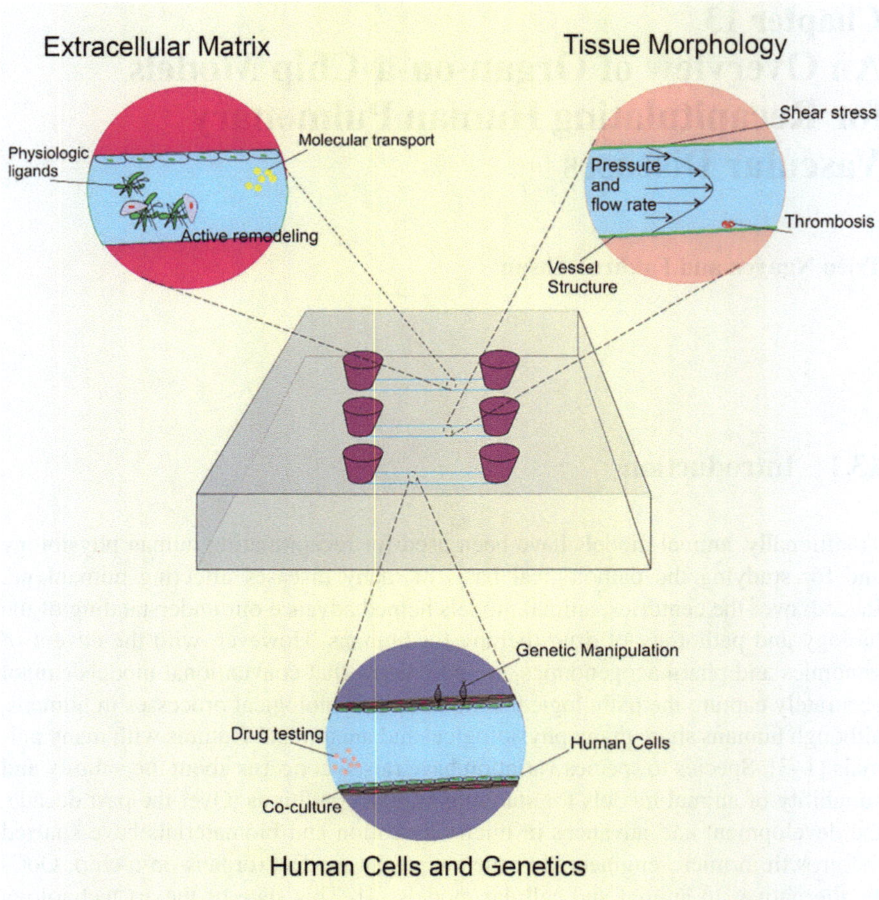


**Fig. 13.1** A microfluidic device for modelling cardiovascular pathophysiology, which can be used for patterning and recapitulating altered tissue morphology and abnormal hemodynamics associated with cardiovascular diseases. (The figure is adapted from Doherty, et al. [4], an open-access journal (Front. Bioeng. Biotechnol))

OoCs can simulate the physiological microenvironment on a miniaturized device in an organ-like fashion and thus have been proposed as a robust alternative experimental tool. This chapter summarizes recent developments surrounding the concept and application OoC in studying biological systems and drug therapy. Here, we briefly described OoC-based models for pulmonary vasculature diseases, especially pulmonary arterial hypertension (PAH). Finally, we explained the future application of OoC-based models that can be deployed to reproduce pulmonary vascular diseases.

## 13.2 Microfluidics and Organ-on-a-Chip

### *13.2.1 Concepts*

Microfluidics is the science of fluid manipulation on a microscale. Due to strong growth in the science surrounding micro, nanofabrication, material and optics, this microengineering-based technology has been used widely in an array of specialities, including biomedical and pharmaceutical sciences, over the past two decades. The surge in microfluidic research stems from the development of lab-on-a-chip (LoC) devices, a technology that may propel a paradigm shift in chemistry, biology and medicine as integrated circuits did in microelectronics [5, 6]. LoCs are microsystems that can encapsulate an entire biological system or a chemical laboratory in a single miniaturized device. This miniaturization was possible because of advances in microengineering to create microfluidic channels with distinct accessories such as filters, valves, mixers, and pumps [7]. LoCs are made of distinct materials such as glass [8, 9] , PDMS, silicon, fused silica [10], and thermal plastics [11], which are compatible with materials of biological origin and do not adversely affect the phenotype of cells or native structure of macromolecules. They have been used to recapitulate various cellular, and molecular cues and organ-like functions and thereby can be engineered to be in tune with the altered physiological conditions that give rise to chronic diseases [12, 13]. Thanks to the state-of-the-art micro-nano fabrication technology that can recreate multicellular structures, tissue-like interfaces, and biological microenvironments in miniaturized devices. Compared with conventional 2D or 3D cultures, OoCs can capture the tissue and organ level functionality with enhanced accuracy, precision, and reproducibility. Thus, these systems can be utilized for studying the underlying mechanisms of disease pathologies and screening both drugs and drug delivery systems [14].

#### Microfluidics in Vascular Biology

The vascular system consists of a network of veins, capillaries and arteries that allow the blood to circulate across the body, which are responsible for performing many functions, including maintaining solutes and water balance between the blood and tissue compartment and adjusting changes in lateral pressure. OoCs can mimic the microphysiological action and three-dimensional (3D) microstructure of human organs more directly and specifically compared to the conventional two-dimensional (2D) cultures and animal models. In addition to supplying nutrients and oxygen to the cultured tissue by perfusing the culture medium, vascularization of OoC can also contribute to the formation of organ-specific microenvironments and microphysiological function by recreating the microvascular with particular barrier function compared to those of in vivo. Microfluidics has emerged as a powerful tool for developing OoC, which can offer precise control over various aspects of the cellular microenvironment, such as a distinct profile of flow patterns, a gradient of various growth factors, and mechanical features of multifaceted biomaterials. These innovative features of OoCs are likely to stimulate major growth in the development of biomimicking vascularized microtissue models.

**Patterning Microvascular Networks**

Various methodologies and materials are used for the fabrication of OoC and construction of vascular networks, including the use of elastomer-based photolithography, hydrogel-based bio-printing and other polymeric devices. First reported by the Whiteside lab, polydimethylsiloxane (PDMS), an elastomer, has been used for the fabrication and rapid prototyping of microfluidic chips [15]. An endothelialized PDMS-based device has been developed for investigating the microvascular interactions in hematologic disease. The device contained channels mimicking the vascular networks [16]. Further, various preparation and fabrication of PDMS-based microfluidics have been summarized in a book edited Marco Rasponi [17]. Although PDMS-based devices have been widely used in academic research labs, they are not amenable to large-scale fabrication. In fact, scale-up and large-scale production have been a significant challenge to the commercialization of PDMS-based OoC models. Thermoplastics such as cyclic olefin copolymer (COC) and cyclic olefin polymer (COP) have been proposed as alternative for commercialization of OoC models [6, 11, 18, 19].

Hydrogels are cross-linked hydrophilic polymer networks that contain a high amount of water content and they have soft textures similar to tissue structures [20]. The culture of cells in a soft network creates a more in vivo-like environment for cell to grow when compared with cells than on hard-plastic like surfaces [21]. Hydrogel has been used as a soft growth matrix for cell cultures to replace hard polymers such as polyester or polycarbonate used as materials in microfluidics and transwells [22, 23]. Jepsen and colleagues developed 3D printed stackable titer plate inserts supporting three interconnected tissue models for drug transport studies [24], in which hydrogel can be cast into the inserts and cells can be cultured in or on the hydrogel.

Moreover, multi-functional scaffolds have recently drawn attention in the field of tissue engineering because of their ability to control cell behavior by providing various cues, including electrical, chemical, and mechanical cues. However, hydrogel-based models and mass fabrication of scaffolds using biocompatible materials remain challenging. Mohanty and co-authors [25] reported a simple 3D printing approach, using polymer casting and supercritical fluid technique to fabricate 3D interpenetrating polymer network (IPN) scaffold of silicon-Poly(2-hydroxyethyl methacrylate)-co-Poly(ethylene glycol) methyl ether acrylate (pHEMA-co-PEGMEA). The pHEMA-co-PEGMEA IPN materials were used to support the growth of human mesenchymal stem cells (hMSC) (Fig. 13.2). By using those scaffolds, the authors reported high cell viability and metabolic activity over a three-week period.

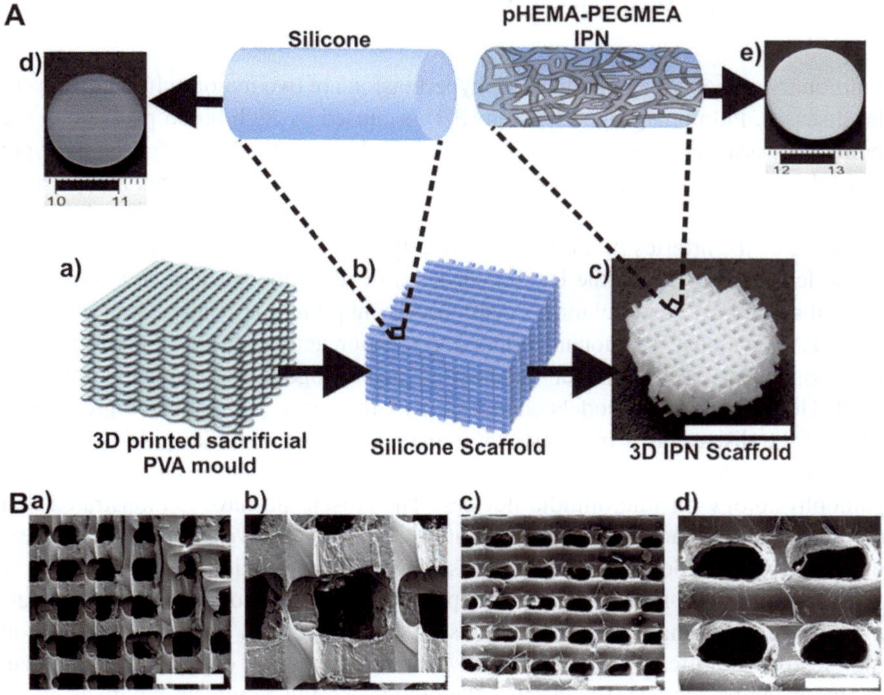


**Fig. 13.2** (**a**) The fabrication process for 3D IPN. (**b**) SEM images of 3D IPN scaffolds. (Reprinted with permission from [25]. Copyright 2016 American Chemical Society)

While PDMS is not suitable for the commercialization of OoC devices because of fabrication cost, material price, and regulatory restrictions [26], polymers such as polystyrene (PS), polycarbonate (PC), polymethyl methacrylate (PMMA), COC and COP can be used in large-scale manufacturing of OoCs. These materials are used to address various needs, such as UV transmittance, water and moisture absorption, thermal resistance, and strong mechanical properties. Of these polymers, PMMA, PS, and PC have been extensively used as the bonding substrate materials for OoC devices because of their solid mechanical property, excellent optical transparency, and reduced auto-fluorescence background [6, 27]. Porous PC membranes are generally sandwiched between microfluidic layers in OoC devices to emulate tissue-tissue interfaces [28]. PS is greatly biocompatible and a good substrate for cell adhesion and growth. COP and COC have excellent optical properties, permitting high-quality fluorescence imaging. They are suitable for applying in experimental settings where elevated temperature is required, such as polymerase chain reaction (PCR) and loop-mediated isothermal amplification (LAMP) [11, 29]. Two common methods that are used for the fabrication OoCs include hot embossing and injection molding. Depending on the size features of the microfluidic devices, the master molds can be fabricated in or outside the cleanrooms [19]. For rapid prototypes, micro-milling or laser micromachining can be used to produce the master molds without the expensive set-ups required in a standard cleanroom.

## 13.3 OoC for Pulmonary Vascular Diseases

Pulmonary embolism and pulmonary hypertension are two major pulmonary vascular diseases. Pulmonary hypertension (PH) is caused by high blood pressure in the pulmonary arteries, which carry blood from the heart to the lungs. In PH, the right heart dangerously enlarges and thus cannot pump blood into the lung, and patients die from right heart failure. Pulmonary embolism occurs due to blood clots blocking branches of the arteries in the lungs, normally following a thrombosis in the veins of the leg. Thrombosis is the leading cause of death in the western countries, with more than half of the population's death involving pulmonary embolism [30].

In PAH patients, pulmonary arteries get stiffer and occluded. Animal models have long been used to better understand the pathogenesis of and therapies for PAH. However, animal models cannot portray the entire pathophysiological spectrum of the disease as it manifests in humans [3]. Recent studies [31–34] suggest that microfluidics and OoC models can be employed for encapsulating the PAH pathophysiology on microfluidic devices. The Ahsan lab have demonstrated that cell-laden multi-channel chips can potentially be used for investigating pathophysiology of PAH and delineating the sex disparity and age-relation in PAH [33, 34].

Pulmonary embolism occurs when a pulmonary vessel is occluded by a circulating blood clot originating from a distant site, such as the legs. This type of occlusion may lead to significant damage to both heart and lung and often death. The size, geometry, and mechanical properties of a blood vessel in conjunction with an endothelial surface can substantially impact blood flow across the pulmonary vasculature. Thus an engineered in vitro replica of 3D microvascular structure can aid in studying many disease processes such as the heterogeneity of endothelial cells, mechanisms of plaque formation and rupture and embolization. Recently, Szulcek and colleagues [35] reported custom-made microfluidic devices for investigating the whole blood-endothelial interactions and blood clot dynamics in real-time. The development of thrombus has been monitored in real-time and characterized by platelet and fibrin deposition. In a separate study, another work, Jain's group [36] reported a tortuosity-powered microfluidic device for assessing thrombosis and antithrombotic therapy. The device was reported to allow accurate measurement of coagulation, platelets, and fibrin contents.

## 13.4 Conclusion

Although it is easy to fabricate and requires less sophisticated equipment compared to glass and silicon microfabrication processes, PDMS-based microfluidic devices for OoC application suffer from several limitations, including mass fabrication and control of structural features of the channels. Cells seeding in sharp corners result in enormous variations in cell physiology [37]. Thus, microfabrication technology produces round-shaped devices, which can be accomplished either by using hexane as a solvent or partially cured PDMS [30].

Current microfluidic platforms for OoC applications mostly use constant pressure to introduce fluidic flow, such as blood flow. A pulsatile flow system should be considered for future studies that would allow studying interactions between fluid flows, shear stress, and tensile strength. In addition to in vitro experimental models, computation models of the microvascular system should be deployed for flow properties and influence stress on vascular systems.

## References


1. Hodge, R.D., et al., *Conserved cell types with divergent features in human versus mouse cortex.* Nature, 2019. **573**(7772): p. 61–68.
2. Wu, Q., et al., *Organ-on-a-chip: recent breakthroughs and future prospects.* BioMedical Engineering OnLine, 2020. **19**(1): p. 9.
3. Campbell, D.R., et al., *Engineering Tissue-Informed Biomaterials to Advance Pulmonary Regenerative Medicine.* Frontiers in Medicine, 2021. **8**: p. 383.
4. Doherty, E.L., et al., *Microfluidic and Organ-on-a-Chip Approaches to Investigate Cellular and Microenvironmental Contributions to Cardiovascular Function and Pathology.* Frontiers in Bioengineering and Biotechnology, 2021. **9**: p. 25.
5. Manz, A., N. Graber, and H.M. Widmer, *Miniaturized total chemical analysis systems: A novel concept for chemical sensing.* Sensors and Actuators B: Chemical, 1990. **1**(1): p. 244–248.
6. Nguyen, T., et al., *Point-of-care devices for pathogen detections: The three most important factors to realise towards commercialization.* TrAC Trends in Analytical Chemistry, 2020. **131**: p. 116004.
7. Reyes, D.R., et al., *Micro Total Analysis Systems. 1. Introduction, Theory, and Technology.* Analytical Chemistry, 2002. **74**(12): p. 2623–2636.
8. Nguyen, T., et al., *Highly enhanced energy conversion from the streaming current by polymer addition.* Lab on a Chip, 2013. **13**(16): p. 3210–3216.
9. Nguyen, T., et al., *Rotary-Atomizer Electric Power Generator.* Physical Review Applied, 2015. **3**(3): p. 034005.
10. Izadi, D., T. Nguyen, and L.J. Lapidus, *Complete Procedure for Fabrication of a Fused Silica Ultrarapid Microfluidic Mixer Used in Biophysical Measurements.* Micromachines, 2017. **8**(1).
11. Nguyen, T., et al., *Optimising the supercritical angle fluorescence structures in polymer microfluidic biochips for highly sensitive pathogen detection: a case study on Escherichia coli.* Lab on a Chip, 2019. **19**(22): p. 3825–3833.
12. Huh, D., et al., *Reconstituting Organ-Level Lung Functions on a Chip.* Science, 2010. **328**(5986): p. 1662–1668.
13. Bhatia, S.N. and D.E. Ingber, *Microfluidic organs-on-chips.* Nature Biotechnology, 2014. **32**(8): p. 760–772.
14. Rothbauer, M., et al., *Tomorrow today: organ-on-a-chip advances towards clinically relevant pharmaceutical and medical in vitro models.* Current Opinion in Biotechnology, 2019. **55**: p. 81–86.
15. Duffy, D.C., et al., *Rapid Prototyping of Microfluidic Systems in Poly(dimethylsiloxane).* Analytical Chemistry, 1998. **70**(23): p. 4974–4984.
16. Myers, D.R., et al., *Endothelialized microfluidics for studying microvascular interactions in hematologic diseases.* J Vis Exp, 2012(64).
17. Rasponi, M., *Organ-on-a-Chip.* 2022: Springer.
18. Vinayaka, A.C., et al., *Pathogen Concentration Combined Solid-Phase PCR on Supercritical Angle Fluorescence Microlens Array for Multiplexed Detection of Invasive Nontyphoidal Salmonella Serovars.* Analytical Chemistry, 2020. **92**(3): p. 2706–2713.

19. Nguyen, T., et al., *A Complete Protocol for Rapid and Low-Cost Fabrication of Polymer Microfluidic Chips Containing Three-Dimensional Microstructures Used in Point-of-Care Devices*. Micromachines, 2019. **10**(9): p. 624.
20. Butcher, D.T., T. Alliston, and V.M. Weaver, *A tense situation: forcing tumour progression*. Nature Reviews Cancer, 2009. **9**(2): p. 108–122.
21. Geiger, B. and A. Bershadsky, *Assembly and mechanosensory function of focal contacts*. Current Opinion in Cell Biology, 2001. **13**(5): p. 584–592.
22. Annabi, N., et al., *Engineered cell-laden human protein-based elastomer*. Biomaterials, 2013. **34**(22): p. 5496–5505.
23. DiMarco, R.L., et al., *Improvement of paracellular transport in the Caco-2 drug screening model using protein-engineered substrates*. Biomaterials, 2017. **129**: p. 152–162.
24. Leth Jepsen, M., et al., *3D Printed Stackable Titer Plate Inserts Supporting Three Interconnected Tissue Models for Drug Transport Studies*. Advanced Biosystems, 2020. **4**(7): p. 1900289.
25. Mohanty, S., et al., *3D Printed Silicone-Hydrogel Scaffold with Enhanced Physicochemical Properties*. Biomacromolecules, 2016. **17**(4): p. 1321–9.
26. Kristiansen, P.M., et al., *Thermoplastic Microfluidics*, in *Organ-on-a-Chip: Methods and Protocols*, M. Rasponi, Editor. 2022, Springer US: New York, NY. p. 39–55.
27. Miller, P.G. and M.L. Shuler, *Design and demonstration of a pumpless 14 compartment microphysiological system*. Biotechnology and bioengineering, 2016. **113**(10): p. 2213–2227.
28. Shah, P., et al., *A microfluidics-based in vitro model of the gastrointestinal human–microbe interface*. Nature communications, 2016. **7**(1): p. 1–15.
29. Ding, C., et al., *Biomedical Application of Functional Materials in Organ-on-a-Chip*. Frontiers in Bioengineering and Biotechnology, 2020. **8**(823).
30. Li, Y., *A microfluidic model for in vitro studies of thromboembolism and thrombolysis*. 2017, University of Toronto (Canada).
31. Wojciak-Stothard, B., et al., *A MICROFLUIDIC CHIP FOR PULMONARY ARTERIAL HYPERTENSION*. 2021.
32. Huisman, J., *Physiological vessel on chip model with integrated flow and oxygen control for in vitro small pulmonary artery studies*. 2021.
33. Trieu Nguyen and F. Ahsan. *A tissue chip device for investigating pediatric pulmonary arterial hypertension (PAH) pathophysiology and developing age-specific therapy*. in *The 25th International Conference on Miniaturized Systems for Chemistry and Life Sciences (μTAS 2021)*. 2021. Palm Spring, California, USA.
34. Al-Hilal, T.A., et al., *Pulmonary-arterial-hypertension (PAH)-on-a-chip: fabrication, validation and application*. Lab Chip, 2020. **20**(18): p. 3334–3345.
35. Manz, X.D., et al., *In Vitro Microfluidic Disease Model to Study Whole Blood-Endothelial Interactions and Blood Clot Dynamics in Real-Time*. J Vis Exp, 2020(159).
36. Luna, D.J., et al., *Tortuosity-powered microfluidic device for assessment of thrombosis and antithrombotic therapy in whole blood*. Sci Rep, 2020. **10**(1): p. 5742.
37. Fisher, A.B., et al., *Endothelial cellular response to altered shear stress*. American Journal of Physiology-Lung Cellular and Molecular Physiology, 2001. **281**(3): p. L529–L533.

# Chapter 14
# Clinical Translation of Engineered Pulmonary Vascular Models

**Yifan Yuan**

## 14.1 Introduction

Pulmonary vascular diseases, including acute respiratory distress syndrome (ARDS), pulmonary arterial hypertension (PAH), capillary leak syndrome, or even COVID-19, are a few causes of death that are increasing over the world [1, 2]. Pulmonary vascular disease is characterized by remodeling and loss of microvessels and is typically attributed to pathological responses in the vascular endothelium. Over the years, numerous preclinical animal models have been developed to elucidate the biological mechanisms underlying the progression of various human pulmonary vascular diseases (reviewed in [3–5] and [6–8]), with the hope of developing new therapies. Although these models capture complex interactions between injury, inflammation, and tissue repair in the lung vasculature, the significant species-specific variation in lung physiology limits their translational potential [4]. A systematic study performed to evaluate how well murine models mimic human diseases [9], demonstrated that the genomic responses to different inflammatory stresses are highly similar in humans, but these responses are not reproduced in the current mouse models. Additionally, numerous studies showed that a large number of drugs that were previously validated in animal models eventually failed during clinical trials, indicating a huge gap between preclinical animal models and human physiology and pathophysiology [10]. Thus, there is a critical need to develop an experimental modeling system that mimics human pulmonary vasculature that can simulate native cellular phenotypes and functions during homeostatic and diseased states.

Y. Yuan (✉)
Department of Medicine (Pulmonary), Department of Anesthesiology, Yale University, New Haven, CT, USA
e-mail: yifan.yuan@yale.edu



C. M. Magin (ed.), *Engineering Translational Models of Lung Homeostasis and Disease*, Advances in Experimental Medicine and Biology 1413,
https://doi.org/10.1007/978-3-031-26625-6_14

The development of translational tissue models, including the pulmonary vascular models, is centered on recapitulating human physiology. The lung vascular niche components, including extracellular matrix (ECM), oxygen tension, cyclic stretch, shear stress, and cell-cell crosstalk signals, are all of pivotal importance for regulating endothelial maturation and maintaining vascular homeostasis. However, their roles in pulmonary vascular tissue models have not been extensively discussed. This chapter will first review the impact of various niche components on pulmonary vascular biology and discuss methods to incorporate them to improve the fidelity of current tissue models. Next, it will highlight the advantages and disadvantages of existing models in studying pulmonary vascular disorders and provide perspectives on their translational potential.

## 14.2 Brief Overview of Pulmonary Vascular Physiology

Lung is the most vascular organ in our body and receives the full cardiac output. At any given time, the pulmonary circulation contains about 10% of the circulating blood volume. To allow efficient flow of this large amount of blood, the overall resistance in pulmonary vasculature needs to remain low as compared to their systemic counterparts. As a result, the pulmonary arterial pressure is low, normally at a level of about 9–16 mmHg. The pulmonary circulation system is highly elastic, with thinner walls and less muscular than vessels of systemic circulation. Any structural changes of the vasculature, including muscularization of arterioles, distributing arteries, and the blockage of microvasculature, could lead to progressive increase in pulmonary vascular resistance and pulmonary arterial hypertension (PAH). Numerous rodent models have been developed to study PAH such as exposure to hypoxia, subcutaneous or intraperitoneal injection of monocrotaline, injection of VEGFR inhibitor Sugen 5416, or the combination of hypoxia and injection of Sugen 5416. More extensive discussion for PAH models could be found in other review articles [11], however, these models do not fully recapitulate the initiation and progression of human vascular disease.

The primary function of the lung is to oxygenate blood and to remove carbon dioxide from systemic circulation. To achieve this, the lung spreads the blood into a very fine layer of intermeshed capillaries within the alveolar walls, allowing extraordinary proximity between perfused blood and inspired gas. This setup allows the compression of nearly 70 $m^2$ of gas-exchanging surface into a total volume of only 5–6 liters [12]. The thin barrier (600 nm–2 μm in thickness) at the gas-exchanging surface or between alveolar and capillary region is called blood-air barrier, which simply contains an epithelial layer, a basement membrane, and an endothelial layer. This barrier not only prevents air bubbles from forming in the blood and regulates blood from entering the alveoli, but also controls the influx of macromolecules, immune cells from the blood capillaries into the alveoli and maintains alveolar homeostasis. The disruption of the blood-air barrier has been a focus in many pulmonary vascular diseases, such as acute respiratory distress syndrome (ARDS)/acute lung injury (ALI), ventilator-induced lung injury, and even COVID-19. For example, dysfunction of pulmonary microvascular endothelial cells increases vascular/alveolar permeability, leading to extravascular leak of protein-rich edema, polymorphonuclear leukocyte influx, microvascular thrombosis, and further lung dysfunction (e.g., ALI, sepsis). Over the years, numerous pre-clinical and in vitro modeling systems have been developed to understand the biological mechanisms causing human ALI/ARDS (reviewed in [3, 4] and [6]). However, currently, there is no single modeling system that could mimic blood-air barrier and recapitulate most of the pathological features of human pulmonary vascular diseases.

## 14.3 Reconstituting Microenvironmental Cues to Improve Model Fidelity

The biochemical and biophysical cues in the pulmonary vascular niche are responsible for maintaining cellular phenotypes and functions (Fig. 14.1). Alterations in the lung vascular niche, such as changes in ECM composition, flow rates and

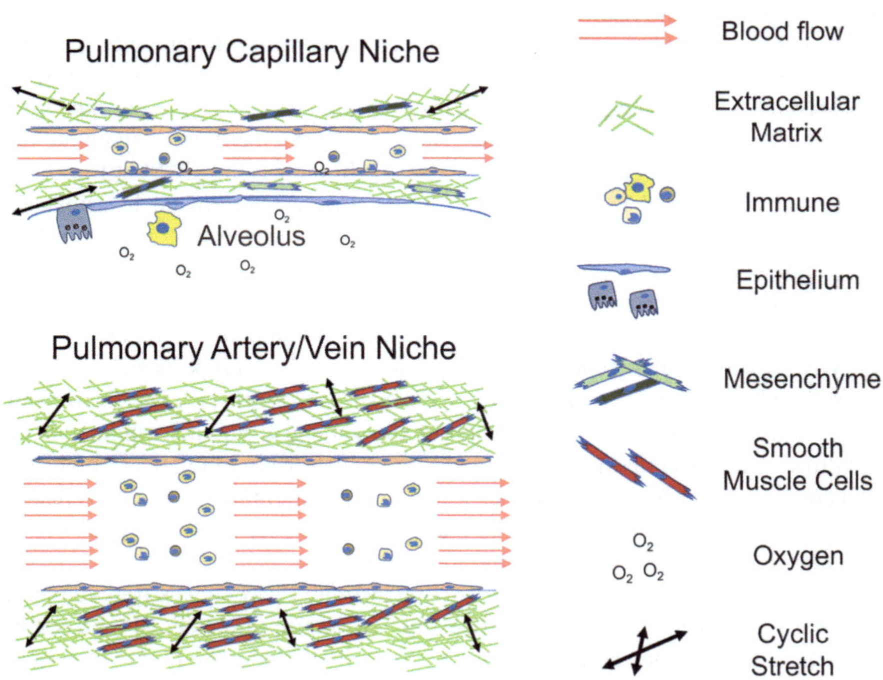


**Fig. 14.1** Key microenvironmental cues in the lung vasculature. In the pulmonary capillary, proper regulation of the niche components such as ECM basement membrane at the blood-air interface, crosstalk with cells in the alveolus and the bloodstream, cyclic stretch, hemodynamics, and oxygen tension are critical to maintaining lung microvascular homeostasis. In the large vessel, the key microenvironmental components are slightly different, and they include thick ECM substrate, crosstalk with smooth muscle cells in the intimal layer and immune cells in the bloodstream, hemodynamics, and mechanical stretch

patterns, cytokine concentrations, ECM substrate stiffness, and oxygen tension, can result in vascular barrier damage, leukocyte adhesion, platelet aggregation, and further disease progression. Under physiological niche conditions such as correct substrate stiffness and stimulation with anti-inflammatory cytokines, the vascular endothelium synthesizes factors (e.g., nitric oxide, prostacyclin, and thrombomodulin) to control thrombosis, barrier function, and vessel wall inflammation. In contrast, under pathophysiological conditions such as the release of pro-inflammatory cytokines TNFα, and IFNγ, the vasculature produces tissue factor, ICAM, and VCAM to promote thrombosis and inflammation and induce vascular barrier breakdown. Here, the impact of various niche components on endothelial phenotypes and functions will be discussed and means to reconstruct the native milieu in vitro to improve the fidelity of current vascular models will be introduced.

## 14.4 ECM Substrates

Extracellular matrix (ECM) is a collection of extracellular molecules that provide structural support and regulate cell adhesion and cell signaling. Both ECM components and their mechanical properties tightly correlate with cellular signaling transduction. For example, a simple coating with vascular basement membrane proteins fibronectin, laminin, and collagen could result in a marked change in the expression and activity of endothelial nitric oxide synthase (eNOS), a critical enzyme for vascular homeostasis, in endothelial cells [13–15]. In the case of ECM stiffness, the stiffer substrate could lead to an increased angiogenic outgrowth, invasion, and neovessel branching and a decreased barrier function through both myosin inhibition and matrix metalloproteinase (MMP)-dependent mechanisms [16]. In the lungs, healthy blood vessels have high compliance and low stiffness, and experience only small systolic pressure increases, with an accompanying relatively large increase in volume and reduced the wall shear stress. These properties also contribute to low flow and pressure pulsatility in downstream vessels [17, 18]. Conversely, stiffer vessels absorb a smaller fraction of the volume from each cardiac stroke, resulting in a substantial increase in pressure, which leads to relatively high flow and pressure pulsatility [19]. This means that matrix stiffening will result in changes in flow and pressure, which will likely lead to endothelial dysfunction as well as vascular remodeling of pulmonary microvessels.

Various methods have been utilized to reconstitute ECM signals in tissue models. Matrigel is the solubilized matrix secreted by mouse sarcoma cells and resembles the native basement membrane ECM microenvironment found in tissues. Various vascular model systems, such as organ chip models [20], organoids [21], and microfluidic-based systems [22] have incorporated Matrigel as a substitute for native ECM. However, there are concerns due to its non-human and cancer cell origin. Studies have then attempted to incorporate hydrogel supplemented with native ECM substrates into the vascular models. A recent study by Palikuqi and colleagues showed that the hydrogel composed of a mixture of laminin, entactin, and type-IV collagen displayed a similar impact as Matrigel in developing a perfusable and plastic vascular plexi [23], demonstrating the potential of using hydrogel for constructing vascular tissue models.

Although various hydrogels show some degrees of recapitulating native ECM, they still fail to reconstitute the structural, compositional, and biological features of native lung vasculature. Whole organ decellularization may open a door to developing a construct that can recapitulate a complete vascular and alveolar structure. We and others previously used quantitative proteomics to examine ECM components remaining after decellularization in rat and human lungs and showed an intact vascular architecture and near-native retention of ECM components such as laminins, proteoglycans, and other basement membrane proteins after decellularization [24–26]. We further developed protocols to seed and culture human epithelial and endothelial cells into the decellularized lungs [27, 28]. Leveraging single-cell RNA-sequencing (scRNAseq), we found that epithelial stem cells or endothelial cells cultured in these decellularized tissue constructs took on phenotypes of their native counterparts [29, 30]. Additionally, to determine the alveolar-capillary barrier function of the repopulated whole lung scaffolds, we developed a monitoring system that enables real-time measurement of microvascular permeability in repopulated whole lung scaffolds [31]. Leveraging these techniques, we established a pulmonary microvascular platform to model acute lung injury by culturing pulmonary microvascular endothelial cells in decellularized lung scaffolds, followed by introducing lipopolysaccharide (LPS) into the bioreactor. Within 24 h of LPS treatment, there was an increase in cell death, and upregulation of inflammation-induced endothelial dysfunction signals. Non-invasive mechanical measurements revealed a loss of barrier function as fluid escaped from the vasculature and into the interstitium or the air sacs [30]. These results suggest that the repopulated lung platform may mimic aspects of endothelial dysfunction during ALI.

Although we show some beneficial impact of the acellular lung matrix on mimicking the native microenvironment and modeling vascular diseases, the caveat is that the current decellularization strategies may not fully retain native matrix composition, including a significant loss of glycosaminoglycans (GAGs) [32–34]. Ren and colleagues have developed a metabolic glycan labeling approach to covalently incorporate click-reactive azide ligands into the native ECM to overcome this issue. The incorporated azides were preserved after decellularization and served as chemoselective ligands for bioconjugation with various substrates such as heparin [35]. This method provides a valuable tool to functionalize the ECM biomaterials and better represent the native ECM microenvironment.

## 14.5 Cell-Cell Crosstalk

In the pulmonary vascular niche, endothelial cells are surrounded by mesenchymal, epithelial, and immune cells, and they work jointly to regulate cellular maturation and maintain vascular homeostasis. Previous studies, including ours, have demonstrated that co-culture of mesenchymal cells, including human lung fibroblast, pericyte, or mesenchymal stem cells, with endothelial cells, could improve vessel growth, stabilization, and maturation in in vitro platforms, including organ chip models, organoid models and decellularized whole lung models [21, 36–38]. Additionally, co-culture with epithelial cells such as type II epithelial cells from the alveolus could improve the maturation in human pulmonary microvascular cells in vitro [39]. While these studies demonstrate the significant benefit of the co-culture systems, it remains questionable whether the cell-cell crosstalk signals from existing platforms represent their native counterparts, partially due to our incomplete understanding of cell populations in the lung. Single-cell transcriptomics (scRNAseq) is an emerging and powerful tool to identify cellular heterogeneity in a multicellular organism at a single cell level. Leveraging this technique, we and others recently determined that the human lung contains at least 58 cell populations, including 15 epithelial, 9 endothelial, 9 mesenchymal, and 25 immune populations [40–42]. With a further understanding of their localization through spatial transcriptomics or even proteomics using platforms such as DBiT-seq [43] or Seq-Scope [44], it would be possible to determine the cellular populations in any given milieu in the human lungs, which could be instructive to engineer co-culture models and recapitulate native cell-cell communications.

Cell-cell paracrine signals are critical to supporting microvascular cellular phenotype and physiological functions. Paracrine signals, including angiopoietin 1 (Ang1), keratin growth factor (KGF), and slit guidance ligand 2 (SLIT2) secreted by lung mesenchymal cells are essential to protect the integrity of the lung microvascular endothelium and prevent the influx of plasma and cells from the bloodstream into the alveolus/airways [45, 46]. Prostaglandin E2 (PGE2) produced by alveolar type II cells has been shown to reduce permeability in human pulmonary microvascular endothelium after being challenged by pharmacological insults [47]. Although the microvascular niche in adult lungs has been studied mechanistically, mostly during development or in response to discrete injuries, such as inhalation injury or lipopolysaccharide injection, we have an incomplete understanding of the paracrine signals that regulate the pulmonary microvasculature and maintain it in a homeostatic state in the adult animal, or human. Leveraging scRNAseq, we recently developed a computational tool, called Connectome, to understand cell-cell crosstalk soluble signals in human distal lungs [42, 48, 49]. We first used scRNAseq to identify cell populations in animal and human lungs and characterized signals coming from alveolar cells that were potentially being sensed by capillary endothelium [42]. We then mapped putative ligand-receptor interactions within and between cell types and assessed the significance of these interactions using the FANTOM5 database [48]. Leveraging this tool in combination with biochemical assays, it is possible to identify important locally acting factors that contribute to a functional human lung microvascular niche, to enable the development of an ex vivo vascular platform mimicking native lung.

## 14.6 Shear Stress

The lung is one of the most highly vascularized organs in the body and has a compliant and high-capacity vascular bed. The endothelium lining the blood vessels is highly sensitive to hemodynamic shear force that acts at the vascular luminal surface in the direction of blood flow (reviewed in [50]). Therefore, alteration of pulmonary hemodynamics is one of the essential factors for pulmonary vascular remodeling. For example, in vivo assessment of wall shear stress (WSS) from pulmonary hypertension (PH) patients reveals that WSS was significantly decreased in the PH cohort as compared to that in controls [51]. This reduced shear stress further induced a cascade of endothelial dysfunction including hyper-proliferation, increased oxidative stress, and inflammation, which worsened the symptoms of disease [52, 53].

Microfluidic systems have been used to construct vascular models and study the impact of shear stresses on vascular cell phenotypes and functions. Unlike the conventional bulk parallel plate flow chambers or cone-and-plate flow devices, microfluidic systems are easily multiplexed, could provide configurable shear stresses, and require a minimal number of cells. For example, Arora and colleagues have established a multiplex microfluidic platform to systematically investigate the impact of different levels of shear forces on induced pluripotent stem cell-derived endothelial cells (iPSC-ECs) differentiation. The multiplex device is comprised of six parallel cell culture chambers, allowing to simultaneously test shear stress ranging from 0.4 to 15 dyne/cm$^2$. Similarly, Rogers and colleagues developed a high throughput microfluidic-based microvascular platform, PREDICT96, to study several perturbations, including fluidic shear stress. This system consists of 96 arrayed bilayer perfusable microfluidic devices containing microvascular endothelial cells and pericytes cultured on opposing sides of a microporous membrane. The microvascular platform is perfused with physiological shear stress and is responsive to multiple perturbations such as barrier disruption, inflammatory stimulation, and fluid shear stress [54]. Such a high throughput system could potentially be used for disease modeling and drug discovery applications.

## 14.7 Cyclic Stretch

Mechanical cyclic stretch (CS) is primarily contributed by alveolar ventilation pressures and vascular perfusion and affects pulmonary vascular functions [55–57]. During physiological conditions, ventilation-generated vascular stretch induces phosphorylation of Akt and eNOS, increases NO production at pulmonary microvessels, and protects vascular integrity from challenges [56, 58]. However, pathophysiological cyclic stretch can be disruptive to the endothelial barrier function and may induce inflammation. Exposure of both pulmonary microvascular and macrovascular endothelium to pathologic 18% CS for 2–96 hours increased thrombin-induced permeability and monolayer disruption. In turn, physiologic 5% CS preconditioning attenuated thrombin-induced paracellular gap formation [57]. Additionally, compared to cells under 5% CS, pulmonary endothelium under pathologic 18% CS increased surface expression of ICAM1 and release of its soluble form (sICAM) as well as pro-inflammatory cytokine IL-8 [55].

Various systems have been used to study the impact of CS on cell biology. Conventionally, the cell culture systems with stretchable polymeric foil were used [57] to contemplate fundamental questions on the impact of uniaxial or biaxial forces on endothelial mechanotransduction. However, they cannot mimic the 3D CS experienced by vessels from cyclically expanding organs, and as such, they are not suitable for studying the cyclic stretch effect on 3D microvessels in the lung. Zeinali and colleagues recently established a dynamic microvascular platform that incorporates perfusion and cyclic strain. They built a 3D perfusable microvascular channel in the middle of a fibrin gel, which could be cyclically bent. This system allows to simultaneously control longitudinal and circumferential strains and study endothelial behavior under different mechanical stimuli. Using this system, they found, unlike the conventional 2D model, cyclic strain in the 3D model could markedly improve vascular barrier function and restore the vascular function challenged by VEGF [59], simulating native conditions. Additionally, the cyclic strain has been incorporated into more complex alveolar models to study various lung diseases. The seminal study by Huh et al. introduced a biomimetic microsystem, called lung-on-a-chip, that reconstitutes the critical functional alveolar-capillary interface of the human lung. They microfabricated the device using poly(dimethylsiloxane) (PDMS) microchannels to form an alveolar-capillary barrier on a thin, porous, flexible PDMS membrane coated with ECM. They cultured human alveolar epithelial cells on one side of the ECM-coated PDMS membrane while human pulmonary microvascular endothelial cells were cultured on the other side. Cyclic vacuum suction was applied to induce cyclic stretching of the adherent cell layers that simulated physiological breathing motions and resultant deformation of the alveolar-capillary barrier [60]. By incorporating cyclic strain in the system, they found that endothelial cell and alveolar cell phenotypes are shifting toward their native counterparts and displayed native-mimicking immune response on the pulmonary vasculature after being challenged with IL-2 [61–63].

## 14.8 Translational Potential of Current Models

From a translational perspective, the ideal model systems should be (1) high fidelity – sufficiently recapitulating human cellular phenotypes and physiological functions during homeostasis and diseased states; (2) Repeatable, Reproducible, Accessible – easily accessible, robust, and repeatable/reproducible across different laboratory settings; (3) High-throughput – scalable to suit manufacturing needs from public or private sectors. Recent development of tissue models on pulmonary vasculature has been centered on organ chips and organoids platforms (Table 14.1). I will review pros and cons of each platform and discuss perspectives for the future development.

**Table 14.1** Examples of engineered pulmonary vascular models

| Engineered models | Methods | Applications | Advantages | Disadvantages |
|---|---|---|---|---|
| Organ chips | Reconstruction of lumina, intimal, medial, adventitial, and perivascular layers of the pulmonary artery | Pulmonary arterial hypertension[64] | Scalable; Recapitulate key biophysical/ mechanical conditions; Relatively cheap | Current models are low-to-moderate throughput; Lack of key biochemical cues in the system |
| | Co-culture human pulmonary vascular endothelial and smooth muscle cells under hypoxia and induced BMPR2 dysfunction in cells | Pulmonary arterial hypertension[65] | | |
| | Lung-on-a-chip model, culture human pulmonary alveolar epithelial cells and microvascular endothelial cells on each side of a stretchable PDMS membrane | Blood-air barrier[60] | | |
| | Added IL2 into microvascular channel in a lung-on-a-chip model | Pulmonary Edema[61] | | |
| | Introduced LPS into microvascular channel in a lung-on-a-chip model | Pulmonary microvascular thrombosis[66] | | |
| | Co-culture HUVECs and human lung fibroblast in a microfluidic device | Vascular function during idiopathic pulmonary fibrosis[67] | | |
| | Second generation lung-on-a-chip model, used a biological membrane made of proteins of the lung ECM, allowed to apply 3D cyclic mechanical strain | Blood-air barrier[68] | | |
| Organoids | Infected human vascular organoids with SARS-CoV-2 | Vascular infection during COVID-19 [69] | Scalable; Recapitulate some biochemical cues; High-throughput | Expensive; Complicated culture procedure; Lack of biophysical cues |

**Table 4.1** (continued)

| Engineered models | Methods | Applications | Advantages | Disadvantages |
|---|---|---|---|---|
| iPSC models | iPSC-derived endothelial cells from BMPR2 mutation carriers | Pulmonary arterial hypertension[70, 71] | Scalable; Great potential for precision medicine; Great for disease with genetic mutations | Complicated culture procedure; Insufficient differentiation protocol |
| Decellularized lung models | Human pulmonary microvascular cells culture in decellularized whole lung scaffolds | Acute lung injury[30] | Recapitulate several key biophysical and biochemical properties | Low-throughput; Complicated culture procedure |

## 14.9 Organ Chips

The organ chip device generally harnesses microfluidic, 3D-bioprinting, and hydrogel technologies and is easy to fabricate, easily accessible, and has great potential for scale- up. This system has shown great promise in mimicking physiological microenvironment at a micro-scale level and modeling different organs and tissues, including lung vasculature, during normal and diseased conditions. Al-Hilal and colleagues have fabricated a device that models the lumina, intimal, medial, adventitial, and perivascular layers of the pulmonary artery. They grew human pulmonary arterial endothelial cells, smooth muscle cells, and adventitial cells in the chip at separate layers, allowing them to study the direct cell-cell communications in the pulmonary artery. They then applied this system to understand the etiology of pulmonary arterial hypertension (PAH). They found that the cells from PAH patients, when grown on the chips, created phenomena similar to the primary pathologies of human PAH, including intimal thickening, muscularization, and arterial remodeling and a sign of endothelial to mesenchymal transition [64]. Various devices have also been developed to model pulmonary capillaries in distal lungs. Zeinali et al. established a pulmonary microvessel-on-a-chip model by co-culturing human endothelial cells with lung fibroblast in a perfusable microchannel. This system allows for studying permeability, vascularized area, and cell-cell interactions. They used this device to understand microvascular remodeling during idiopathic pulmonary fibrosis (IPF) and the impact of nintedanib, a medication for the treatment of IPF, on vascular protection. They found that nintedanib improves vessel permeability and inhibits vascular remodeling in the chip, consistent with in vivo studies [67]. Hul et al. introduced a more complicated model that used a microfluidic-based system to model the blood-air barrier in vitro, called lung-on-a-chip (LOC). This system consists of an upper alveolar epithelial and a bottom microvascular endothelial culture chamber separated by a microporous stretchable PDMS membrane. This initial version of LOC allows studying breath motion in the alveolar compartment and shear forces in the microvascular compartment [60]. This system has been widely used to model various pulmonary disorders, including pulmonary edema and thrombosis, which are key pathological features of human ARDS/ALI [61, 66]. Although the LOC system recapitulates some critical physical cues such as shear stress, cyclic stretch, and stiffness, it fails to reconstitute other biochemical and biophysical features of lung vasculature, such as cell-cell crosstalk signals, oxygen tension, and correct ECM substrate, which are critical for intracellular regulation that I have discussed in previous sections. Thus, further efforts would be made to improve the fidelity of current models that could better resemble the native microenvironment.

## 14.10 Organoids

The organoids are tiny, self-assembled 3D tissues/organs derived from stem cells. This miniature organ system could resemble several key physiological features of their native counterparts [72–74], enabling the study of organ biology in a dish. In 2019, a seminal study by Wimmer et al. introduced the first human blood vessel organoids from pluripotent stem cells. In this organoid, the stem cell-derived endothelial and perivascular cells could self-assemble into capillary networks covered by the basement membrane. After transplantation, they could form stable and perfusable vascular branches, including arteries, arterioles, and venules [21]. This system has been used to model various vascular disorders, including diabetes mellitus and COVID-19 [21, 69]. However, there are many hurdles to overcome before translating this technique into clinics. First, most organoids (reviewed in [75, 76]) to date still only represent cellular phenotypes during the fetal stage rather than the adult stage, limiting their potential to study organ physiology or pathophysiology, especially for chronic diseases. Second, there are huge variations in culture protocols, and the results are hardly comparable between different laboratories, limiting their potential for scale-up. Third, mechanical properties such as hemodynamics and cyclic stretch are poorly studied in organoids. Thus, although organoids are a great candidate to model human organ biology, significant work will be conducted to improve the fidelity of this system to study human diseases, including pulmonary vascular diseases.

## 14.11 Conclusion

The development of in vitro native-mimicking model systems offers a way to bridge the gap between preclinical animal models and human physiology. Various systems were developed to recapitulate key microenvironmental features of the pulmonary vasculature, and to elucidate the biological mechanisms during the development and diseases. However, from a translational standpoint, there are many obstacles to overcome: (1) very few validation studies were performed to ensure the fidelity of engineered models, possibly due to our deficient information on cellular phenotypes in human tissues. Multi-omics is a powerful tool for understanding transcriptomics and proteomics in multicellular organisms at a single-cell level. Various databases on human cell atlas have been developed in recent years [40, 77, 78]. Incorporating multi-omics may enable the establishment of standards for validating engineered models. (2) Non-human or cancer-derived biological products such as cell lines, fetal bovine serum, and Matrigel were still commonly used in various model systems, limiting their translational potential. Instead, primary cells, human serum or chemically defined medium, and native ECM should be used. (3) Critical microenvironmental cues such as native ECM, correct cell-cell communications, and oxygen tensions are still not incorporated into various systems. Therefore, reconstruction of pulmonary vasculature is still in its infancy, and extensive research on vascular niche regulation, validation, and scalability is needed to improve their translational potential.

## References


1. Huertas A, Guignabert C, Barbera JA, Bartsch P, Bhattacharya J, Bhattacharya S, et al. Pulmonary vascular endothelium: the orchestra conductor in respiratory diseases: Highlights from basic research to therapy. Eur Respir J. 2018;51(4).
2. Ackermann M, Verleden SE, Kuehnel M, Haverich A, Welte T, Laenger F, et al. Pulmonary Vascular Endothelialitis, Thrombosis, and Angiogenesis in Covid-19. N Engl J Med. 2020;383(2):120–8.
3. Aeffner F, Bolon B, Davis IC. Mouse Models of Acute Respiratory Distress Syndrome: A Review of Analytical Approaches, Pathologic Features, and Common Measurements. Toxicol Pathol. 2015;43(8):1074–92.
4. Viola H, Chang J, Grunwell JR, Hecker L, Tirouvanziam R, Grotberg JB, et al. Microphysiological systems modeling acute respiratory distress syndrome that capture mechanical force-induced injury-inflammation-repair. APL Bioeng. 2019;3(4):041503.
5. Sztuka K, Jasinska-Stroschein M. Animal models of pulmonary arterial hypertension: A systematic review and meta-analysis of data from 6126 animals. Pharmacol Res. 2017;125(Pt B):201–14.
6. Mahida RY, Scott A, Parekh D, Lugg ST, Belchamber KBR, Hardy RS, et al. Assessment of Alveolar Macrophage Dysfunction Using an in vitro Model of Acute Respiratory Distress Syndrome. Front Med (Lausanne). 2021;8:737859.
7. Park JG, Pino PA, Akhter A, Alvarez X, Torrelles JB, Martinez-Sobrido L. Animal Models of COVID-19: Transgenic Mouse Model. Methods Mol Biol. 2022;2452:259–89.
8. Singh DK, Cole J, Escobedo RA, Alfson KJ, Singh B, Lee TH, et al. Animal Models of COVID-19: Nonhuman Primates. Methods Mol Biol. 2022;2452:227–58.
9. Seok J, Warren HS, Cuenca AG, Mindrinos MN, Baker HV, Xu W, et al. Genomic responses in mouse models poorly mimic human inflammatory diseases. Proc Natl Acad Sci U S A. 2013;110(9):3507–12.
10. Ledford H. Translational research: 4 ways to fix the clinical trial. Nature. 2011;477(7366):526–8.
11. Stenmark KR, Meyrick B, Galie N, Mooi WJ, McMurtry IF. Animal models of pulmonary arterial hypertension: the hope for etiological discovery and pharmacological cure. Am J Physiol Lung Cell Mol Physiol. 2009;297(6):L1013–32.
12. Leiby KL, Raredon MSB, Niklason LE. Bioengineering the Blood-gas Barrier. Compr Physiol. 2020;10(2):415–52.
13. Viji RI, Kumar VB, Kiran MS, Sudhakaran PR. Modulation of endothelial nitric oxide synthase by fibronectin. Mol Cell Biochem. 2009;323(1–2):91–100.
14. Kumar VB, Viji RI, Kiran MS, Sudhakaran PR. Negative modulation of eNOS by laminin involving post-translational phosphorylation. J Cell Physiol. 2009;219(1):123–31.
15. Yuan Y, Khan S, Stewart DJ, Courtman DW. Engineering blood outgrowth endothelial cells to optimize endothelial nitric oxide synthase and extracellular matrix production for coating of blood contacting surfaces. Acta Biomater. 2020;109:109–20.
16. Bordeleau F, Mason BN, Lollis EM, Mazzola M, Zanotelli MR, Somasegar S, et al. Matrix stiffening promotes a tumor vasculature phenotype. Proc Natl Acad Sci U S A. 2017;114(3):492–7.
17. Anayiotos AS, Jones SA, Giddens DP, Glagov S, Zarins CK. Shear stress at a compliant model of the human carotid bifurcation. J Biomech Eng. 1994;116(1):98–106.

18. Perktold K, Rappitsch G. Computer simulation of local blood flow and vessel mechanics in a compliant carotid artery bifurcation model. J Biomech. 1995;28(7):845–56.
19. Tan W, Madhavan K, Hunter KS, Park D, Stenmark KR. Vascular stiffening in pulmonary hypertension: cause or consequence? (2013 Grover Conference series). Pulm Circ. 2014;4(4):560–80.
20. Andrique L, Recher G, Alessandri K, Pujol N, Feyeux M, Bon P, et al. A model of guided cell self-organization for rapid and spontaneous formation of functional vessels. Sci Adv. 2019;5(6):eaau6562.
21. Wimmer RA, Leopoldi A, Aichinger M, Wick N, Hantusch B, Novatchkova M, et al. Human blood vessel organoids as a model of diabetic vasculopathy. Nature. 2019;565(7740):505–10.
22. Polacheck WJ, Kutys ML, Tefft JB, Chen CS. Microfabricated blood vessels for modeling the vascular transport barrier. Nat Protoc. 2019;14(5):1425–54.
23. Palikuqi B, Nguyen DT, Li G, Schreiner R, Pellegata AF, Liu Y, et al. Adaptable haemodynamic endothelial cells for organogenesis and tumorigenesis. Nature. 2020;585(7825):426–32.
24. Calle EA, Hill RC, Leiby KL, Le AV, Gard AL, Madri JA, et al. Targeted proteomics effectively quantifies differences between native lung and detergent-decellularized lung extracellular matrices. Acta Biomater. 2016;46:91–100.
25. Balestrini JL, Gard AL, Gerhold KA, Wilcox EC, Liu A, Schwan J, et al. Comparative biology of decellularized lung matrix: Implications of species mismatch in regenerative medicine. Biomaterials. 2016;102:220–30.
26. Li Q, Uygun BE, Geerts S, Ozer S, Scalf M, Gilpin SE, et al. Proteomic analysis of naturally-sourced biological scaffolds. Biomaterials. 2016;75:37–46.
27. Yuan Y, Engler AJ, Raredon MS, Le A, Baevova P, Yoder MC, et al. Epac agonist improves barrier function in iPSC-derived endothelial colony forming cells for whole organ tissue engineering. Biomaterials. 2019;200:25–34.
28. Leiby KL, Yuan Y, Ng R, Raredon MSB, Adams TS, Baevova P, et al. <em>In vitro</em> engineering of the lung alveolus. bioRxiv. 2022:2022.03.13.484143.
29. Greaney AM, Adams TS, Brickman Raredon MS, Gubbins E, Schupp JC, Engler AJ, et al. Platform Effects on Regeneration by Pulmonary Basal Cells as Evaluated by Single-Cell RNA Sequencing. Cell Rep. 2020;30(12):4250–65.e6.
30. Yuan Y, Leiby KL, Greaney AM, Raredon MSB, Qian H, Schupp JC, et al. A Pulmonary Vascular Model From Endothelialized Whole Organ Scaffolds. Front Bioeng Biotechnol. 2021;9:760309.
31. Engler AJ, Raredon MSB, Le AV, Yuan Y, Oczkowicz YA, Kan EL, et al. Non-invasive and real-time measurement of microvascular barrier in intact lungs. Biomaterials. 2019;217:119313.
32. Uhl FE, Zhang F, Pouliot RA, Uriarte JJ, Rolandsson Enes S, Han X, et al. Functional role of glycosaminoglycans in decellularized lung extracellular matrix. Acta Biomater. 2020;102:231–46.
33. Tsuchiya T, Balestrini JL, Mendez J, Calle EA, Zhao L, Niklason LE. Influence of pH on extracellular matrix preservation during lung decellularization. Tissue Eng Part C Methods. 2014;20(12):1028–36.
34. Petersen TH, Calle EA, Colehour MB, Niklason LE. Matrix composition and mechanics of decellularized lung scaffolds. Cells Tissues Organs. 2012;195(3):222–31.
35. Ren X, Evangelista-Leite D, Wu T, Rajab TK, Moser PT, Kitano K, et al. Metabolic glycan labeling and chemoselective functionalization of native biomaterials. Biomaterials. 2018;182:127–34.
36. Kosyakova N, Kao DD, Figetakis M, Lopez-Giraldez F, Spindler S, Graham M, et al. Differential functional roles of fibroblasts and pericytes in the formation of tissue-engineered microvascular networks in vitro. NPJ Regen Med. 2020;5:1.
37. Doi R, Tsuchiya T, Mitsutake N, Nishimura S, Matsuu-Matsuyama M, Nakazawa Y, et al. Transplantation of bioengineered rat lungs recellularized with endothelial and adipose-derived stromal cells. Sci Rep. 2017;7(1):8447.

38. Ren X, Moser PT, Gilpin SE, Okamoto T, Wu T, Tapias LF, et al. Engineering pulmonary vasculature in decellularized rat and human lungs. Nat Biotechnol. 2015;33(10):1097–102.
39. Barnthaler T, Maric J, Platzer W, Konya V, Theiler A, Hasenohrl C, et al. The Role of PGE2 in Alveolar Epithelial and Lung Microvascular Endothelial Crosstalk. Sci Rep. 2017;7(1):7923.
40. Travaglini KJ, Nabhan AN, Penland L, Sinha R, Gillich A, Sit RV, et al. A molecular cell atlas of the human lung from single-cell RNA sequencing. Nature. 2020;587(7835):619–25.
41. Gillich A, Zhang F, Farmer CG, Travaglini KJ, Tan SY, Gu M, et al. Capillary cell-type specialization in the alveolus. Nature. 2020;586(7831):785–9.
42. Schupp JC, Adams TS, Cosme C, Jr., Raredon MSB, Yuan Y, Omote N, et al. Integrated Single-Cell Atlas of Endothelial Cells of the Human Lung. Circulation. 2021;144(4):286–302.
43. Liu Y, Yang M, Deng Y, Su G, Enninful A, Guo CC, et al. High-Spatial-Resolution Multi-Omics Sequencing via Deterministic Barcoding in Tissue. Cell. 2020;183(6):1665–81.e18.
44. Cho CS, Xi J, Si Y, Park SR, Hsu JE, Kim M, et al. Microscopic examination of spatial transcriptome using Seq-Scope. Cell. 2021;184(13):3559–72.e22.
45. Lee JW, Gupta N, Serikov V, Matthay MA. Potential application of mesenchymal stem cells in acute lung injury. Expert Opin Biol Ther. 2009;9(10):1259–70.
46. London NR, Zhu W, Bozza FA, Smith MC, Greif DM, Sorensen LK, et al. Targeting Robo4-dependent Slit signaling to survive the cytokine storm in sepsis and influenza. Sci Transl Med. 2010;2(23):23ra19.
47. Katsura H, Sontake V, Tata A, Kobayashi Y, Edwards CE, Heaton BE, et al. Human Lung Stem Cell-Based Alveolospheres Provide Insights into SARS-CoV-2-Mediated Interferon Responses and Pneumocyte Dysfunction. Cell Stem Cell. 2020;27(6):890–904.e8.
48. Raredon MSB, Adams TS, Suhail Y, Schupp JC, Poli S, Neumark N, et al. Single-cell connectomic analysis of adult mammalian lungs. Sci Adv. 2019;5(12):eaaw3851.
49. Raredon MSB, Yang J, Garritano J, Wang M, Kushnir D, Schupp JC, et al. Computation and visualization of cell-cell signaling topologies in single-cell systems data using Connectome. Sci Rep. 2022;12(1):4187.
50. Davies PF. Hemodynamic shear stress and the endothelium in cardiovascular pathophysiology. Nat Clin Pract Cardiovasc Med. 2009;6(1):16–26.
51. Schafer M, Kheyfets VO, Schroeder JD, Dunning J, Shandas R, Buckner JK, et al. Main pulmonary arterial wall shear stress correlates with invasive hemodynamics and stiffness in pulmonary hypertension. Pulm Circ. 2016;6(1):37–45.
52. Siasos G, Sara JD, Zaromytidou M, Park KH, Coskun AU, Lerman LO, et al. Local Low Shear Stress and Endothelial Dysfunction in Patients With Nonobstructive Coronary Atherosclerosis. J Am Coll Cardiol. 2018;71(19):2092–102.
53. Milovanova T, Chatterjee S, Manevich Y, Kotelnikova I, Debolt K, Madesh M, et al. Lung endothelial cell proliferation with decreased shear stress is mediated by reactive oxygen species. Am J Physiol Cell Physiol. 2006;290(1):C66–76.
54. Rogers MT, Gard AL, Gaibler R, Mulhern TJ, Strelnikov R, Azizgolshani H, et al. A high-throughput microfluidic bilayer co-culture platform to study endothelial-pericyte interactions. Sci Rep. 2021;11(1):12225.
55. Tian Y, Gawlak G, O'Donnell JJ, 3rd, Mambetsariev I, Birukova AA. Modulation of Endothelial Inflammation by Low and High Magnitude Cyclic Stretch. PLoS One. 2016;11(4):e0153387.
56. Hu Z, Xiong Y, Han X, Geng C, Jiang B, Huo Y, et al. Acute mechanical stretch promotes eNOS activation in venous endothelial cells mainly via PKA and Akt pathways. PLoS One. 2013;8(8):e71359.
57. Birukova AA, Rios A, Birukov KG. Long-term cyclic stretch controls pulmonary endothelial permeability at translational and post-translational levels. Exp Cell Res. 2008;314(19):3466–77.
58. Kuebler WM, Uhlig U, Goldmann T, Schael G, Kerem A, Exner K, et al. Stretch activates nitric oxide production in pulmonary vascular endothelial cells in situ. Am J Respir Crit Care Med. 2003;168(11):1391–8.
59. Zeinali S, Thompson EK, Gerhardt H, Geiser T, Guenat OT. Remodeling of an in vitro microvessel exposed to cyclic mechanical stretch. APL Bioeng. 2021;5(2):026102.

60. Huh D, Matthews BD, Mammoto A, Montoya-Zavala M, Hsin HY, Ingber DE. Reconstituting organ-level lung functions on a chip. Science. 2010;328(5986):1662–8.
61. Huh D, Leslie DC, Matthews BD, Fraser JP, Jurek S, Hamilton GA, et al. A human disease model of drug toxicity-induced pulmonary edema in a lung-on-a-chip microdevice. Sci Transl Med. 2012;4(159):159ra47.
62. Bai H, Si L, Jiang A, Belgur C, Zhai Y, Plebani R, et al. Mechanical control of innate immune responses against viral infection revealed in a human lung alveolus chip. Nat Commun. 2022;13(1):1928.
63. Si L, Bai H, Rodas M, Cao W, Oh CY, Jiang A, et al. A human-airway-on-a-chip for the rapid identification of candidate antiviral therapeutics and prophylactics. Nat Biomed Eng. 2021;5(8):815–29.
64. Al-Hilal TA, Keshavarz A, Kadry H, Lahooti B, Al-Obaida A, Ding Z, et al. Pulmonary-arterial-hypertension (PAH)-on-a-chip: fabrication, validation and application. Lab Chip. 2020;20(18):3334–45.
65. Wojciak-Stothard B, Ainscough A, Smith T, Rhodes C, Fellows A, Howard L, et al. A MICROFLUIDIC CHIP FOR PULMONARY ARTERIAL HYPERTENSION. Research Square; 2021.
66. Jain A, Barrile R, van der Meer AD, Mammoto A, Mammoto T, De Ceunynck K, et al. Primary Human Lung Alveolus-on-a-chip Model of Intravascular Thrombosis for Assessment of Therapeutics. Clin Pharmacol Ther. 2018;103(2):332–40.
67. Zeinali S, Bichsel CA, Hobi N, Funke M, Marti TM, Schmid RA, et al. Human microvasculature-on-a chip: anti-neovasculogenic effect of nintedanib in vitro. Angiogenesis. 2018;21(4):861–71.
68. Stucki AO, Stucki JD, Hall SR, Felder M, Mermoud Y, Schmid RA, et al. A lung-on-a-chip array with an integrated bio-inspired respiration mechanism. Lab Chip. 2015;15(5):1302–10.
69. Monteil V, Kwon H, Prado P, Hagelkruys A, Wimmer RA, Stahl M, et al. Inhibition of SARS-CoV-2 Infections in Engineered Human Tissues Using Clinical-Grade Soluble Human ACE2. Cell. 2020;181(4):905–13.e7.
70. Sa S, Gu M, Chappell J, Shao NY, Ameen M, Elliott KA, et al. Induced Pluripotent Stem Cell Model of Pulmonary Arterial Hypertension Reveals Novel Gene Expression and Patient Specificity. Am J Respir Crit Care Med. 2017;195(7):930–41.
71. Gu M, Shao NY, Sa S, Li D, Termglinchan V, Ameen M, et al. Patient-Specific iPSC-Derived Endothelial Cells Uncover Pathways that Protect against Pulmonary Hypertension in BMPR2 Mutation Carriers. Cell Stem Cell. 2017;20(4):490–504.e5.
72. Hu H, Gehart H, Artegiani B, C LO-I, Dekkers F, Basak O, et al. Long-Term Expansion of Functional Mouse and Human Hepatocytes as 3D Organoids. Cell. 2018;175(6):1591–606.e19.
73. Takasato M, Er PX, Chiu HS, Maier B, Baillie GJ, Ferguson C, et al. Kidney organoids from human iPS cells contain multiple lineages and model human nephrogenesis. Nature. 2016;536(7615):238.
74. Dye BR, Hill DR, Ferguson MA, Tsai YH, Nagy MS, Dyal R, et al. In vitro generation of human pluripotent stem cell derived lung organoids. Elife. 2015;4.
75. Clevers H. Modeling Development and Disease with Organoids. Cell. 2016;165(7):1586–97.
76. Tanaka Y, Cakir B, Xiang Y, Sullivan GJ, Park IH. Synthetic Analyses of Single-Cell Transcriptomes from Multiple Brain Organoids and Fetal Brain. Cell Rep. 2020;30(6):1682–9.e3.
77. Delorey TM, Ziegler CGK, Heimberg G, Normand R, Yang Y, Segerstolpe A, et al. COVID-19 tissue atlases reveal SARS-CoV-2 pathology and cellular targets. Nature. 2021;595(7865):107–13.
78. Vieira Braga FA, Kar G, Berg M, Carpaij OA, Polanski K, Simon LM, et al. A cellular census of human lungs identifies novel cell states in health and in asthma. Nat Med. 2019;25(7):1153–63.

# Part V
# Engineering and Modeling the Interface Between Medical Devices and the Lung

# Chapter 15
# Extracorporeal Membrane Oxygenation: Set-up, Indications, and Complications

**Anna Niroomand, Franziska Olm, and Sandra Lindstedt**

Extracorporeal membrane oxygenation (ECMO) is a form of support given to patients experiencing lethal failure of their cardiac and/or pulmonary systems. Given the possibility of substantial complications following ECMO initiation, the choice to manage a patient with the circuit must be taken only when the degree of organ failure is severe enough that other medical management is insufficient, and the patient faces the threat of mortality. ECMO as a device set-up relies on a pump to facilitate the flow of blood which is removed via the drainage cannula and

A. Niroomand
Wallenberg Center for Molecular Medicine, Lund University, Lund, Sweden

Department of Clinical Sciences, Lund University, Lund, Sweden

Lund Stem Cell Center, Lund University, Lund, Sweden

Rutgers Robert University, New Brunswick, NJ, USA

F. Olm
Wallenberg Center for Molecular Medicine, Lund University, Lund, Sweden

Department of Clinical Sciences, Lund University, Lund, Sweden

Lund Stem Cell Center, Lund University, Lund, Sweden

S. Lindstedt (✉)
Department of Cardiothoracic Surgery and Transplantation, Skåne University Hospital, Lund, Sweden

Wallenberg Center for Molecular Medicine, Lund University, Lund, Sweden

Department of Clinical Sciences, Lund University, Lund, Sweden

Lund Stem Cell Center, Lund University, Lund, Sweden
e-mail: sandra.lindstedt@med.lu.se



C. M. Magin (ed.), *Engineering Translational Models of Lung Homeostasis and Disease*, Advances in Experimental Medicine and Biology 1413,
https://doi.org/10.1007/978-3-031-26625-6_15

subsequently oxygenated through an oxygenator also called a membrane lung. Ultimately, the blood is then sent back through the return cannula once it has been filtered, warmed, or cooled.

After an interlude following a demonstration of no survival benefit to ECMO in its earlier years, there was a period of stagnation in the system's implementation in adult clinics [1]. Management of patients on ECMO at the time had not been refined and complications relating to experience and technique facing those patients led to high mortality rates and cast doubt on the utility of ECMO [2]. Experience with ECMO was a contributing factor to its low use: a survey sent out in 2004 to Australian intensivists reported that their lack of staff and skills was the most common reason behind not providing ECMO [3]. The emergence of the H1N1 epidemic and the seminal CESAR and ANZ ECMO trials in 2009 launched a resurgence in ECMO as centers gained more experience. Within the CESAR trial, patients with severe acute respiratory failure were divided between management with ECMO or conventional methods. Survival without disability was found to be improved among those on ECMO [4]. The study helped to bolster confidence in this form of extracorporeal circulation as an effective management choice in these severe cases. These results were also backed by the Australia and New Zealand ANZ ECMO study comparing influenza-associated acute respiratory distress syndrome (ARDS) patients with and without ECMO [4, 5]. As experience has risen for centers utilizing ECMO with growing case numbers and increasing confidence managing cases on ECMO, outcomes for these patients have and continue to improve.

The Extracorporeal Life Support Organization (ELSO) has since its foundation in 1989 maintained a registry of ECMO cases, allowing for a retrospective look into the distribution of use with respect to geography, ECMO mode, and within patient populations. ECMO was initially dominated by neonatology in 1990 with over 80% of all reported cases having an indication for neonatal pulmonary use [6]. With the results of the aforementioned trials and growing experience across centers globally, the use of ECMO in indications within adult patients surged. For the complete years of 2016 to 2021, adult cases made up 77% of the registry. Within that allotment of cases, pulmonary indications comprised 46% of those adults on ECMO, cardiac indications the next 41%, and extracorporeal cardiopulmonary resuscitation (ECPR) the last 12.5% [7]. The uptick in use across the globe can be accounted for by a number of factors, including the advent of technology which has improved the circuity, such as more durable membranes [8]. Additionally, expansion of the product to new horizons, such as potentially portable devices has enlarged its use-case. Of course, increasing familiarity with the technology and the improved outcomes contributes to the heightened faith that ECMO has an invaluable spot in modern medicine as a tool for management of cardiac and pulmonary failure. This is bolstered by the broad use of ECMO, as the system can be a bridge to a number of destinations, including transplantation, decision, and recovery.

## 15.1 Introduction to Modes of ECMO

ECMO modes are delineated by their points of access and return, dividing the types primarily into veno-arterial ECMO (VA-ECMO) and veno-venous ECMO (VV-ECMO). Venous blood is drained from the systemic circulation, pressured with a pump where it can then pass through a membrane oxygenator, heat exchanger, and return either into the venous circulation (VV-ECMO) or arterial circulation (VA-ECMO). Other less common derivatives on this system include venovenous-arterial (VVA)-ECMO and veno-arterial-venous (VAV)-ECMO which are variants that expand on the dual-cannula setup to include another cannula for either an additional point of access or return.

### *15.1.1 Veno-Venous ECMO*

VV-ECMO is indicated in cases of respiratory failure when cardiac function is adequate, and the risk of patient mortality is high despite optimal conventional respiratory therapy [9]. Importantly, VV-ECMO does not provide hemodynamic support. To this end, it has been increasingly used in cases of severe ARDS secondary to multiple etiologies, such as aspiration, pneumonia, barotrauma, or pneumonitis [10]. Additionally, the application of VV-ECMO has been used to bridge patients to lung transplantation, an indication detailed in a later section.

To initiate VV-ECMO, both the draining and returning cannulas are placed within venous circulation with the choice of cannula location depending on the strategy. Commonly, drainage is through the inferior vena cava (IVC) with access from the femoral vein while return is through the jugular vein to the superior vena cava (SVC) to the right atrium [10, 11] (Fig. 15.1a). The return cannula can alternatively be placed through the femoral vein in the context of a femorofemoral setup. Choice of cannula size is dependent on patient anatomy as well as the required blood flow. In a dual-lumen approach, a single cannula is placed into the internal jugular vein and the two lumens are oriented to allow for drainage through the IVC and return into the right atrium through the other lumen (Fig. 15.1a). Proper placement of the dual-lumen catheter is particularly important to prevent recirculation in which oxygenated blood is returned to the right atrium but enters the drainage cannula rather than the right ventricle. In the case of recirculation, there is relative systemic hypoxemia given the now decreasing proportion of oxygenated circulating blood as that oxygenated blood from the return cannula is feeding back into the drainage cannula instead of the systemic circulation [12–14]. This can be prevented by placing the dual-lumen cannula such that the return flow is directed to the tricuspid valve, although doing so and maintaining said positioning may be difficult, especially when factoring in practicalities such as patient movement. To attempt placement of the cannula, transesophageal or transthoracic echocardiography or fluoroscopy can guide cannulation [13]. The use of concurrent imaging to guide

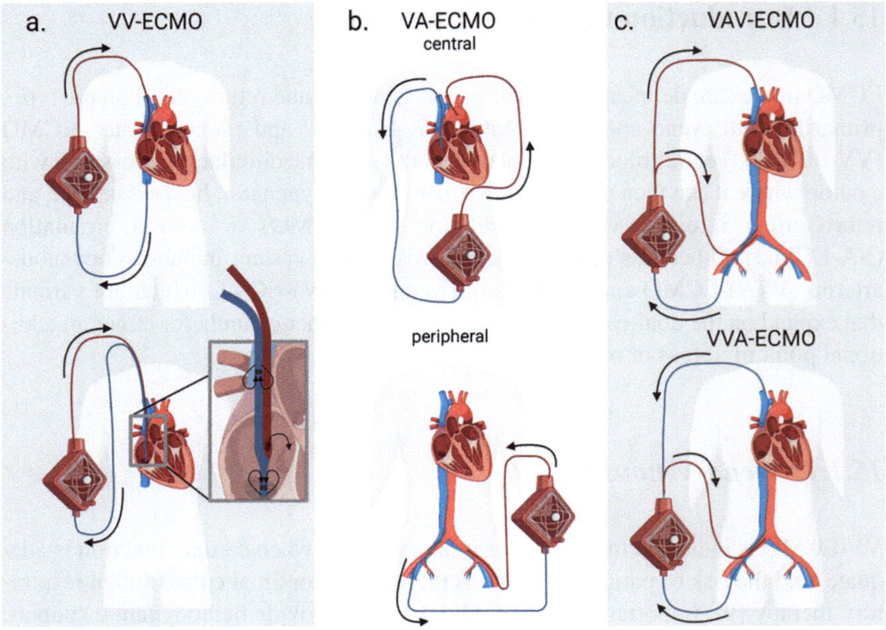


**Fig. 15.1** ECMO can be set up through the veno-venous configuration (**a**) with two cannulas (*top*) or with a dual-lumen cannula (*bottom*). Alternatively, in the veno-arterial configuration (**b**), the cannulas can either be centrally placed (*top*) or peripherally through groin access (*bottom*). Special arrangements of ECMO involving more cannulas (**c**) include the veno-arterial-venous configuration (*top*) and the veno-veno-arterial configuration (*bottom*). Created with biorender.com

placement helps to reduce the risk of vessel or cardiac injury [15–19]. This is also an opportunity for technological innovation to use methods such as ultrasound dilution and mixed venous oxygen saturation to measure the extent of recirculation [20–22]. The benefits of placing a dual-lumen single catheter would include reducing the risk of bleeding as well as increasing the possibility of ambulation while continuing ECMO therapy. Early ambulation on ECMO has been studied for feasibility and safety with the goal of improving functional recovery and independence [23–25]. It should be noted that success of ambulatory physical rehabilitation comes only with the assistance and availability of a trained multidisciplinary team [26].

### *15.1.2 Veno-Arterial ECMO*

In the event of hemodynamic compromise, VA-ECMO can be employed to supply both respiratory and circulatory support to a patient. Indications for VA-ECMO initiation include cardiogenic shock, encompassing conditions such as acute myocardial infarction, myocarditis, acute or chronic heart failure, post-cardiotomy syndrome, as well as refractory ventricular arrythmia and cardiac arrest [8, 15].

Patients receiving extracorporeal cardiopulmonary resuscitation have shown improved survival compared to conventional CPR [27, 28] and the utility of VA-ECMO has been explored following out-of-hospital and emergency department cardiac arrests [29]. After placement on VA-ECMO, patients can then be bridged to recovery, transplantation, or further circulatory support. Patients must be considered carefully prior to ECMO initiation and relative contraindications include severe aortic regurgitation [30].

To discuss VA-ECMO set-up, the category must first be further subdivided between peripheral and central techniques. In peripheral VA-ECMO, the femoral vein is cannulated to drain the right atrium. Following oxygenation and decarboxylation by the system, the blood returns to the femoral artery, traveling retrograde up the aorta [31]. Alternatively, with central VA-ECMO, the right atrium or pulmonary artery is directly cannulated and then the return likewise is a direct cannulation of either the left atrium, aorta, or left ventricle [31, 32] (Fig. 15.1b).

Reviewing peripheral VA-ECMO inherently leads to a discussion of the effects of retrograde flow on both left ventricular afterload and left ventricular end-diastolic pressure in which an increase in both could potentially lead to pulmonary edema [33–35]. As the flow may also affect the left ventricular ejection, the stasis of blood within that chamber could result in thrombosis with downstream consequences [36–38]. This has prompted the adjunct use of alternative devices. Techniques to combat this effect are entitled “left ventricular (LV) venting” and include Impella placement and the use of intra-aortic balloon pumps (IABPs) [31, 39, 40]. Improved outcomes including lower mortality rate have been observed in tangential use of IABPs with VA-ECMO relative to ECMO alone [38]. Context should be considered, however, as LV venting may be more clinically useful in patients who have minimal to no cardiac contractility relative to patients with maintained function [41].

Another consideration in VA-ECMO is the so-called “Harlequin” or “North-South” syndrome which occurs in the peripheral configuration. Differential hypoxia between upper and lower limbs develops due to retrograde flow from cannulation of the femoral artery. The retrograde-moving oxygenated blood mixes with deoxygenated blood ejected from a heart with impaired lung function and can consequently result in oxygenated blood failing to reach the coronary arteries and the brain [42, 43]. The phenomenon can be counteracted by the insertion of another cannula to create a venoarterial venous (VAV)-ECMO setup or by a switch to a central VA-ECMO system [30]. The occurrence of such differential hypoxia should be suspected if the oxygen measures differ between the radial and femoral arterial samples.

In peripheral VA-ECMO, there is also an issue of maintaining adequate perfusion in the extremities to avoid ischemia of the leg. Rates of limb ischemia range depending on the study from 10% to 50% in VA-ECMO cases [44–47]. Those at greater risk for developing ischemia include populations in which the femoral artery for cannulation may be smaller: in women, younger patients, patients with peripheral arterial disease and when larger cannulas are utilized [48, 49]. To this end, insertion of an additional distal perfusion catheter (DPC) into the superficial femoral artery can help relieve the risk of limb ischemia (Fig. 15.2a). The DPC has been reported for use both prophylactically at the time of cannulation to prevent ischemia

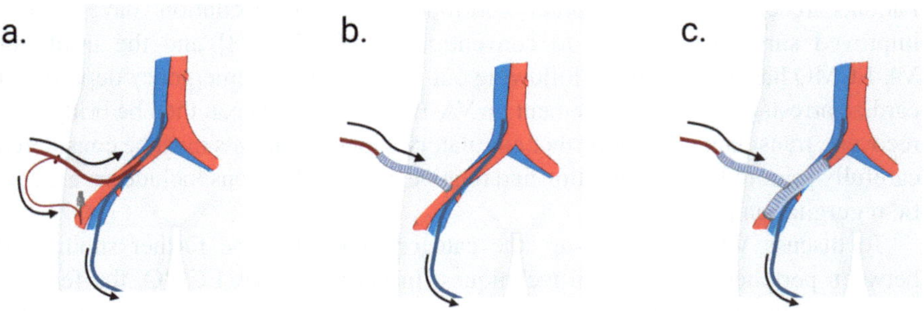


**Fig. 15.2** (**a**) Techniques to avoid or treat ischemia of the leg include the use of a distal perfusion catheter; (**b**) as well as the implementation of end-to-side chimney grafts; (**c**) or T-grafts. Created with biorender.com

as well as after the onset of clinical ischemia to salvage the limb [48]. Other techniques have been reported, including cannulation of the ipsilateral posterior tibial artery [50]. Lack of a DPC was a risk factor in one study of VA-ECMO for the development of distal limb ischemia and the study recommended making DLCs a mandatory component of peripheral VA-ECMO [51]. Even with additional DLC, clinically significant limb ischemia may occur and could require serious measures to correct the damage, including compartment fasciotomy and amputation [52].

An additional technique employed to avoid distal ischemia involves using a graft to the artery to circumvent placing the cannula directly into the artery lumen which could lead to obstruction of the lumen. This includes the use of either an end-to-side chimney graft or T-graft [53–56] (Fig. 15.2b, c). The use of a bidirectional cannula, as is currently being tested with cardiopulmonary bypass (CPB), presents as another solution to the dilemma of limb ischemia [57]. This set-up with a side hole for antegrade perfusion was successful in facilitating distal flow with adequate systemic pressures but has yet to be extended beyond CPB and to ECMO.

### *15.1.3 VVA- and VAV-ECMO*

As briefly mentioned, the canonical set-ups of VV- and VA-ECMO can be augmented with the addition of a third cannula. In VVA-ECMO, another cannula is deployed such that the second drainage cannula within an existing VA-ECMO set-up drains blood flow from the right atrium or ventricle [58] (Fig. 15.1c). The addition of the second drainage cannula aims to reduce the incidence of differential hypoxia.

Another form of triple cannulation is VAV-ECMO which seeks to combine the advantages of VA- and VV-ECMO by supporting both the respiratory and circulatory systems (Fig. 15.1c). The additional cannula is a return one so that there is arterial outflow both to the right atrium and through the femoral artery towards the aorta. The potential benefit of this alternative configuration to either VA- or

VV-ECMO would be the ability to initiate respiratory support when failure occurs secondary to cardiac failure or vice versa (cardiac failure secondary to respiratory failure). The studies of either VAV- or VVA-ECMO are limited with a need for investigation beyond feasibility to demonstrate improvement in outcomes [59–61]. Some data including a retrospective review of 23 patients did not demonstrate survival advantage or complication reduction associated with VVA-ECMO use [62]. There in addition exists a study of successful treatment of cerebral hypoxemia in patients on VA-ECMO using VAV-ECMO [61].

### 15.1.4 ECCO$_2$R ECMO

In extracorporeal carbon dioxide removal (ECCO$_2$R), the use of reduced blood and sweep gas flow has emerged for the clearance of carbon dioxide while also allowing for "ultraprotective ventilation" or other configuration modifications [63]. ECCO$_2$R aims to remove carbon dioxide rather than focusing on oxygenating the blood. This specialized version of ECLS is being trialed in patients with ARDS to allow for ventilation with a markedly reduced tidal volume of 3–4 mL/kg as well as in patients with chronic obstructive pulmonary disease (COPD) [64–66]. "Ultraprotective ventilation" settings at such low tidal volumes with low plateau pressures ≤25 cmH2O are thought to reduce the risk of ventilator-induce lung injury but increase the incidence of respiratory acidosis [67]. ECCO$_2$R complements the lungs' function of removing carbon dioxide to cope with the respiratory acidosis that may result from the ultraprotective ventilation. The SUPERNOVA study assesses ECCO$_2$R in moderate ARDS, finding that the design of the system was feasible and reporting a 73% survival rate at day 28 with 62% alive at hospital discharge [68].

### 15.1.5 Other Components of the ECMO System

Generally, the blood pumps utilized in ECMO can be divided between the occlusive roller pumps and the nonocclusive axial and centrifugal pumps. Nonocclusive pumps are preferred as roller pumps may lead to emboli formation and mechanical stress [15]. Centrifugal pumps are widely used now that lower resistance oxygenators are employed allowing for small priming volumes and reduced hemolysis [69, 70]. These changes in design translated to minimized circuit volume and connections which in turn reduce the area of blood to machine interface and decrease the points at which turbulent flow and thus thrombi can occur [71]. Other components of the system not discussed in this chapter include heat exchangers, line pressure measurers, ultrasonic flow detectors, and oxyhemoglobin saturation measures.

## 15.2 Thrombosis and Bleeding

Thromboembolic complications are a common occurrence on ECMO, due to the exposure of blood to the ECMO circuit. The coagulation cascade is known to be activated by the interaction of blood with artificial surfaces [72]. Contact with the surfaces also results in coagulation and fibrinolytic pathway activation as well as complement-mediated inflammatory responses. A patient's own underlying disease can have a relative contribution to thrombogenesis, and the lack of or reduction in cardiac function in patients on VA-ECMO makes it important to maintain blood flow to prevent stasis in the heart [69]. The emboli that are formed may manifest as severe complications, such as embolic stroke. The incidence of embolic brain infarction ranges in estimates from 1.7% to 15% [73–76]. Thrombus formation is known to occur at the oxygenator and consequently the pressure including any drops in this part of the ECMO circuit should be monitored [77]. In VA-ECMO, if a clot is formed at the oxygenator or pump, when the blood is returned to arterial circulation, that emboli will then travel systemically. Rapid increases in D-dimer can signify failure of the oxygenator and circuit change [78].

To prevent the sequelae of thromboemboli, patients are placed on an anticoagulative regimen, most commonly with unfractionated heparin. Treatment with bivalirudin and argatroban is also possible, with a relative advantage over heparin of not relying on antithrombin III for an anticoagulative effect [79]. Under the current ELSO guidelines, a bolus of 50–100 units/kg of heparin should be given at the time of cannulation followed by a continuous infusion throughout the duration of ECMO use [9]. This infusion rate should be sufficient to keep the activated clotting time at 1.5 times the normal time with measures taken every hour [9]. Other measures of anticoagulation like activated partial thromboplastin time (aPTT), anti-factor Xa levels, thromboelastography, and the absolute value of the heparin dose are used to varying degrees based on the center. Within patients treated with ECMO, there is no standard dose of heparin [9]. The introduction of heparin-coated cannulas was instrumental in decreasing the blood-surface interaction, considering the degree of blood interface with both the tubing and the oxygenator [80, 81]. The use of other coatings, such as phosphorylcholine and polymethoxyethyl acrylate has also been studied and put into use [82–86]. With the use of heparin comes the need to monitor for heparin induced thrombocytopenia (HIT), which while rare can result in a prothrombotic state. Development of HIT would mean withdrawal of heparin use in the patient, which could be mitigated by use of other anticoagulants such as bivalirudin and argatroban [87]. Lepirudin, a direct thrombin inhibitor, has been used in ECMO patients with HIT [88, 89]. There is also the added complication of increasing heparin resistance over time as antithrombin levels decrease, with antithrombin supplementation being a controversial choice [90].

Inherent to the administration of anticoagulation is the risk of bleeding. Factors which increase the likelihood of bleeding include platelet activation associated with shear stress, an acquired coagulopathy due to exposure to artificial surfaces, and systemic inflammation which accompany the conditions that had indicated the use of ECMO, such as cardiac arrest and cardiogenic shock [32]. There is also always the possibility of bleeding from the access site of the cannula. The configuration of the ECMO set-up may also play a role, with the risk of bleeding found to be lower in peripheral VA-ECMO compared to central [91].

The burden of bleeding in ECMO may be substantial, and as reported in the use of ECMO in the ANZ ECMO study, the main cause of mortality there was hemorrhage [5]. In a study of the ELSO registry, bleeding events associated with mortality included intracranial, pulmonary, and gastrointestinal hemorrhage [92]. The patients who require ECMO are typically critically ill and thus may have impaired hemostasis. Thrombocytopenia can also be found in up to 50% of ICU patients and is associated with increased bleeding [93–95]. Practices may vary, but there are reports describing a target platelet count greater than $100x10^9/L$ [96]. It is of course important to also consider the role that disseminated intravascular coagulation (DIC) may contribute in the event of development within a patient on ECMO. As platelets and coagulation factors are consumed, there can be severe bleeding and thrombosis [97]. In the occurrence of bleeding, anticoagulative therapy would need to be held or reduced, which has been shown to be safe in some reports for up to three days [98, 99]. In one study of eight patients with peripherally-inserted ECMO, hemostasis following a bleed was successfully established through interventions including ultrasound-guided thrombin injection, *n*-butyl cyanoacrylate, stent grafts and coil embolization [100]. In these cases, the bleeding was a complication of the percutaneous cannulation and interventional methods were pursued as the patients were medically unsuitable for surgical revision.

Other vascular complications that ECMO patients may face include the development of pseudoaneurysms at the access site, arterial dissections, and hematomas [101]. For pseudoaneurysms, following confirmation with an ultrasound, management can include either thrombin injection or surgical intervention depending on the size [102]. Symptomatic dissections require a stent and for hematomas, the anticoagulation must be revised along with possible need for transfusion or in severe cases, embolization or surgery [101]. To avoid bleeding during decannulation of percutaneous cannulation, closure devices have been tested for increased efficacy and complication avoidance [103–105].

## 15.3 Infection

Infection can occur with ECMO, including bacteremia and sepsis with longer duration of support correlating to higher incidence of infection. It is estimated that within two weeks of ECMO initiation, 53% of adults acquire an infection and that mortality is as high as 60% in patients with infection [8]. Sources of infection include both the access site as well as urinary tract due to prolonged indwelling catheter use and surgical wound infections [32]. Contributing to this burden of infection is the risk inherent to the machinery of ECMO and the use of multiple cannulation sites, ventilators, and sometimes an open chest. The configuration of the ECMO system itself can introduce risk, with rates of infection higher in VA-ECMO relative to VV-ECMO [106–109]. Additionally, beyond the risk due to mechanical and logistical components of ECMO, the effect which ECMO initiation may have on the innate and adaptive immune system have also been reported to impact infection [110]. The underlying condition of the patient prior to ECMO is also a factor, with higher sequential organ failure assessment (SOFA) scores acting as a risk factor in infection as a complication [108, 111].The longer a patient remains on ECMO, the higher the likelihood of infection development [112].

The consequent rate of infection observed in ECMO studies ranges, with values from 9% to 65% [113–115]. The most common infection was that of the bloodstream with a prevalence of 3–18% of cases and an incidence of 2.98–20.55 episodes per 1000 days [112]. Other reports have noted lower respiratory tract infections occurring at an incidence of 24.4 episodes/1000 days and infections at the surgical site in 0.6–14.7% of cases [112, 116, 117]. Relative rates of infection appear to be higher in adults on ECMO than children [106, 107].

A documented infection after ECMO must be managed with the appropriate antibiotics, but the pharmacokinetics of these drugs when given in the context of extracorporeal support is an area in need of further research [112]. Changes such as increased volume of distribution in vancomycin and cefotaxime and reduced clearance of gentamicin have been noted [118–120]. In evaluating patients for the development of infection and fever as a reflection of that infection, it is important to remain cognizant that the ECMO circuit is maintaining body temperature with heating or cooling of the outflow. In looking to reduce infections, double-lumen cannulation in VV-ECMO has lower risk of infection development, which could be a consideration during ECMO initiation [106].

## 15.4 Inflammatory Response

When ECMO is initiated, there is a known inflammatory reaction that takes place owing to the contact of patient blood with the circuit, activating coagulation and inflammatory pathways [121–124]. With activation of factor XII in the coagulation cascade occurring within minutes of ECMO initiation [125], the intrinsic pathway is quickly initiated and ultimately thrombin is activated. Thrombin is not only important in coagulation but increases neutrophil activation and leads to expression of pro-inflammatory cytokines [126–129]. Other mediators of inflammation within the ECMO system include platelets and the complement system [124, 130–133]. It is thought, however, that the use of heparin in more recent circuits is tied to reduced complement activation [134]. Activated complement component C3b was shown to reach its peak level within 60 minutes of ECMO initiation but ultimately fell to a baseline by two days, an important consideration when taking into account the longer duration of patients on ECMO [124].

In response to complement, the endothelium can become activated, up-regulating P-selectin [135]. The rise in pro-inflammatory cytokines further promotes endothelial activation with the downstream effect of adhesion and migration of activated neutrophils, the infiltration of which is associated with end-organ damage in ECMO [122, 136]. Neutrophil activation rises within thirty minutes of ECMO initiation, reaching peak levels over hours, and declining with their activation resulting in neutrophil degranulation and enzyme release [136–138].

Cytokines, which are key mediators in inflammation with widespread effects, have also been demonstrated to rise within ECMO. TNF-α in a porcine model of ECMO has been shown to be released from gut mast cells and associated in clinical studies of neonates with lower survival [122, 138, 139]. IL-6, another potent proinflammatory cytokine, is elevated within ECMO [140–144].

In another porcine model of lipopolysaccharide (LPS)-induced acute lung injury, there were significantly greater levels of proinflammatory cytokines IL-6, IL-12 and TNF-α in those who were also placed on ECMO compared to those without ECMO [145]. Monitoring the rate of inflammation as it develops with the use of ECMO may then become an important clinical measure. Exhaled breath particles which originate from the respiratory tract lining fluid of the lung could be used as a surrogate to understand the extent of lung damage as it develops, as particle flow rate was found to be increased in ECMO-treated porcine compared to those without. These findings were corroborated in human patients who also experienced increasing particle flow rate as their condition deteriorated on ECMO [145]. Bedside measures which correspond to the underlying state of the lung could become more important as the prevalence of ECMO increases.

Importantly, IL-6 has correlated with survival within patients with survivors demonstrating lowering levels after reaching a peak while non-survivors experienced maintained elevations in the cytokine [146]. It would follow accordingly that therapies which aim to reduce circulating cytokines could theoretically benefit patients experiencing high levels of inflammation, such as persistently elevated IL-6. Trials using devices such as cytokine adsorbers are currently being investigated. In a single-center randomized control of cytokine adsorption in COVID-19 pneumonia placed in line with the ECMO circuit for 72 hours, levels of IL-6 were not significantly decreased and there was a negative effect on survival, leading to a caution of use of the device within the first days of ECMO support in a COVID-19 population [147]. Another single-center study of severe ARDS with VV-ECMO found that cytokine adsorption was associated with reduced morality and reduced vasopressor need [148]. Other studies with the device in line with dialysis circuits have found reduced IL-6 and improvement in mortality with use of cytokine adsorption devices in septic patients [149–151]. It appears then, that not only is there a need for more studies with larger patient cohorts to gather evidence, but also a need to study timing of treatment. It is plausible that this could be a case where cytokine removal would be the right treatment if given at the right time, but potentially harmful if started at the wrong one. There may also be a relationship with the method of administration, as it could be placed in line with ECMO, dialysis or as a standalone circuit which may impact the efficacy and outcome.

## 15.5 Bridge to Transplant

With the current lack of available donor organs and prolonged waitlist times a potential transplant recipient may face, the need for techniques to help a patient survive their wait is great. For lung transplants, 54% of adults in 2019 spent between three months and two years waiting for a donor organ [152]. As the patient's condition may deteriorate while waiting for a transplant, "bridging to transplant" with ECMO is considered a viable option for helping a critically ill patient survive until the procedure.

Importantly within this subsection of ECMO use, mortality rates in this patient subpopulation have been tied to the ECMO volume of the lung transplant center. Those centers with more time having utilized ECMO and those with greater experience appear to observe improved outcomes. High volume centers have noted survival rates in their bridged population of 70–90% at one-year post-transplantation [153–155]. In a meta-analysis of center volume, low volume was associated with post-transplant hazard [156]. High volume centers have been able to report similar survival rates post-transplantation in their patients who are bridged to transplant to those who do not receive ECMO prior to transplantation [157]. When considering the effect that experience may have as a center gains practice with the technique, the analysis of outcomes over "eras" within a high-volume center highlighted improvement over time [158]. The mounting success of programs with a bridge-to-transplant program has resulted in consideration of transplanting a greater breadth of patients as one study found that while the bridged patients had lower lung allocation scores than non-bridged patients, they had no difference in survival [159].

The use of ambulatory ECMO has shown a benefit of preventing deconditioning in this patient population. Loss of strength in the intensive care unit is associated with increased morbidity and mortality [160]. In patients bridged to transplant, ambulation during ECMO was an independent predictor for survival [158, 159]. The International Society for Heart and Lung Transplantation's 2021 consensus statement remarks that when placing lung transplant candidates on extracorporeal life support, they should preferably be awake and mobilized [161].

## 15.6 Conclusion

As ECMO gains a stronger foothold as a critical tool in the management of severe to fatal respiratory and cardiac failure, the exploration of important factors related to its use – from the potential for thrombosis and bleeding to the extent it elicits an inflammatory and infectious response – will continue to help our understanding of best practices. Importantly, protocols regarding the management of conditions like heparin dosing and antibiotic use should be analyzed for standardization. With the use of ECMO across many different patient populations, such as those with cardiac arrest and those awaiting lung transplant, these settings will need customization for the specific patients whom it seeks to manage. Globally, the use of ECMO in centers is likely to grow and with it, improvements in patient mortality and morbidity.

**Acknowledgements** The authors gratefully acknowledge funding received from the Wallenberg Molecular Medicine Foundation and the Knut and Alice Wallenberg Foundation. The authors thank Dr. Martin Stenlo for his contributions in the editing process.

## References

1. Zapol WM, Snider MT, Hill JD, et al. Extracorporeal membrane oxygenation in severe acute respiratory failure. A randomized prospective study. *JAMA*. Nov 16 1979;242(20):2193–6. doi:https://doi.org/10.1001/jama.242.20.2193
2. Chalwin RP, Moran JL, Graham PL. The role of extracorporeal membrane oxygenation for treatment of the adult respiratory distress syndrome: review and quantitative analysis. *Anaesth Intensive Care*. Mar 2008;36(2):152–61. doi:https://doi.org/10.1177/0310057X0803600203
3. Hastings SL, Pellegrino VA, Preovolos A, Salamonsen RF. Survey of adult extracorporeal membrane oxygenation (ECMO) practice and attitudes among Australian and New Zealand intensivists. *Crit Care Resusc*. Mar 2008;10(1):46.
4. Peek GJ, Mugford M, Tiruvoipati R, et al. Efficacy and economic assessment of conventional ventilatory support versus extracorporeal membrane oxygenation for severe adult respiratory failure (CESAR): a multicentre randomised controlled trial. *Lancet*. Oct 17 2009;374(9698):1351–63. doi:https://doi.org/10.1016/S0140-6736(09)61069-2
5. Australia, New Zealand Extracorporeal Membrane Oxygenation Influenza I, Davies A, et al. Extracorporeal Membrane Oxygenation for 2009 Influenza A(H1N1) Acute Respiratory Distress Syndrome. *JAMA*. Nov 4 2009;302(17):1888–95. doi:https://doi.org/10.1001/jama.2009.1535
6. Quintel M, Bartlett RH, Grocott MPW, et al. Extracorporeal Membrane Oxygenation for Respiratory Failure. *Anesthesiology*. May 2020;132(5):1257–1276. doi:https://doi.org/10.1097/ALN.0000000000003221
7. Organization ELS. ELSO Live Registry Dashboard of ECMO Patient Data. Accessed May 31, 2022. https://www.elso.org/Registry/ELSOLiveRegistryDashboard.aspx
8. Guglin M, Zucker MJ, Bazan VM, et al. Venoarterial ECMO for Adults: JACC Scientific Expert Panel. *J Am Coll Cardiol*. Feb 19 2019;73(6):698–716. doi:https://doi.org/10.1016/j.jacc.2018.11.038
9. Organization ELS. General Guideline for all ECLS Cases. Accessed May 31, 2022. https://www.elso.org/Portals/0/ELSO%20Guidelines%20General%20All%20ECLS%20Version%201_4.pdf
10. Banfi C, Pozzi M, Siegenthaler N, et al. Veno-venous extracorporeal membrane oxygenation: cannulation techniques. *J Thorac Dis*. Dec 2016;8(12):3762–3773. doi:https://doi.org/10.21037/jtd.2016.12.88
11. Pooboni SK, Gulla KM. Vascular access in ECMO. *Indian J Thorac Cardiovasc Surg*. Apr 2021;37(Suppl 2):221–231. doi:https://doi.org/10.1007/s12055-020-00999-w
12. Lindholm JA. Cannulation for veno-venous extracorporeal membrane oxygenation. *J Thorac Dis*. Mar 2018;10(Suppl 5):S606–S612. doi:https://doi.org/10.21037/jtd.2018.03.101
13. Abrams D, Bacchetta M, Brodie D. Recirculation in venovenous extracorporeal membrane oxygenation. *ASAIO J*. Mar–Apr 2015;61(2):115–21. doi:https://doi.org/10.1097/MAT.0000000000000179
14. Xie A, Yan TD, Forrest P. Recirculation in venovenous extracorporeal membrane oxygenation. *J Crit Care*. Dec 2016;36:107–110. doi:https://doi.org/10.1016/j.jcrc.2016.05.027

15. Kwak J, Majewski MB, Jellish WS. Extracorporeal Membrane Oxygenation: The New Jack-of-All-Trades? *J Cardiothorac Vasc Anesth*. Jan 2020;34(1):192–207. doi:https://doi.org/10.1053/j.jvca.2019.09.031
16. Kessler A, Coker B, Townsley M, Zaky A. Extracorporeal Membrane Oxygenator Rotational Cannula Catastrophe: A Role of Echocardiography in Rescue. *J Cardiothorac Vasc Anesth*. Jun 2016;30(3):720–4. doi:https://doi.org/10.1053/j.jvca.2015.07.010
17. Banayan JM, Barry A, Chaney MA. Right Ventricular Rupture during Insertion of an Avalon Elite((R)) Catheter. *J Cardiothorac Vasc Anesth*. Aug 2016;30(4):e34–5. doi:https://doi.org/10.1053/j.jvca.2015.12.022
18. Tarola CL, Nagpal AD. Internal Jugular Vein Avulsion Complicating Dual-Lumen VV-ECMO Cannulation: An Unreported Complication of Avalon Cannulas. *Can J Cardiol*. Dec 2016;32(12):1576 e5–1576 e6. doi:https://doi.org/10.1016/j.cjca.2016.02.049
19. Hirose H, Yamane K, Marhefka G, Cavarocchi N. Right ventricular rupture and tamponade caused by malposition of the Avalon cannula for venovenous extracorporeal membrane oxygenation. *J Cardiothorac Surg*. Apr 20 2012;7:36. doi:https://doi.org/10.1186/1749-8090-7-36
20. Krivitski N, Galyanov G, Gehron JM, Bandorski D, Boning A. New noninvasive methodology to measure cardiac output in veno-venous extracorporeal membrane oxygenation patients. *Perfusion*. May 2020;35(1_suppl):73–80. doi:https://doi.org/10.1177/0267659120908507
21. Walker JL, Gelfond J, Zarzabal LA, Darling E. Calculating mixed venous saturation during veno-venous extracorporeal membrane oxygenation. *Perfusion*. Sep 2009;24(5):333–9. doi:https://doi.org/10.1177/0267659109354790
22. van Heijst AF, van der Staak FH, de Haan AF, et al. Recirculation in double lumen catheter veno-venous extracorporeal membrane oxygenation measured by an ultrasound dilution technique. *ASAIO J*. Jul–Aug 2001;47(4):372–6. doi:https://doi.org/10.1097/00002480-200107000-00015
23. Investigators E-PS, International EN. Early mobilisation during extracorporeal membrane oxygenation was safe and feasible: a pilot randomised controlled trial. *Intensive Care Med*. May 2020;46(5):1057–1059. doi:https://doi.org/10.1007/s00134-020-05994-8
24. Ko Y, Cho YH, Park YH, et al. Feasibility and Safety of Early Physical Therapy and Active Mobilization for Patients on Extracorporeal Membrane Oxygenation. *ASAIO J*. Sep–Oct 2015;61(5):564–8. doi:https://doi.org/10.1097/MAT.0000000000000239
25. Abrams D, Javidfar J, Farrand E, et al. Early mobilization of patients receiving extracorporeal membrane oxygenation: a retrospective cohort study. *Crit Care*. Feb 27 2014;18(1):R38. doi:https://doi.org/10.1186/cc13746
26. Wells CL, Forrester J, Vogel J, Rector R, Tabatabai A, Herr D. Safety and Feasibility of Early Physical Therapy for Patients on Extracorporeal Membrane Oxygenator: University of Maryland Medical Center Experience. *Crit Care Med*. Jan 2018;46(1):53–59. doi:https://doi.org/10.1097/CCM.0000000000002770
27. Chen YS, Lin JW, Yu HY, et al. Cardiopulmonary resuscitation with assisted extracorporeal life-support versus conventional cardiopulmonary resuscitation in adults with in-hospital cardiac arrest: an observational study and propensity analysis. *Lancet*. Aug 16 2008;372(9638):554–61. doi:https://doi.org/10.1016/S0140-6736(08)60958-7
28. Maekawa K, Tanno K, Hase M, Mori K, Asai Y. Extracorporeal cardiopulmonary resuscitation for patients with out-of-hospital cardiac arrest of cardiac origin: a propensity-matched study and predictor analysis. *Crit Care Med*. May 2013;41(5):1186–96. doi:https://doi.org/10.1097/CCM.0b013e31827ca4c8
29. Johnson NJ, Acker M, Hsu CH, et al. Extracorporeal life support as rescue strategy for out-of-hospital and emergency department cardiac arrest. *Resuscitation*. Nov 2014;85(11):1527–32. doi:https://doi.org/10.1016/j.resuscitation.2014.08.028
30. Lorusso R, Shekar K, MacLaren G, et al. ELSO Interim Guidelines for Venoarterial Extracorporeal Membrane Oxygenation in Adult Cardiac Patients. *ASAIO J*. Aug 1 2021;67(8):827–844. doi:https://doi.org/10.1097/MAT.0000000000001510

31. Combes A, Price S, Slutsky AS, Brodie D. Temporary circulatory support for cardiogenic shock. *Lancet*. Jul 18 2020;396(10245):199–212. doi:https://doi.org/10.1016/S0140-6736(20)31047-3
32. Tsangaris A, Alexy T, Kalra R, et al. Overview of Veno-Arterial Extracorporeal Membrane Oxygenation (VA-ECMO) Support for the Management of Cardiogenic Shock. *Front Cardiovasc Med*. 2021;8:686558. doi:https://doi.org/10.3389/fcvm.2021.686558
33. Kawashima D, Gojo S, Nishimura T, et al. Left ventricular mechanical support with Impella provides more ventricular unloading in heart failure than extracorporeal membrane oxygenation. *ASAIO J*. May–Jun 2011;57(3):169–76. doi:https://doi.org/10.1097/MAT.0b013e31820e121c
34. Uriel N, Sayer G, Annamalai S, Kapur NK, Burkhoff D. Mechanical Unloading in Heart Failure. *J Am Coll Cardiol*. Jul 31 2018;72(5):569–580. doi:https://doi.org/10.1016/j.jacc.2018.05.038
35. Ostadal P, Mlcek M, Kruger A, et al. Increasing venoarterial extracorporeal membrane oxygenation flow negatively affects left ventricular performance in a porcine model of cardiogenic shock. *J Transl Med*. Aug 15 2015;13:266. doi:https://doi.org/10.1186/s12967-015-0634-6
36. Petroni T, Harrois A, Amour J, et al. Intra-aortic balloon pump effects on macrocirculation and microcirculation in cardiogenic shock patients supported by venoarterial extracorporeal membrane oxygenation*. *Crit Care Med*. Sep 2014;42(9):2075–82. doi:https://doi.org/10.1097/CCM.0000000000000410
37. Schrage B, Burkhoff D, Rubsamen N, et al. Unloading of the Left Ventricle During Venoarterial Extracorporeal Membrane Oxygenation Therapy in Cardiogenic Shock. *JACC Heart Fail*. Dec 2018;6(12):1035–1043. doi:https://doi.org/10.1016/j.jchf.2018.09.009
38. Russo JJ, Aleksova N, Pitcher I, et al. Left Ventricular Unloading During Extracorporeal Membrane Oxygenation in Patients With Cardiogenic Shock. *J Am Coll Cardiol*. Feb 19 2019;73(6):654–662. doi:https://doi.org/10.1016/j.jacc.2018.10.085
39. Schrage B, Becher PM, Bernhardt A, et al. Left Ventricular Unloading Is Associated With Lower Mortality in Patients With Cardiogenic Shock Treated With Venoarterial Extracorporeal Membrane Oxygenation: Results From an International, Multicenter Cohort Study. *Circulation*. Dec 2020;142(22):2095–2106. doi:https://doi.org/10.1161/CIRCULATIONAHA.120.048792
40. Brechot N, Demondion P, Santi F, et al. Intra-aortic balloon pump protects against hydrostatic pulmonary oedema during peripheral venoarterial-extracorporeal membrane oxygenation. *Eur Heart J Acute Cardiovasc Care*. Feb 2018;7(1):62–69. doi:https://doi.org/10.1177/2048872617711169
41. Fuhrman BP, Hernan LJ, Rotta AT, Heard CM, Rosenkranz ER. Pathophysiology of cardiac extracorporeal membrane oxygenation. *Artif Organs*. Nov 1999;23(11):966–9. doi:https://doi.org/10.1046/j.1525-1594.1999.06484.x
42. Falk L, Sallisalmi M, Lindholm JA, et al. Differential hypoxemia during venoarterial extracorporeal membrane oxygenation. *Perfusion*. Apr 2019;34(1_suppl):22–29. doi:https://doi.org/10.1177/0267659119830513
43. Kato J, Seo T, Ando H, Takagi H, Ito T. Coronary arterial perfusion during venoarterial extracorporeal membrane oxygenation. *J Thorac Cardiovasc Surg*. Mar 1996;111(3):630–6. doi:https://doi.org/10.1016/s0022-5223(96)70315-x
44. Gander JW, Fisher JC, Reichstein AR, et al. Limb ischemia after common femoral artery cannulation for venoarterial extracorporeal membrane oxygenation: an unresolved problem. *J Pediatr Surg*. Nov 2010;45(11):2136–40. doi:https://doi.org/10.1016/j.jpedsurg.2010.07.005
45. Jaski BE, Lingle RJ, Overlie P, et al. Long-term survival with use of percutaneous extracorporeal life support in patients presenting with acute myocardial infarction and cardiovascular collapse. *ASAIO J*. Nov–Dec 1999;45(6):615–8. doi:https://doi.org/10.1097/00002480-199911000-00018
46. Buckley E, Sidebotham D, McGeorge A, Roberts S, Allen SJ, Beca J. Extracorporeal membrane oxygenation for cardiorespiratory failure in four patients with pandemic H1N1 2009

influenza virus and secondary bacterial infection. *Br J Anaesth*. Mar 2010;104(3):326–9. doi:https://doi.org/10.1093/bja/aep396
47. Zangrillo A, Landoni G, Biondi-Zoccai G, et al. A meta-analysis of complications and mortality of extracorporeal membrane oxygenation. *Crit Care Resusc*. Sep 2013;15(3):172–8.
48. Lamb KM, DiMuzio PJ, Johnson A, et al. Arterial protocol including prophylactic distal perfusion catheter decreases limb ischemia complications in patients undergoing extracorporeal membrane oxygenation. *J Vasc Surg*. Apr 2017;65(4):1074–1079. doi:https://doi.org/10.1016/j.jvs.2016.10.059
49. Lamb KM, Hirose H, Cavarocchi NC. Preparation and technical considerations for percutaneous cannulation for veno-arterial extracorporeal membrane oxygenation. *J Card Surg*. Mar 2013;28(2):190–2. doi:https://doi.org/10.1111/jocs.12058
50. Spurlock DJ, Toomasian JM, Romano MA, Cooley E, Bartlett RH, Haft JW. A simple technique to prevent limb ischemia during veno-arterial ECMO using the femoral artery: the posterior tibial approach. *Perfusion*. Mar 2012;27(2):141–5. doi:https://doi.org/10.1177/0267659111430760
51. Kaufeld T, Beckmann E, Ius F, et al. Risk factors for critical limb ischemia in patients undergoing femoral cannulation for venoarterial extracorporeal membrane oxygenation: Is distal limb perfusion a mandatory approach? *Perfusion*. Sep 2019;34(6):453–459. doi:https://doi.org/10.1177/0267659119827231
52. Yau P, Xia Y, Shariff S, et al. Factors Associated with Ipsilateral Limb Ischemia in Patients Undergoing Femoral Cannulation Extracorporeal Membrane Oxygenation. *Ann Vasc Surg*. Jan 2019;54:60–65. doi:https://doi.org/10.1016/j.avsg.2018.08.073
53. Calderon D, El-Banayosy A, Koerner MM, Reed AB, Aziz F. Modified T-Graft for Extracorporeal Membrane Oxygenation in a Patient with Small-Caliber Femoral Arteries. *Tex Heart Inst J*. Dec 2015;42(6):537–9. doi:https://doi.org/10.14503/THIJ-14-4728
54. Jackson KW, Timpa J, McIlwain RB, et al. Side-arm grafts for femoral extracorporeal membrane oxygenation cannulation. *Ann Thorac Surg*. Nov 2012;94(5):e111–2. doi:https://doi.org/10.1016/j.athoracsur.2012.05.064
55. Foley PJ, Morris RJ, Woo EY, et al. Limb ischemia during femoral cannulation for cardiopulmonary support. *J Vasc Surg*. Oct 2010;52(4):850–3. doi:https://doi.org/10.1016/j.jvs.2010.05.012
56. Read R, St Cyr J, Tornabene S, Whitman G. Improved cannulation method for extracorporeal membrane oxygenation. *Ann Thorac Surg*. Oct 1990;50(4):670–1. doi:https://doi.org/10.1016/0003-4975(90)90219-v
57. Marasco SF, Tutungi E, Vallance SA, et al. A Phase 1 Study of a Novel Bidirectional Perfusion Cannula in Patients Undergoing Femoral Cannulation for Cardiac Surgery. *Innovations (Phila)*. Mar/Apr 2018;13(2):97–103. doi:https://doi.org/10.1097/IMI.0000000000000489
58. Napp LC, Kuhn C, Hoeper MM, et al. Cannulation strategies for percutaneous extracorporeal membrane oxygenation in adults. *Clin Res Cardiol*. Apr 2016;105(4):283–96. doi:https://doi.org/10.1007/s00392-015-0941-1
59. Kustermann J, Gehrmann A, Kredel M, Wurmb T, Roewer N, Muellenbach RM. [Acute respiratory distress syndrome and septic cardiomyopathy: successful application of veno-venoarterial extracorporeal membrane oxygenation]. *Anaesthesist*. Aug 2013;62(8):639–43. Akutes Lungenversagen und septische Kardiomyopathie: Erfolgreicher Einsatz der veno-venoarteriellen extrakorporalen Membranoxygenierung. doi:https://doi.org/10.1007/s00101-013-2213-7
60. Ius F, Sommer W, Tudorache I, et al. Veno-veno-arterial extracorporeal membrane oxygenation for respiratory failure with severe haemodynamic impairment: technique and early outcomes. *Interact Cardiovasc Thorac Surg*. Jun 2015;20(6):761–7. doi:https://doi.org/10.1093/icvts/ivv035
61. Moravec R, Neitzel T, Stiller M, et al. First experiences with a combined usage of veno-arterial and veno-venous ECMO in therapy-refractory cardiogenic shock patients with cerebral hypoxemia. *Perfusion*. May 2014;29(3):200–9. doi:https://doi.org/10.1177/0267659113502832

62. Werner NL, Coughlin M, Cooley E, et al. The University of Michigan Experience with Veno-Venoarterial Hybrid Mode of Extracorporeal Membrane Oxygenation. *ASAIO J.* Sep–Oct 2016;62(5):578–83. doi:https://doi.org/10.1097/MAT.0000000000000405
63. Bein T, Aubron C, Papazian L. Focus on ECMO and ECCO2R in ARDS patients. *Intensive Care Med.* Sep 2017;43(9):1424–1426. doi:https://doi.org/10.1007/s00134-017-4882-1
64. Fanelli V, Ranieri MV, Mancebo J, et al. Feasibility and safety of low-flow extracorporeal carbon dioxide removal to facilitate ultra-protective ventilation in patients with moderate acute respiratory distress sindrome. *Crit Care.* Feb 10 2016;20:36. doi:https://doi.org/10.1186/s13054-016-1211-y
65. Sklar MC, Beloncle F, Katsios CM, Brochard L, Friedrich JO. Extracorporeal carbon dioxide removal in patients with chronic obstructive pulmonary disease: a systematic review. *Intensive Care Med.* Oct 2015;41(10):1752–62. doi:https://doi.org/10.1007/s00134-015-3921-z
66. Inal V, Efe S. Extracorporeal carbon dioxide removal (ECCO2R) in COPD and ARDS patients with severe hypercapnic respiratory failure. A retrospective case-control study. *Turk J Med Sci.* Aug 30 2021;51(4):2127–2135. doi:https://doi.org/10.3906/sag-2012-151
67. Fanelli V, Costamagna A, Ranieri VM. Extracorporeal support for severe acute respiratory failure. *Semin Respir Crit Care Med.* Aug 2014;35(4):519–27. doi:https://doi.org/10.1055/s-0034-1383866
68. Combes A, Fanelli V, Pham T, Ranieri VM, European Society of Intensive Care Medicine Trials G, the "Strategy of Ultra-Protective lung ventilation with Extracorporeal CORfN-OmtsAi. Feasibility and safety of extracorporeal CO2 removal to enhance protective ventilation in acute respiratory distress syndrome: the SUPERNOVA study". *Intensive Care Med.* May 2019;45(5):592–600. doi:https://doi.org/10.1007/s00134-019-05567-4
69. Murphy DA, Hockings LE, Andrews RK, et al. Extracorporeal membrane oxygenation-hemostatic complications. *Transfus Med Rev.* Apr 2015;29(2):90–101. doi:https://doi.org/10.1016/j.tmrv.2014.12.001
70. Horton AaB, Warwick. Pump-induced haemolysis: is the constrained vortex pump better or worse than the roller pump? *Perfusion.* 1992 1992;7(2):103–108. doi:https://doi.org/10.1177/026765919200700204
71. Reul HM, Akdis M. Blood pumps for circulatory support. *Perfusion.* Jul 2000;15(4):295–311. doi:https://doi.org/10.1177/026765910001500404
72. Urlesberger B, Zobel G, Zenz W, et al. Activation of the clotting system during extracorporeal membrane oxygenation in term newborn infants. *J Pediatr.* Aug 1996;129(2):264–8. doi:https://doi.org/10.1016/s0022-3476(96)70252-4
73. Xie A, Lo P, Yan TD, Forrest P. Neurologic Complications of Extracorporeal Membrane Oxygenation: A Review. *J Cardiothorac Vasc Anesth.* Oct 2017;31(5):1836–1846. doi:https://doi.org/10.1053/j.jvca.2017.03.001
74. Omar HR, Mirsaeidi M, Shumac J, Enten G, Mangar D, Camporesi EM. Incidence and predictors of ischemic cerebrovascular stroke among patients on extracorporeal membrane oxygenation support. *J Crit Care.* Apr 2016;32:48–51. doi:https://doi.org/10.1016/j.jcrc.2015.11.009
75. Lorusso R, Barili F, Mauro MD, et al. In-Hospital Neurologic Complications in Adult Patients Undergoing Venoarterial Extracorporeal Membrane Oxygenation: Results From the Extracorporeal Life Support Organization Registry. *Crit Care Med.* Oct 2016;44(10):e964–72. doi:https://doi.org/10.1097/CCM.0000000000001865
76. Iacobelli R, Fletcher-Sandersjoo A, Lindblad C, Keselman B, Thelin EP, Broman LM. Predictors of brain infarction in adult patients on extracorporeal membrane oxygenation: an observational cohort study. *Sci Rep.* Feb 15 2021;11(1):3809. doi:https://doi.org/10.1038/s41598-021-83157-5
77. Zanatta P, Forti A, Bosco E, et al. Microembolic signals and strategy to prevent gas embolism during extracorporeal membrane oxygenation. *J Cardiothorac Surg.* Feb 4 2010;5:5. doi:https://doi.org/10.1186/1749-8090-5-5

78. Lubnow M, Philipp A, Dornia C, et al. D-dimers as an early marker for oxygenator exchange in extracorporeal membrane oxygenation. *J Crit Care*. Jun 2014;29(3):473 e1–5. doi:https://doi.org/10.1016/j.jcrc.2013.12.008
79. Trigonis R, Smith N, Porter S, et al. Efficacy of Bivalirudin for Therapeutic Anticoagulation in COVID-19 Patients Requiring ECMO Support. *J Cardiothorac Vasc Anesth*. Feb 2022;36(2):414–418. doi:https://doi.org/10.1053/j.jvca.2021.10.026
80. Gu YJ, van Oeveren W, Akkerman C, Boonstra PW, Huyzen RJ, Wildevuur CR. Heparin-coated circuits reduce the inflammatory response to cardiopulmonary bypass. *Ann Thorac Surg*. Apr 1993;55(4):917–22. doi:https://doi.org/10.1016/0003-4975(93)90117-z
81. Sakiyama-Elbert SE. Incorporation of heparin into biomaterials. *Acta Biomater*. Apr 2014;10(4):1581–7. doi:https://doi.org/10.1016/j.actbio.2013.08.045
82. MacLaren G, Combes A, Bartlett RH. Contemporary extracorporeal membrane oxygenation for adult respiratory failure: life support in the new era. *Intensive Care Med*. Feb 2012;38(2):210–20. doi:https://doi.org/10.1007/s00134-011-2439-2
83. Muntean W. Coagulation and anticoagulation in extracorporeal membrane oxygenation. *Artif Organs*. Nov 1999;23(11):979–83. doi:https://doi.org/10.1046/j.1525-1594.1999.06451.x
84. Sohn N, Marcoux J, Mycyk T, Krahn J, Meng Q. The impact of different biocompatible coated cardiopulmonary bypass circuits on inflammatory response and oxidative stress. *Perfusion*. Jul 2009;24(4):231–7. doi:https://doi.org/10.1177/0267659109351218
85. de Vroege R, Huybregts R, van Oeveren W, et al. The impact of heparin-coated circuits on hemodynamics during and after cardiopulmonary bypass. *Artif Organs*. Jun 2005;29(6):490–7. doi:https://doi.org/10.1111/j.1525-1594.2005.29083.x
86. de Vroege R, van Oeveren W, van Klarenbosch J, et al. The impact of heparin-coated cardiopulmonary bypass circuits on pulmonary function and the release of inflammatory mediators. *Anesth Analg*. Jun 2004;98(6):1586–1594. doi:https://doi.org/10.1213/01.ANE.0000114551.64123.79
87. Natt B, Hypes C, Basken R, Malo J, Kazui T, Mosier J. Suspected Heparin-Induced Thrombocytopenia in Patients Receiving Extracorporeal Membrane Oxygenation. *J Extra Corpor Technol*. Mar 2017;49(1):54–58.
88. Dager WE, Gosselin RC, Yoshikawa R, Owings JT. Lepirudin in heparin-induced thrombocytopenia and extracorporeal membranous oxygenation. *Ann Pharmacother*. Apr 2004;38(4):598–601. doi:https://doi.org/10.1345/aph.1D436
89. Balasubramanian SK, Tiruvoipati R, Chatterjee S, Sosnowski A, Firmin RK. Extracorporeal membrane oxygenation with lepirudin anticoagulation for Wegener's granulomatosis with heparin-induced thrombocytopenia. *ASAIO J*. Jul–Aug 2005;51(4):477–9. doi:https://doi.org/10.1097/01.mat.0000169123.21946.31
90. Chlebowski MM, Baltagi S, Carlson M, Levy JH, Spinella PC. Clinical controversies in anticoagulation monitoring and antithrombin supplementation for ECMO. *Crit Care*. Jan 20 2020;24(1):19. doi:https://doi.org/10.1186/s13054-020-2726-9
91. Raffa GM, Kowalewski M, Brodie D, et al. Meta-Analysis of Peripheral or Central Extracorporeal Membrane Oxygenation in Postcardiotomy and Non-Postcardiotomy Shock. *Ann Thorac Surg*. Jan 2019;107(1):311–321. doi:https://doi.org/10.1016/j.athoracsur.2018.05.063
92. Nunez JI, Gosling AF, O'Gara B, et al. Bleeding and thrombotic events in adults supported with venovenous extracorporeal membrane oxygenation: an ELSO registry analysis. *Intensive Care Med*. Feb 2022;48(2):213–224. doi:https://doi.org/10.1007/s00134-021-06593-x
93. Vanderschueren S, De Weerdt A, Malbrain M, et al. Thrombocytopenia and prognosis in intensive care. *Crit Care Med*. Jun 2000;28(6):1871–6. doi:https://doi.org/10.1097/00003246-200006000-00031
94. Crowther MA, Cook DJ, Meade MO, et al. Thrombocytopenia in medical-surgical critically ill patients: prevalence, incidence, and risk factors. *J Crit Care*. Dec 2005;20(4):348–53. doi:https://doi.org/10.1016/j.jcrc.2005.09.008

95. Williamson DR, Albert M, Heels-Ansdell D, et al. Thrombocytopenia in critically ill patients receiving thromboprophylaxis: frequency, risk factors, and outcomes. *Chest*. Oct 2013;144(4):1207–1215. doi:https://doi.org/10.1378/chest.13-0121
96. Bembea MM, Annich G, Rycus P, Oldenburg G, Berkowitz I, Pronovost P. Variability in anticoagulation management of patients on extracorporeal membrane oxygenation: an international survey. *Pediatr Crit Care Med*. Feb 2013;14(2):e77–84. doi:https://doi.org/10.1097/PCC.0b013e31827127e4
97. Levi M, Ten Cate H. Disseminated intravascular coagulation. *N Engl J Med*. Aug 19 1999;341(8):586–92. doi:https://doi.org/10.1056/NEJM199908193410807
98. Chung YS, Cho DY, Sohn DS, et al. Is Stopping Heparin Safe in Patients on Extracorporeal Membrane Oxygenation Treatment? *ASAIO J*. Jan/Feb 2017;63(1):32–36. doi:https://doi.org/10.1097/MAT.0000000000000442
99. Wood KL, Ayers B, Gosev I, et al. Venoarterial-Extracorporeal Membrane Oxygenation Without Routine Systemic Anticoagulation Decreases Adverse Events. *Ann Thorac Surg*. May 2020;109(5):1458–1466. doi:https://doi.org/10.1016/j.athoracsur.2019.08.040
100. Zheng L, Kim PH, Shin JH, et al. Interventional treatment of bleeding complications due to percutaneous cannulation for peripheral extracorporeal membrane oxygenation. *Diagn Interv Imaging*. Jun 2019;100(6):337–345. doi:https://doi.org/10.1016/j.diii.2019.01.003
101. Pillai AK, Bhatti Z, Bosserman AJ, Mathew MC, Vaidehi K, Kalva SP. Management of vascular complications of extra-corporeal membrane oxygenation. *Cardiovasc Diagn Ther*. Jun 2018;8(3):372–377. doi:https://doi.org/10.21037/cdt.2018.01.11
102. Mishra A, Rao A, Pimpalwar Y. Ultrasound Guided Percutaneous Injection of Thrombin: Effective Technique for Treatment of Iatrogenic Femoral Pseudoaneurysms. *J Clin Diagn Res*. Apr 2017;11(4):TC04–TC06. doi:https://doi.org/10.7860/JCDR/2017/25582.9512
103. Hassan MF, Lawrence M, Lee D, Velazco J, Martin C, Reddy R. Simplified percutaneous VA ECMO decannulation using the MANTA vascular closure device: Initial US experience. *J Card Surg*. Jan 2020;35(1):217–221. doi:https://doi.org/10.1111/jocs.14308
104. Montero-Cabezas JM, van der Meer RW, van der Kley F, et al. Percutaneous Decannulation of Femoral Venoarterial ECMO Cannulas Using MANTA Vascular Closure Device. *Can J Cardiol*. Jun 2019;35(6):796 e9–796 e11. doi:https://doi.org/10.1016/j.cjca.2019.02.010
105. Hwang JW, Yang JH, Sung K, et al. Percutaneous removal using Perclose ProGlide closure devices versus surgical removal for weaning after percutaneous cannulation for venoarterial extracorporeal membrane oxygenation. *J Vasc Surg*. Apr 2016;63(4):998–1003 e1. doi:https://doi.org/10.1016/j.jvs.2015.10.067
106. Vogel AM, Lew DF, Kao LS, Lally KP. Defining risk for infectious complications on extracorporeal life support. *J Pediatr Surg*. Dec 2011;46(12):2260–4. doi:https://doi.org/10.1016/j.jpedsurg.2011.09.013
107. Bizzarro MJ, Conrad SA, Kaufman DA, Rycus P, Extracorporeal Life Support Organization Task Force on Infections EMO. Infections acquired during extracorporeal membrane oxygenation in neonates, children, and adults. *Pediatr Crit Care Med*. May 2011;12(3):277–81. doi:https://doi.org/10.1097/PCC.0b013e3181e28894
108. Aubron C, Cheng AC, Pilcher D, et al. Infections acquired by adults who receive extracorporeal membrane oxygenation: risk factors and outcome. *Infect Control Hosp Epidemiol*. Jan 2013;34(1):24–30. doi:https://doi.org/10.1086/668439
109. Coffin SE, Bell LM, Manning M, Polin R. Nosocomial infections in neonates receiving extracorporeal membrane oxygenation. *Infect Control Hosp Epidemiol*. Feb 1997;18(2):93–6. doi:https://doi.org/10.1086/647561
110. Frerou A, Lesouhaitier M, Gregoire M, et al. Venoarterial extracorporeal membrane oxygenation induces early immune alterations. *Crit Care*. Jan 6 2021;25(1):9. doi:https://doi.org/10.1186/s13054-020-03444-x
111. Pieri M, Agracheva N, Fumagalli L, et al. Infections occurring in adult patients receiving mechanical circulatory support: the two-year experience of an Italian National Referral

Tertiary Care Center. *Med Intensiva*. Oct 2013;37(7):468–75. doi:https://doi.org/10.1016/j.medin.2012.08.009
112. Biffi S, Di Bella S, Scaravilli V, et al. Infections during extracorporeal membrane oxygenation: epidemiology, risk factors, pathogenesis and prevention. *Int J Antimicrob Agents*. Jul 2017;50(1):9–16. doi:https://doi.org/10.1016/j.ijantimicag.2017.02.025
113. Lo Coco V, Lorusso R, Raffa GM, et al. Clinical complications during veno-arterial extracorporeal membrane oxigenation in post-cardiotomy and non post-cardiotomy shock: still the achille's heel. *J Thorac Dis*. Dec 2018;10(12):6993–7004. doi:https://doi.org/10.21037/jtd.2018.11.103
114. Schmidt M, Brechot N, Hariri S, et al. Nosocomial infections in adult cardiogenic shock patients supported by venoarterial extracorporeal membrane oxygenation. *Clin Infect Dis*. Dec 2012;55(12):1633–41. doi:https://doi.org/10.1093/cid/cis783
115. Bouadma L, Mourvillier B, Deiler V, et al. A multifaceted program to prevent ventilator-associated pneumonia: impact on compliance with preventive measures. *Crit Care Med*. Mar 2010;38(3):789–96. doi:https://doi.org/10.1097/CCM.0b013e3181ce21af
116. Biancari F, Perrotti A, Dalen M, et al. Meta-Analysis of the Outcome After Postcardiotomy Venoarterial Extracorporeal Membrane Oxygenation in Adult Patients. *J Cardiothorac Vasc Anesth*. Jun 2018;32(3):1175–1182. doi:https://doi.org/10.1053/j.jvca.2017.08.048
117. Rupprecht L, Lunz D, Philipp A, Lubnow M, Schmid C. Pitfalls in percutaneous ECMO cannulation. *Heart Lung Vessel*. 2015;7(4):320–6.
118. Sherwin J, Heath T, Watt K. Pharmacokinetics and Dosing of Anti-infective Drugs in Patients on Extracorporeal Membrane Oxygenation: A Review of the Current Literature. *Clin Ther*. Sep 2016;38(9):1976–94. doi:https://doi.org/10.1016/j.clinthera.2016.07.169
119. Amaker RD, DiPiro JT, Bhatia J. Pharmacokinetics of vancomycin in critically ill infants undergoing extracorporeal membrane oxygenation. *Antimicrob Agents Chemother*. May 1996;40(5):1139–42. doi:https://doi.org/10.1128/AAC.40.5.1139
120. Ahsman MJ, Wildschut ED, Tibboel D, Mathot RA. Pharmacokinetics of cefotaxime and desacetylcefotaxime in infants during extracorporeal membrane oxygenation. *Antimicrob Agents Chemother*. May 2010;54(5):1734–41. doi:https://doi.org/10.1128/AAC.01696-09
121. Wang S, Krawiec C, Patel S, et al. Laboratory Evaluation of Hemolysis and Systemic Inflammatory Response in Neonatal Nonpulsatile and Pulsatile Extracorporeal Life Support Systems. *Artif Organs*. Sep 2015;39(9):774–81. doi:https://doi.org/10.1111/aor.12466
122. Mc IRB, Timpa JG, Kurundkar AR, et al. Plasma concentrations of inflammatory cytokines rise rapidly during ECMO-related SIRS due to the release of preformed stores in the intestine. *Lab Invest*. Jan 2010;90(1):128–39. doi:https://doi.org/10.1038/labinvest.2009.119
123. Mildner RJ, Taub N, Vyas JR, et al. Cytokine imbalance in infants receiving extracorporeal membrane oxygenation for respiratory failure. *Biol Neonate*. 2005;88(4):321–7. doi:https://doi.org/10.1159/000087630
124. Vallhonrat H, Swinford RD, Ingelfinger JR, et al. Rapid activation of the alternative pathway of complement by extracorporeal membrane oxygenation. *ASAIO J*. Jan–Feb 1999;45(1):113–4. doi:https://doi.org/10.1097/00002480-199901000-00025
125. Wendel HP, Scheule AM, Eckstein FS, Ziemer G. Haemocompatibility of paediatric membrane oxygenators with heparin-coated surfaces. *Perfusion*. Jan 1999;14(1):21–8. doi:https://doi.org/10.1177/026765919901400104
126. Kaplanski G, Fabrigoule M, Boulay V, et al. Thrombin induces endothelial type II activation in vitro: IL-1 and TNF-alpha-independent IL-8 secretion and E-selectin expression. *J Immunol*. Jun 1 1997;158(11):5435–41.
127. Zimmerman GA, McIntyre TM, Prescott SM. Thrombin stimulates the adherence of neutrophils to human endothelial cells in vitro. *J Clin Invest*. Dec 1985;76(6):2235–46. doi:https://doi.org/10.1172/JCI112232
128. Levy JH, Tanaka KA. Inflammatory response to cardiopulmonary bypass. *Ann Thorac Surg*. Feb 2003;75(2):S715–20. doi:https://doi.org/10.1016/s0003-4975(02)04701-x

129. Szaba FM, Smiley ST. Roles for thrombin and fibrin(ogen) in cytokine/chemokine production and macrophage adhesion in vivo. *Blood.* Feb 1 2002;99(3):1053–9. doi:https://doi.org/10.1182/blood.v99.3.1053
130. Cheung PY, Sawicki G, Salas E, Etches PC, Schulz R, Radomski MW. The mechanisms of platelet dysfunction during extracorporeal membrane oxygenation in critically ill neonates. *Crit Care Med.* Jul 2000;28(7):2584–90. doi:https://doi.org/10.1097/00003246-200007000-00067
131. Graulich J, Sonntag J, Marcinkowski M, et al. Complement activation by in vivo neonatal and in vitro extracorporeal membrane oxygenation. *Mediators Inflamm.* Apr 2002;11(2):69–73. doi:https://doi.org/10.1080/09629350220131908
132. Moen O, Fosse E, Braten J, et al. Roller and centrifugal pumps compared in vitro with regard to haemolysis, granulocyte and complement activation. *Perfusion.* Mar 1994;9(2):109–17. doi:https://doi.org/10.1177/026765919400900205
133. Bergman P, Friberg, G., Liu, B., Al-Khaja, N., Belboul, A., Heideman, M., Mellgren, G., & Roberts, D. Blood cell rheologic deterioration by complement activation during experimental prolonged perfusion with membrane oxygenation. *Perfusion.* 1992;7(1):13–19. doi:https://doi.org/10.1177/026765919200700104
134. Hein E, Munthe-Fog L, Thiara AS, Fiane AE, Mollnes TE, Garred P. Heparin-coated cardiopulmonary bypass circuits selectively deplete the pattern recognition molecule ficolin-2 of the lectin complement pathway in vivo. *Clin Exp Immunol.* Feb 2015;179(2):294–9. doi:https://doi.org/10.1111/cei.12446
135. Perkins GD, Nathani N, McAuley DF, Gao F, Thickett DR. In vitro and in vivo effects of salbutamol on neutrophil function in acute lung injury. *Thorax.* Jan 2007;62(1):36–42. doi:https://doi.org/10.1136/thx.2006.059410
136. Kruger P, Saffarzadeh M, Weber AN, et al. Neutrophils: Between host defence, immune modulation, and tissue injury. *PLoS Pathog.* Mar 2015;11(3):e1004651. doi:https://doi.org/10.1371/journal.ppat.1004651
137. Graulich J, Walzog B, Marcinkowski M, et al. Leukocyte and endothelial activation in a laboratory model of extracorporeal membrane oxygenation (ECMO). *Pediatr Res.* Nov 2000;48(5):679–84. doi:https://doi.org/10.1203/00006450-200011000-00021
138. Fortenberry JD, Bhardwaj V, Niemer P, Cornish JD, Wright JA, Bland L. Neutrophil and cytokine activation with neonatal extracorporeal membrane oxygenation. *J Pediatr.* May 1996;128(5 Pt 1):670–8. doi:https://doi.org/10.1016/s0022-3476(96)80133-8
139. Plotz FB, van Oeveren W, Bartlett RH, Wildevuur CR. Blood activation during neonatal extracorporeal life support. *J Thorac Cardiovasc Surg.* May 1993;105(5):823–32.
140. Rungatscher A, Tessari M, Stranieri C, et al. Oxygenator Is the Main Responsible for Leukocyte Activation in Experimental Model of Extracorporeal Circulation: A Cautionary Tale. *Mediators Inflamm.* 2015;2015:484979. doi:https://doi.org/10.1155/2015/484979
141. Adrian K, Mellgren K, Skogby M, Friberg LG, Mellgren G, Wadenvik H. Cytokine release during long-term extracorporeal circulation in an experimental model. *Artif Organs.* Oct 1998;22(10):859–63. doi:https://doi.org/10.1046/j.1525-1594.1998.06121.x
142. Shi J, Chen Q, Yu W, et al. Continuous renal replacement therapy reduces the systemic and pulmonary inflammation induced by venovenous extracorporeal membrane oxygenation in a porcine model. *Artif Organs.* Mar 2014;38(3):215–23. doi:https://doi.org/10.1111/aor.12154
143. Yimin H, Wenkui Y, Jialiang S, et al. Effects of continuous renal replacement therapy on renal inflammatory cytokines during extracorporeal membrane oxygenation in a porcine model. *J Cardiothorac Surg.* Apr 29 2013;8:113. doi:https://doi.org/10.1186/1749-8090-8-113
144. S J, S J, C Q, et al. In-line hemofiltration minimized extracorporeal membrane oxygenation-related inflammation in a porcine model. *Perfusion.* Nov 2014;29(6):526–33. doi:https://doi.org/10.1177/0267659114529320
145. Stenlo M, Silva IAN, Hyllen S, et al. Monitoring lung injury with particle flow rate in LPS- and COVID-19-induced ARDS. *Physiol Rep.* Jul 2021;9(13):e14802. doi:https://doi.org/10.14814/phy2.14802

146. Risnes I, Wagner K, Ueland T, Mollnes T, Aukrust P, Svennevig J. Interleukin-6 may predict survival in extracorporeal membrane oxygenation treatment. *Perfusion*. May 2008;23(3):173–8. doi:https://doi.org/10.1177/0267659108097882
147. Supady A, Weber E, Rieder M, et al. Cytokine adsorption in patients with severe COVID-19 pneumonia requiring extracorporeal membrane oxygenation (CYCOV): a single centre, open-label, randomised, controlled trial. *Lancet Respir Med*. Jul 2021;9(7):755–762. doi:https://doi.org/10.1016/S2213-2600(21)00177-6
148. Rieder M, Duerschmied D, Zahn T, et al. Cytokine Adsorption in Severe Acute Respiratory Failure Requiring Veno-Venous Extracorporeal Membrane Oxygenation. *ASAIO J*. Mar 1 2021;67(3):332–338. doi:https://doi.org/10.1097/MAT.0000000000001302
149. Friesecke S, Stecher SS, Gross S, Felix SB, Nierhaus A. Extracorporeal cytokine elimination as rescue therapy in refractory septic shock: a prospective single-center study. *J Artif Organs*. Sep 2017;20(3):252–259. doi:https://doi.org/10.1007/s10047-017-0967-4
150. Kogelmann K, Jarczak D, Scheller M, Druner M. Hemoadsorption by CytoSorb in septic patients: a case series. *Crit Care*. Mar 27 2017;21(1):74. doi:https://doi.org/10.1186/s13054-017-1662-9
151. Mehta Y, Mehta C, Kumar A, et al. Experience with hemoadsorption (CytoSorb((R))) in the management of septic shock patients. *World J Crit Care Med*. Jan 31 2020;9(1):1–12. doi:https://doi.org/10.5492/wjccm.v9.i1.1
152. Valapour M, Lehr CJ, Skeans MA, et al. OPTN/SRTR 2019 Annual Data Report: Lung. *Am J Transplant*. Feb 2021;21 Suppl 2:441–520. doi:https://doi.org/10.1111/ajt.16495
153. Hoetzenecker K, Donahoe L, Yeung JC, et al. Extracorporeal life support as a bridge to lung transplantation-experience of a high-volume transplant center. *J Thorac Cardiovasc Surg*. Mar 2018;155(3):1316–1328 e1. doi:https://doi.org/10.1016/j.jtcvs.2017.09.161
154. Kukreja J, Tsou S, Chen J, et al. Risk Factors and Outcomes of Extracorporeal Membrane Oxygenation as a Bridge to Lung Transplantation. *Semin Thorac Cardiovasc Surg*. Winter 2020;32(4):772–785. doi:https://doi.org/10.1053/j.semtcvs.2020.05.008
155. Hakim AH, Ahmad U, McCurry KR, et al. Contemporary Outcomes of Extracorporeal Membrane Oxygenation Used as Bridge to Lung Transplantation. *Ann Thorac Surg*. Jul 2018;106(1):192–198. doi:https://doi.org/10.1016/j.athoracsur.2018.02.036
156. Hayes D, Jr., Tobias JD, Tumin D. Center Volume and Extracorporeal Membrane Oxygenation Support at Lung Transplantation in the Lung Allocation Score Era. *Am J Respir Crit Care Med*. Aug 1 2016;194(3):317–26. doi:https://doi.org/10.1164/rccm.201511-2222OC
157. Halpern AL, Kohtz PD, Helmkamp L, et al. Improved Mortality Associated With the Use of Extracorporeal Membrane Oxygenation. *Ann Thorac Surg*. Aug 2019;108(2):350–357. doi:https://doi.org/10.1016/j.athoracsur.2019.03.057
158. Benazzo A, Schwarz S, Frommlet F, et al. Twenty-year experience with extracorporeal life support as bridge to lung transplantation. *J Thorac Cardiovasc Surg*. Jun 2019;157(6):2515–2525 e10. doi:https://doi.org/10.1016/j.jtcvs.2019.02.048
159. Tipograf Y, Salna M, Minko E, et al. Outcomes of Extracorporeal Membrane Oxygenation as a Bridge to Lung Transplantation. *Ann Thorac Surg*. May 2019;107(5):1456–1463. doi:https://doi.org/10.1016/j.athoracsur.2019.01.032
160. Garcia JP, Iacono A, Kon ZN, Griffith BP. Ambulatory extracorporeal membrane oxygenation: a new approach for bridge-to-lung transplantation. *J Thorac Cardiovasc Surg*. Jun 2010;139(6):e137–9. doi:https://doi.org/10.1016/j.jtcvs.2009.12.021
161. Leard LE, Holm AM, Valapour M, et al. Consensus document for the selection of lung transplant candidates: An update from the International Society for Heart and Lung Transplantation. *J Heart Lung Transplant*. Nov 2021;40(11):1349–1379. doi:https://doi.org/10.1016/j.healun.2021.07.005

# Chapter 16
# Current and Future Engineering Strategies for ECMO Therapy

**Deniz A. Bölükbas and Sinem Tas**

## 16.1 Introduction

Extracorporeal membrane oxygenation (ECMO) is a life-saving therapy for patients who suffer from cardiac and/or respiratory failure where the gas exchange function of the lung is impaired. The first model of ECMO was developed in the 1950s as a modified form of cardiopulmonary bypass. In the 1970s, the first successful use of ECMO in adults and neonates with respiratory failure was reported. Since its development, the use of ECMO has steadily increased in clinics from neonates to pediatric and adult patients as a last resort in acute and chronic cases [1, 2]. ECMO is an expensive therapy that requires considerable expertise and sophisticated equipment. Nevertheless, it was successfully exploited to treat patients with Swine Flu in 2009 and during the COVID-19 outbreak, as a last line therapy if mechanical ventilation fails [2, 3].

A typical ECMO circuit consists of vascular access, tubing, pump, gas exchange unit (membrane oxygenator) and interface. ECMO replaces the gas exchange function of the lungs by pumping venous blood from a large central vein to a membrane oxygenator outside of the body. The oxygenator consists of hollow fiber membranes where blood flows across the fiber bundle and gas flows within the lumen of the individual fibers. A thin gas permeable membrane allows the diffusion of oxygen from the lumen side into blood and carbon dioxide from the blood into the gas for

D. A. Bölükbas · S. Tas (✉)
Wallenberg Center for Molecular Medicine, Lund University, Lund, Sweden

Lund Stem Cell Center, Lund University, Lund, Sweden

Department of Clinical Sciences, Lund University, Lund, Sweden

Department of Cardiothoracic Surgery and Transplantation, Skåne University Hospital, Lund, Sweden



C. M. Magin (ed.), *Engineering Translational Models of Lung Homeostasis and Disease*, Advances in Experimental Medicine and Biology 1413,
https://doi.org/10.1007/978-3-031-26625-6_16

disposal. In a membrane lung, the main focus is on achieving adequate levels of oxygenation of the deoxygenated blood. This, being 250 to 300 ml of oxygen/min for an average adult at a resting state, is much more challenging than the removal of carbon dioxide [4]. Thus, all devices on the market have a large membrane surface area to enable this amount of oxygen transfer. In addition to the characteristics of the membrane oxygenator (e.g., surface area or oxygenator size), blood flow and hemoglobin concentration should be carefully planned not only to match adequate levels of oxygenation but also to better management of ECMO [4].

ECMO membranes are made of highly hydrophobic polymers such as silicone, polypropylene (PP) and polymethylpentene (PMP). Non-wetting ECMO membranes have gas-filled pores that allow fast and efficient mass transfer between gas and liquid phases. However, the downside of these hydrophobic surfaces is that they are attractive for protein adsorption. Thus, the incompatibility of ECMO circuits (including tubing, cannulas, and membrane oxygenator) with blood is the main reason for inflammation and bleeding complications. In addition to blood incompatibility, non-physiological flow conditions in the cannula and the pump can activate the platelets, the coagulation cascade and lead to thrombus growth within the circuit [5, 6]. Large foreign surface area of the membrane oxygenator represents a high risk for thrombus formation especially at the edges and corners where the velocities are the lowest. Additionally, higher shear rates due to high blood flow velocities in the cannulas and pumps can cause hemolysis [7, 8]. To prevent blood coagulation and thrombus formation, anticoagulants are given to the patients, which can induce bleeding complications and contribute to the high morbidity and mortality seen in ECMO patients [6].

While ECMO can be life-saving, it is not a first-option therapy because of the high risk of mortality. Over the years, ECMO mortality remained unchanged, despite the patient population becoming at increasingly higher risk of mortality due to severity of diseases and associated comorbidities. An increased experience and refined treatment strategies have contributed to improve the outcomes in ECMO patients; the technological advances in ECMO circuit design improved the overall safety, ease of use, accessibility and potentially contributed to decreased morbidity and mortality. This chapter will provide an overview of the requirements for blood oxygenation and technological advancements in the ECMO circuit, as well as emerging strategies to minimize the complications for patients undergoing ECMO therapy.

## 16.2 Blood Oxygen

The oxygen in blood is composed of the amount that is bound to hemoglobin and the amount that is dissolved in plasma [4]. These together define the oxygen content of the blood and is the most important factor for tissue oxygenation. In fact, the amount of oxygen that is delivered to tissue is measured by the oxygen content in arterial blood multiplied by the blood flow, termed as $DO_2$.

At rest, $DO_2$ is controlled to be five times more than the oxygen consumption of the body, termed as $VO_2$. A drop in the $DO_2$ compared to $VO_2$ results in more oxygen extraction from the blood flow, leaving less oxygen in venous blood, to maintain physiologic metabolism rates. If the $DO_2$ becomes less than two times $VO_2$, aerobic metabolism is hampered. This results in anaerobic metabolism with lactic acid production which can lead to supply dependency systemic acidosis and organ failure [9]. ECMO therapy allows for control of $DO_2$ independent of the native lung function. It is aimed to keep the $DO_2$ levels close to five times the $VO_2$ levels as seen in healthy adults [4].

## 16.3 Advances in ECMO Circuit

Although individual centers customize their ECMO circuits to best fit their patient population needs, there has been a significant amount of effort put into the design of compact ECMO to reduce extrinsic surface area of blood contact, priming volumes and reduceconnectors and stopcocks to eliminate potential sites of flow disturbance.

### *16.3.1 Cannula and Circuit Tubing*

ECMO circuit has undergone an important improvement in terms of biocompatibility. However, efficacy of ECMO is greatly dependent on adequate blood flow in the circuit. Therefore, there have been significant design changes in ECMO circuits to improve the outcome of ECMO. The design of cannulas has been focused on maximizing blood flow while minimizing the damage on blood, having the least possible coagulation pathway activation. The cannula and the circuit tubing are the first site of blood contact with a foreign surface. The use of cannula and/or tubing with a surface coating is essential for preventing fibrin shear and thrombus formation. Even a small thrombus can significantly affect the blood flow. This effect is especially critical if the thrombus is positioned at a narrower part of the cannula or at a drainage hole since these areas occlude a large portion of the blood volume. Biocompatible polymers such as polyurethane, silicone and polyvinylchloride (PVC) have been commonly used to manufacture cannulas and circuit tubing [10, 11]. Although they are made of biocompatible materials, these materials have drawbacks that contribute to the complications related with ECMO. These materials are hydrophobic and make the surface attractive for protein adsorption which eventually provokes coagulation cascade. Current trends in ECMO management move towards prevention of blood clotting without anticoagulants. Therefore, several surface coatings have been applied to enhance the hemocompatibility of these materials. The most commonly used surface coating for ECMO circuits is heparin surface coatings [10, 11]. The main advantage of this material is its anticoagulation properties. Heparin lowers the degree of activation of the hemostatic system, prevents platelet adhesion and activation while preserving platelet counts. Despite the advantages, heparin causes the development of osteopenia and thrombocytopenia [12]. Importantly, heparin-based coatings are not always covalently attached to the surface [13], thus, can leach into the blood, resulting in extremely high heparin concentration in the blood. This can cause severe bleeding complications, which can be lethal [11].

As an alternative, biomimetic surfaces that closely resemble the natural endothelium have been designed. Zwitterionic polymers have been used to develop endothelium-mimicking surfaces [10, 11]. The use of zwitterionic polymer is inspired by the external surface of the cell membrane that has zwitterionic phosphorylcholine (PPC) group on the outer side [11, 14, 15]. PPC coating possesses repeating positive and negative regions on the surface resulting in strong hydration forces, prohibiting the direct adsorption on top of proteins, thereby creating a layer between the components of the human blood and the artificial surface.

Additionally, cannulas with different lengths, shape, with/without drainage holes, and double lumen have been developed to compare and determine the best choice for each situation [16–19].

For example, the large cannulation strategy is in general recommended to maximize blood flow [18]. On the other hand, small arterial cannulation strategy reduces the possibility of lower limb ischemia in peripheral venoarterial extracorporeal membrane oxygenation (VA ECMO) [17]. Moreover, cannulas with drainage holes with different hole size, angles, and spacing between the wholes can affect the blood flow dynamics and eventually the thrombosis potential [16].

### *16.3.2 Pumps*

Roller or centrifugal pumps have been the most common types of pumps used in ECMO circuits [20]. Although more economic, roller pumps have adverse effects on the flow due to mechanical forces and pressure on the circuit tubing. Additionally, shear stress caused by the roller-head contributes to hemolysis. To minimize the risk of rupture caused by extreme positive pressures, the tubing length is extended with an increased need for pressure monitoring. Large circuit size with longer tubing leads to elevated exposure to extracorporeal units and it affects the priming volume required proportionally [21].

The newest centrifugal pumps with a magnetic drive can perform in smaller circuits, hence requiring less priming volume and resulting in less amount of hemolysis. Mostly due to these attributes, the number of smaller centrifugal pumps used in ECMO circuits has been on the rise [22]. The drawbacks of this type include the risk for cavitation or air formation due to high negative pressure on the venous side. Although a bladder reservoir may be used to minimize these risks, there is no consensus on whether that benefit outweighs the risks associated with increased surface exposure.

### *16.3.3 Membrane Oxygenator*

The first model of ECMO oxygenators used flat sheet polyethylene, ethylcellulose membranes that could provide the required gas exchange [23]. However, these two membranes had significant drawbacks. Manufacturing techniques introduced pin-hole defects on the polyethylene membranes that allowed oxygen bubbles to enter the blood channel. Ethylcellulose membranes allowed water to seep into the gas channel. Additionally, their brittleness made them difficult to support mechanically. Later, silicone films were used to minimize the material related defects [23, 24]. In the mid-1980s, there were significant changes in ECMO oxygenator design. ECMO oxygenators adopted the hollow fiber membrane after their success in hemodialysis [24]. The first large scale hollow fiber oxygenators developed with extraluminal cross-flow of blood. High gas exchange capacity was achieved due to luminal flow where blood passed through the lumen side and sweep gas flowed through a jacket surrounding the fibers. The circular cross-section maximized the surface to volume ratio, resulting in high gas exchange capacity. Oxygen and carbon dioxide were exchanged in the oxygenator based on partial pressure differences.

To achieve high gas exchange capacity, the membranes should be hydrophobic to prevent membrane wetting and blood plasma leakage. Therefore, current oxygenators are made of hydrophobic PP and PMP membranes [23]. The hydrophobic nature of the membranes makes the surface attractive for protein fouling which leads to blood clots formation. Moreover, adsorption of proteins on membrane surface degrades the performance and lifetime of the membrane, eventually necessitating the change of oxygenator which puts patients life at risk [25]. Antithrombogenic and antifouling membranes have been developed to minimize the blood clotting [10, 11, 13]. Thus, various commercial coatings are available from the oxygenator manufacturers. Heparin is also the most commonly used coating material in the ECMO membranes due to its anticoagulation effect [10, 26]. As alternative strategies to circumvent the side effects associated with heparin, membranes with coatings of albumin, polyethylene glycol, phosphorylcholine were developed [10]. Although newer oxygenators using surface coatings have demonstrated some improvements in hemocompatibility compared to prior devices, further improvement is needed to prevent the protein and platelet adhesion on the membrane surface to improve the performance of ECMO.

## 16.4 Experimental Strategies

The contact of blood with extracorporeal surfaces can lead to the adverse effects, thus can worsen the patient condition [27]. Therefore, recent research efforts have primarily focused on improving hemocompatibility of the surface by following ways to reduce incidence of clotting, bleeding and minimize reliance on anticoagulants: (i) development of surface coatings, (ii) endothelialization. On the other hand, microfluidic oxygenator technologies have emerged as an alternative to overcome some of the principal drawbacks of the current ECMO circuits.

## 16.5 Membrane Surface Coatings

To date, several alternative materials have been utilized as antithrombotic coatings to prevent the adverse effects of passing blood through an ECMO circuit. The mechanism of action of antithrombotic coatings can be divided into two categories: bioactive coatings and biopassive coatings [10, 11].

### *16.5.1 Bioactive Coatings*

Bioactive coatings have been traditionally applied to the surface of the blood contacting devices. The main objective of the bioactive coatings is inhibition of coagulation factors and accumulation of platelets. Heparin is the most commonly used bioactive coating on the market due to its clinically proven benefits in several medical device applications. Examples of products engineered with heparin coatings are Cortiva (Medtronic), Hepaface (Sorin), and Rheoparin (Xenios) [10]. Despite the presence of bioactive coated ECMO circuits on the market, there have been continuous efforts focused on the improvement of the performance of heparin-based coatings as well as development of new coating materials. The major issues of the commercially available heparin coatings are the loss of antithrombin binding effect and leaching of heparin from the surface [13, 26]. To overcome these issues, the strategies of immobilization of heparin on long chain dialkyl groups via ionic interactions, and heparin coupled polyethylene glycol (PEG) have been developed [26, 28–32]. These coatings exhibited improved anti-thrombogenicity and long-term durability without diminishing gas exchange capacity. In vivo, endothelial cells can inhibit platelet activation via multiple pathways and release of nitric oxide (NO) [33]. Therefore, NO-releasing polymer surfaces have been developed and shown ability to prevent platelet activation and thrombus formation in in vivo studies with healthy animal models [34–37]. However, only a finite amount of NO can be released from these surfaces, which limits the duration of activity, thus further advancements are needed for the clinical translation of this approach.

### *16.5.2 Biopassive Coatings*

Inspired by the cell membrane, biopassive coatings aim to inhibit the attachment of proteins on membrane surface by minimizing the interaction between the blood and the foreign surface [10]. The potential of zwitterionic polymers has been recognized and applied on the surfaces of membranes to improve the anti-fouling properties. Zwitterionic polymers possess an equal number of both negatively and positively charged groups which allow them to maintain overall electrical neutrality. The anti-fouling properties associated with zwitterionic polymers is attributed to the formation of a hydration layer on the surface of the biomaterial, forming a physical and energetic barrier that hinders unspecific protein adhesion. 2-ethacryloyloxyethyl phosphorylcholine (MPC), sulfobetamine (SB), and carboxybetamine (CB)-based polymers are the most commonly used zwitterionic polymers to develop novel coatings [10]. MPC [35, 38–40], SB [41], CB [42], or their block-copolymer have been grafted or coated on ECMO membrane surfaces to improve the hemocompatibility and antifouling properties (Fig. 16.1) of the membranes. Although there have been promising early in vitro studies with zwitterionic polymers, it has not yet been tested in larger trials. Besides zwitterionic polymers, PEG is the most commonly used non-ionic polymer to create biopassive surfaces. Presence of PEG coating increased the hydrophilicity of the membrane surface thereby decreasing the susceptibility of the membrane fouling [5, 10, 13, 32].

## 16.6 Endothelialization of ECMO Membrane: Biohybrid Approach

One emerging approach is the use of endothelialization of the membrane surface (biohybrid) which aims to recapitulate the in vivo function of vascular endothelial cells (ECs) to yield a biocompatible and bioresponsive surface, dynamically inhibiting platelet deposition and activation. ECs sit as a non-thrombogenic monolayer in the endothelium, with their anticoagulation and anti-proliferation features in vivo [11]. Experimental work for endothelialization of ECMO membranes has been conducted both as in vitro pre-endothelialization and in vivo endothelial progenitor cell (EPC)-based endothelialization.

In vitro pre-endothelialization has been mostly based on autologous ECs seeded on surfaces in contact with blood. This has been particularly explored for coating of stents to prevent restenosis and thrombus formation [43]. To increase the efficiency of endothelialization, the hydrophilicity of the biomaterial needs to be enhanced. Several groups have designed coatings to promote EC adhesion and growth in vitro. Some of the earliest attempts were based on dip coating of extracellular matrix (ECM) proteins such as collagen, gelatin, Arginylglycylaspartic acid (RGD) peptide sequence of fibronectin [44–46]. However, some of these proteins are thrombogenic, and thus blood clotting can occur in case of premature EC layer formation or loss of EC layer under blood flow. Therefore, it is crucial to monitor cell growth and validate the stability of the endothelial monolayer under flow conditions over time (Fig. 16.2).

Recently, researchers introduced the idea of deposition of titanium dioxide ($TiO_2$) coating on membrane surfaces by pulsed vacuum cathodic arc plasma which

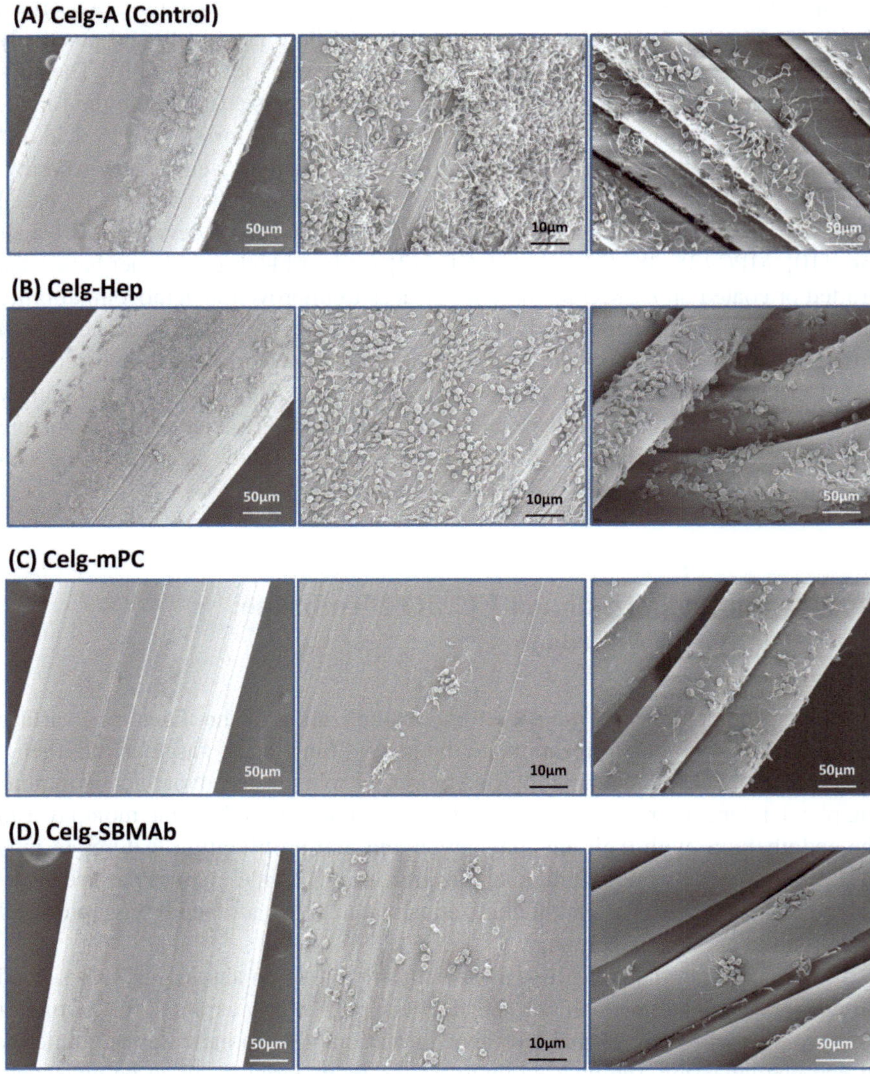


**Fig. 16.1** Scanning electron micrographs of membranes after citrated ovine blood contact for 3 h for (**a**) aminated polypropylene control (Celg-A) and (**b**) heparin-modified (Celg-Hep), (**c**) mPC-modified (Celg-mPC), and (**d**) SBMAb block copolymer-modified (Celg-SBMAb) hollow fibers (left, middle columns). Binding fibers are seen in the right column. (Reprinted with permission from Ref. [39]. Copyright 2014 American Chemical Society)

enables the formation of stable, confluent EC monolayer [47]. However, this process is relatively costly in comparison to other coatings.

Major drawbacks of in vitro endothelialization approaches are that these require long culture times established only in select institutions and unavailable for acute demand [48]. An alternative approach is based on circulating EPC-based

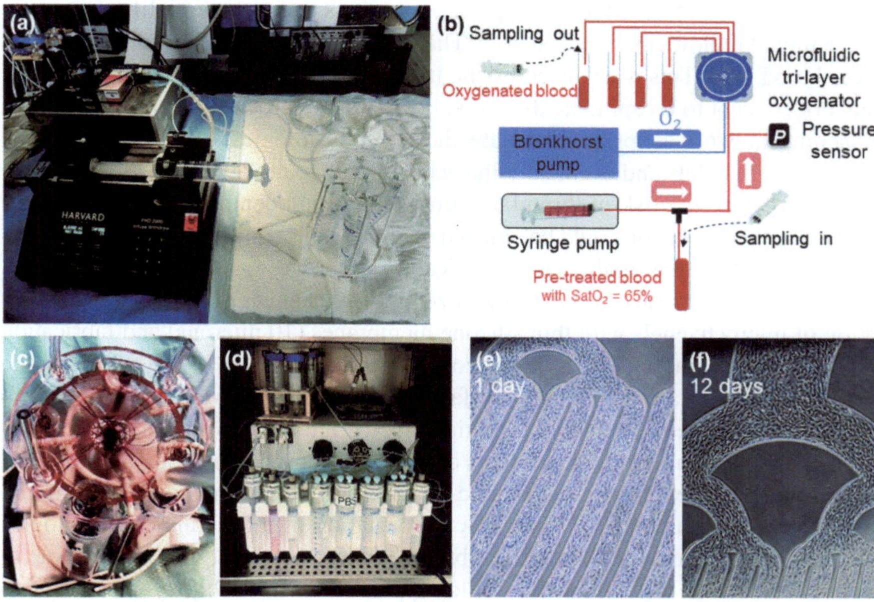


**Fig. 16.2** An in vitro endothelialized microfluidic approach with high gas exchange efficiency and compatibility. (**a**) An image of the whole setup and (**b**) the corresponding scheme describing each element. (**c**) An image of the single trilayer unit oxygenator recorded during the oxygenation/decarbonation experiment. (**d**) An image of the microfluidic flow control system (MFCS™, Fluigent) installed inside the incubator during the experiment for endothelialization. Optical images recorded during endothelialization in the branching injection parts of one single-layer device at different times after (**e**) one day and (**f**) 12 days. (Reproduced from Ref. [51] with permission from the Royal Society of Chemistry)

endothelialization of the membranes. EPCs are mononuclear cells located in the bone marrow with low numbers circulating in the peripheral blood. They can differentiate into mature ECs and play a pivotal role in vascular repair and reendothelialization [49]. For EPCs to adhere and differentiate into ECs on artificial surfaces, materials should be engineered with regard to their surface characteristics and charge. Specifically, these materials are often surface modified with EPC-homing and differentiation factors by immobilization of cell-specific antibodies, proteins, or peptide sequences as well as micro- or nano-patterning. Although this technology seems promising and emerging, endothelial cell proliferation in vivo is low and further studies are needed to evaluate its feasibility in clinical translation [43].

## 16.7 Miniaturization of ECMO Circuit

Despite the emerging novel biomaterials and surface coatings, there has been limited progress in the designing of alternative membrane oxygenators. Microfluidic approaches have emerged as an alternative technology to generate membrane oxygenators [50]. Microfluidic oxygenators possess major potential benefits over conventional hollow fiber oxygenators. They can offer uniform, non-turbulent, low-shear blood flow. This is achieved by the introduction of a branching microchannel network concept to recapitulate the blood flow in the microcirculation [50–53]. The branching network helps to eliminate the non-physiological forces that damage blood cells, platelets and eventually the activation of coagulation and inflammatory responses. Thus, the risk of thrombus formation, thromboemboli, and the need for systemic anticoagulation could be reduced. Another major advantage of microfluidic designs includes the ability to significantly improve gas exchange capacity. Microfluidic membrane oxygenators realize high gas exchange capacity due to separation of microchannels with thin silicone membranes (10 microns) and fabrication of circular flow patterns [54, 55]. Moreover, superior gas transfer efficiency can be achieved by using a much smaller surface area and priming volume compared to standard membrane oxygenators. Additionally, the reduction of blood contact with foreign surfaces in microfluidic designs can offer improved hemocompatibility.

Despite the potential of microfluidic oxygenators demonstrated in in vitro and in vivo studies in animal models, clinical translation of microfluidic technologies remains a challenge. None of the membrane oxygenator prototypes are yet able to accommodate large blood flow that could support a human adult [50]. This is due to the difficulties in upscaling of these devices and the cost of manufacturing.

## 16.8 Conclusion

ECMO is a lifesaving therapy for patients with lung or cardiac failure as well as for chronically ill patients awaiting a lung transplant, replacing the lung function outside of the body. While lifesaving, ECMO has 50% mortality due to blood incompatibility of current ECMO circuits. However, improvements in material and membrane technologies have made the ECMO circuit and components more compact and compatible, allowing its use across a wide variety of patients from neonates to adults.

The hemocompatibility of membrane materials determines the lifetime of the membrane oxygenator. Membrane fouling and eventually the drop in gas exchange capacity leads to frequent replacement of the oxygenator, resulting in high treatment cost and operational difficulties. Importantly, changing the oxygenator during ECMO treatment puts the patients' life at a high risk. Several surface modification strategies have been applied to improve the oxygenator performance. Although there are several manufacturers that produce ECMO oxygenators with coatings, they can still not address the clinical needs fully. Thus, recent scientific efforts have been heavily focused on modification of membrane materials for long-term operation. The concept of endothelialization of membrane surfaces has emerged as a biohybrid approach. The goal of endothelization is to mimic the natural bioresponsive endothelium. Healthy vascular endothelium is covered with a monolayer of ECs, which provides the natural antithrombotic property to blood vessels. Several coatings have been designed and tested for ECs attachment. However, there is an urgent need of coatings that support rapid endothelialization ex vivo or in vivo to meet the acute clinical demand. Also, the stability of EC monolayer under flow conditions should be optimized.

Recent advances in advanced manufacturing and fabrication techniques have already demonstrated their potential to generate microfluidic oxygenators. In principle, microfluidic oxygenators can address many of the limitations of conventional ECMO oxygenators since they are designed to mimic the flow properties of vascular microcirculation. Currently, the microfluidic oxygenator technology is limited to small scale prototypes due to limitations in manufacturing, scalability as well as cost effectiveness. This approach has demonstrated promising results in vitro, however, testing in large animal models in parallel to their scaling up process will be critical prior to their translation to the clinic.

One major area which is under-researched so far is the testing of ECMO circuits with diseased conditions both in vitro and in vivo. The clinically used ECMO membranes are made with a "one-size-fits-all-mentality" for all diseases. Furthermore, they have been optimized and mainly tested using healthy blood samples and in healthy animals. However, emerging evidence suggests that different diseases have markedly different inflammatory properties and thrombus potential. This clearly indicates the urgent need of new membrane coatings that are designed by considering the disease and patient specific needs.

To improve the clinical outcomes of ECMO, further developments will require interdisciplinary work between different scientific fields, clinical experts, and industry. However, it is already clear that advances in material sciences and manufacturing approaches are poised to play a critical role in the creation of safe and efficient ECMO circuits that will meet both patient and clinical needs and might also expedite the development of portable artificial lung devices.

## References

1. Patel, B., Chatterjee, S., Davignon, S. and Herlihy, J., (2019) Extracorporeal membrane oxygenation as rescue therapy for severe hypoxemic respiratory failure. Journal of Thoracic Disease, 11(S14), pp. S1688–S1697
2. Kilic A (2018) Extracorporeal Membrane Oxygenation. In: Vasan RS, Sawyer DB (eds) Encyclopedia of Cardiovascular Research and Medicine. Elsevier, Oxford, pp. 281–284
3. Bertini P, Guarracino F, Falcone M, Nardelli P, Landoni G, Nocci M, Paternoster G, (2022) ECMO in COVID-19 Patients: A Systematic Review and Meta-analysis. Journal of Cardiothoracic and Vascular Anesthesia 36:2700–2706.
4. Bartlett, R., (2016) Physiology of Gas Exchange During ECMO for Respiratory Failure. Journal of Intensive Care Medicine, 32(4), pp. 243–248
5. Maul, T. M., Massicotte, M. P., Wearden, P. D. (2016), 'ECMO Biocompatibility: Surface Coatings, Anticoagulation, and Coagulation Monitoring', in M. S. Firstenberg (ed.), Extracorporeal Membrane Oxygenation – Advances in Therapy, IntechOpen
6. Sniderman, J., Monagle, P., Annich, G. and MacLaren, G., 2020. Hematologic concerns in extracorporeal membrane oxygenation. Research and Practice in Thrombosis and Haemostasis, 4(4), pp. 455–468

7. Lehle K, Philipp A, Zeman F, Lunz D, Lubnow M, Wendel H-P, Göbölös L, Schmid C, Müller T (2015) Technical-Induced Hemolysis in Patients with Respiratory Failure Supported with Veno-Venous ECMO – Prevalence and Risk Factors. PLOS ONE 10:e0143527
8. Köhne, I., (2020) Haemolysis induced by mechanical circulatory support devices: unsolved problems. Perfusion, 35(6), 474–483
9. Vincent JL (2005) DO2/VO2 relationships. In: Pinsky MR, Payen D (eds) Functional Hemodynamic Monitoring. Springer Berlin Heidelberg, Berlin, Heidelberg, pp. 251–258
10. Zhang, M., Pauls, J., Bartnikowski, N., Haymet, A., Chan, C., Suen, J., Schneider, B., Ki, K., Whittaker, A., Dargusch, M. and Fraser, J., 2021. Anti-thrombogenic Surface Coatings for Extracorporeal Membrane Oxygenation: A Narrative Review. ACS Biomaterials Science & Engineering, 7(9), 4402–4419
11. Ontaneda, A. and Annich, G., (2018) Novel Surfaces in Extracorporeal Membrane Oxygenation Circuits. Frontiers in Medicine, 5
12. Walenga JM, Bick RL (1998) Heparin-Induced Thrombocytopenia, Paradoxical Thromboembolism, And Other Side Effects Of Heparin Therapy. Medical Clinics of North America 82:635–658.
13. He T, He J, Wang Z, Cui Z (2021) Modification strategies to improve the membrane hemocompatibility in extracorporeal membrane oxygenator (ECMO). Advanced Composites and Hybrid Materials 4:847–864
14. Iwasaki Y, Ishihara K (2012) Cell membrane-inspired phospholipid polymers for developing medical devices with excellent biointerfaces. Science and Technology of Advanced Materials 13:064101
15. Watanabe J, Ishihara K (2008) Establishing ultimate biointerfaces covered with phosphorylcholine groups. Colloids and Surfaces B: Biointerfaces 65:155–165
16. Vatani, A., Liao, S., Burrell, A., Carberry, J., Azimi, M., Steinseifer, U., Arens, J., Soria, J., Pellegrino, V., Kaye, D. and Gregory, S., (2021) Improved Drainage Cannula Design to Reduce Thrombosis in Veno-Arterial Extracorporeal Membrane Oxygenation. ASAIO Journal, Publish Ahead of Print.
17. Kim, J., Cho, Y., Sung, K., Park, T., Lee, G., Lee, J., Song, Y., Hahn, J., Choi, J., Choi, S., Gwon, H. and Yang, J., (2019) Impact of Cannula Size on Clinical Outcomes in Peripheral Venoarterial Extracorporeal Membrane Oxygenation. ASAIO Journal, 65(6), 573–579
18. Strunina, S., Hozman, J. and Ostadal, P., 2018. The peripheral cannulas in extracorporeal life support. Biomedical Engineering / Biomedizinische Technik, 64(2), 127–133
19. Modine, T., Vincent, F., Delhaye, C. and Van Belle, E., (2020) A dedicated Y-shaped percutaneous ECMO cannula for femoral 2-in-1 vascular access during high-risk procedures. Catheterization and Cardiovascular Interventions, 97(5), 959–961
20. Rehder KJ, Turner DA, Bonadonna D, Walczak RJ, Rudder RJ, Cheifetz IM (2012) Technological advances in extracorporeal membrane oxygenation for respiratory failure. Expert Review of Respiratory Medicine 6:377–384.
21. Dalton HJ (2011) Extracorporeal Life Support: Moving at the Speed of Light. Respiratory Care 56:1445.
22. Khan S, Vasavada R, Qiu F, Kunselman A, Ündar A (2011) Extracorporeal life support systems: alternative vs. conventional circuits. Perfusion 26:191–198.
23. Evseev AK, Zhuravel SV, Alentiev A. Yu, Goroncharovskaya IV, Petrikov SS (2019) Membranes in Extracorporeal Blood Oxygenation Technology. Membranes and Membrane Technologies 1:201–211.
24. Yeager T, Roy S (2017) Evolution of Gas Permeable Membranes for Extracorporeal Membrane Oxygenation. Artificial Organs 41:700–709.
25. Lehle K, Philipp A, Gleich O, Holzamer A, Müller T, Bein T, Schmid C (2008) Efficiency in Extracorporeal Membrane Oxygenation—Cellular Deposits on Polymethypentene Membranes Increase Resistance to Blood Flow and Reduce Gas Exchange Capacity. ASAIO Journal 54:612–617

26. Biran R, Pond D (2017) Heparin coatings for improving blood compatibility of medical devices. Advanced Drug Delivery Reviews 112:12–23.
27. Olson SR, Murphree CR, Zonies D, Meyer AD, Mccarty OJT, Deloughery TG, Shatzel JJ (2021) Thrombosis and Bleeding in Extracorporeal Membrane Oxygenation (ECMO) Without Anticoagulation: A Systematic Review. ASAIO Journal 67: 290–296
28. Nishinaka T, Tatsumi E, Taenaka Y, Katagiri N, Ohnishi H, Shioya K, Fukuda T, Oshikawa M, Sato K, Tsukiya T, Homma A, Takewa Y, Takano H, Sato M, Kashiwabara S, Tanaka H, Sakai K, Matsuda T (2002) At Least Thirty-Four Days of Animal Continuous Perfusion by a Newly Developed Extracorporeal Membrane Oxygenation System without Systemic Anticoagulants. Artificial Organs 26:548–551
29. Nishinaka T, Tatsumi E, Katagiri N, Ohnishi H, Mizuno T, Shioya K, Tsukiya T, Homma A, Kashiwabara S, Tanaka H, Sato M, Taenaka Y (2007) Up to 151 days of continuous animal perfusion with trivial heparin infusion by the application of a long-term durable antithrombogenic coating to a combination of a seal-less centrifugal pump and a diffusion membrane oxygenator. Journal of Artificial Organs 10:240–244.
30. Zhang M, Chan CHH, Pauls JP, Semenzin C, Ainola C, Peng H, Fu C, Whittaker AK, Heinsar S, Fraser JF (2022) Investigation of heparin-loaded poly(ethylene glycol)-based hydrogels as anti-thrombogenic surface coatings for extracorporeal membrane oxygenation. Journal of Materials Chemistry B.
31. Wang W, Zheng Z, Huang X, Fan W, Yu W, Zhang Z, Li L, Mao C (2017) Hemocompatibility and oxygenation performance of polysulfone membranes grafted with polyethylene glycol and heparin by plasma-induced surface modification. Journal of Biomedical Materials Research Part B: Applied Biomaterials 105:1737–1746.
32. Abednejad AS, Amoabediny G, Ghaee A (2013) Surface Modification of Polypropylene Blood Oxygenator Membrane by Poly Ethylene Glycol Grafting. Advanced Materials Research 816–817:459–463.
33. Hamilos, M., Petousis, S. and Parthenakis, F., (2018) Interaction between platelets and endothelium: from pathophysiology to new therapeutic options. Cardiovascular Diagnosis and Therapy, 8(5), pp. 568–580
34. Jeakle MM, Major TC, Meyerhoff ME, Bartlett RH (2020) Comparison of Diazeniumdiolated Dialkylhexanediamines as Nitric Oxide Release Agents on Nonthrombogenicity in an Extracorporeal Circulation Model. ACS Applied Bio Materials 3:466–476.
35. Brisbois EJ, Handa H, Major TC, Bartlett RH, Meyerhoff ME (2013) Long-term nitric oxide release and elevated temperature stability with S-nitroso-N-acetylpenicillamine (SNAP)-doped Elast-eon E2As polymer. Biomaterials 34:6957–6966.
36. Hopkins SP, Pant J, Goudie MJ, Schmiedt C, Handa H (2018) Achieving Long-Term Biocompatible Silicone via Covalently Immobilized S-Nitroso-N-acetylpenicillamine (SNAP) That Exhibits 4 Months of Sustained Nitric Oxide Release. ACS Applied Materials & Interfaces 10:27316–27325.
37. Major TC, Brant DO, Burney CP, Amoako KA, Annich GM, Meyerhoff ME, Handa H, Bartlett RH (2011) The hemocompatibility of a nitric oxide generating polymer that catalyzes S-nitrosothiol decomposition in an extracorporeal circulation model. Biomaterials 32:5957–5969.
38. Wang Y-B, Gong M, Yang S, Nakashima K, Gong Y-K (2014) Hemocompatibility and film stability improvement of crosslinkable MPC copolymer coated polypropylene hollow fiber membrane. Journal of Membrane Science 452:29–36.
39. Ye S-H, Arazawa DT, Zhu Y, Shankarraman V, Malkin AD, Kimmel JD, Gamble LJ, Ishihara K, Federspiel WJ, Wagner WR (2015) Hollow Fiber Membrane Modification with Functional Zwitterionic Macromolecules for Improved Thromboresistance in Artificial Lungs. Langmuir 31:2463–2471.
40. Huang X, Wang W, Zheng Z, Fan W, Mao C, Shi J, Li L (2016) Surface monofunctionalized polymethyl pentene hollow fiber membranes by plasma treatment and hemocompatibility modification for membrane oxygenators. Applied Surface Science 362:355–363.

41. Malkin AD, Ye S-H, Lee EJ, Yang X, Zhu Y, Gamble LJ, Federspiel WJ, Wagner WR (2018) Development of zwitterionic sulfobetaine block copolymer conjugation strategies for reduced platelet deposition in respiratory assist devices. Journal of Biomedical Materials Research Part B: Applied Biomaterials 106:2681–2692.
42. Ukita R, Wu K, Lin X, Carleton NM, Naito N, Lai A, Do-Nguyen CC, Demarest CT, Jiang S, Cook KE (2019) Zwitterionic poly-carboxybetaine coating reduces artificial lung thrombosis in sheep and rabbits. Acta Biomaterialia 92:71–81.
43. Goh, E., Wong, E., Farhatnia, Y., Tan, A. and Seifalian, A., (2014) Accelerating in Situ Endothelialisation of Cardiovascular Bypass Grafts. International Journal of Molecular Sciences, 16(1), pp. 597–627
44. Klein S, Hesselmann F, Djeljadini S, Berger T, Thiebes AL, Schmitz-Rode T, Jockenhoevel S, Cornelissen CG (2020) EndOxy: Dynamic Long-Term Evaluation of Endothelialized Gas Exchange Membranes for a Biohybrid Lung. Annals of Biomedical Engineering 48:747–756.
45. Hellmann A, Klein S, Hesselmann F, Djeljadini S, Schmitz-Rode T, Jockenhoevel S, Cornelissen CG, Thiebes AL (2020) EndOxy: Mid-term stability and shear stress resistance of endothelial cells on PDMS gas exchange membranes. Artificial Organs 44:E419–E433.
46. Pflaum, M., Dahlmann, J., Engels, L., Naghilouy-Hidaji, H., Adam, D., Zöllner, J., Otto, A., Schmeckebier, S., Martin, U., Haverich, A., Olmer, R. and Wiegmann, B., (2021) Towards Biohybrid Lung: Induced Pluripotent Stem Cell Derived Endothelial Cells as Clinically Relevant Cell Source for Biologization. Micromachines, 12:981
47. Pflaum M, Kühn-Kauffeldt M, Schmeckebier S, Dipresa D, Chauhan K, Wiegmann B, Haug RJ, Schein J, Haverich A, Korossis S (2017) Endothelialization and characterization of titanium dioxide-coated gas-exchange membranes for application in the bioartificial lung. Acta Biomaterialia 50:510–521.
48. Liu, T., Liu, S., Zhang, K., Chen, J. and Huang, N., (2013) Endothelialization of implanted cardiovascular biomaterial surfaces: The development from in vitro to in vivo Journal of Biomedical Materials Research Part A, 102(10), pp. 3754–3772
49. Pang, J., Farhatnia, Y., Godarzi, F., Tan, A., Rajadas, J., Cousins, B. and Seifalian, A., (2015) In situ Endothelialization: Bioengineering Considerations to Translation. Small, 11(47), pp. 6248–6264
50. Astor TL, Borenstein JT (2022) The microfluidic artificial lung: Mimicking nature's blood path design to solve the biocompatibility paradox. Artificial Organs 46:1227–1239.
51. Lachaux J, Hwang G, Arouche N, Naserian S, Harouri A, Lotito V, Casari C, Lok T, Menager JB, Issard J, Guihaire J, Denis CV, Lenting PJ, Barakat AI, Uzan G, Mercier O, Haghiri-Gosnet A-M (2021) A compact integrated microfluidic oxygenator with high gas exchange efficiency and compatibility for long-lasting endothelialization. Lab on a Chip 21:4791–4804.
52. Santos JA, Gimbel AA, Peppas A, Truslow JG, Lang DA, Sukavaneshvar S, Solt D, Mulhern TJ, Markoski A, Kim ES, Hsiao JC-M, Lewis DJ, Harjes DI, DiBiasio C, Charest JL, Borenstein JT (2021) Design and construction of three-dimensional physiologically-based vascular branching networks for respiratory assist devices. Lab on a Chip 21:4637–4651.
53. Santos, J., Vedula, E., Lai, W., Isenberg, B., Lewis, D., Lang, D., Sutherland, D., Roberts, T., Harea, G., Wells, C., Teece, B., Karandikar, P., Urban, J., Risoleo, T., Gimbel, A., Solt, D., Leazer, S., Chung, K., Sukavaneshvar, S., Batchinsky, A. and Borenstein, J., (2021) Toward Development of a Higher Flow Rate Hemocompatible Biomimetic Microfluidic Blood Oxygenator. Micromachines, 12(8), p. 888
54. Dabaghi M, Saraei N, Fusch G, Rochow N, Brash JL, Fusch C, Ravi Selvaganapathy P (2019) An ultra-thin, all PDMS-based microfluidic lung assist device with high oxygenation capacity. Biomicrofluidics 13:034116.
55. Kniazeva T, Hsiao JC, Charest JL, Borenstein JT (2011) A microfluidic respiratory assist device with high gas permeance for artificial lung applications. Biomedical Microdevices 13:315–323.

# Correction to: Lung Development in a Dish: Models to Interrogate the Cellular Niche and the Role of Mechanical Forces in Development

**Brea Chernokal, Cailin R. Gonyea, and Jason P. Gleghorn**

## Correction to: Chapter 3 in: C. M. Magin (ed.), *Engineering Translational Models of Lung Homeostasis and Disease*, Advances in Experimental Medicine and Biology 1413, https://doi.org/10.1007/978-3-031-26625-6_3

Chapter 3 "Lung Development in a Dish: Models to Interrogate the Cellular Niche and the Role of Mechanical Forces in Development" was previously published as open access. It has now been changed to non-open access. The book has also been updated with these changes.

The updated original version of this chapter can be found at
https://doi.org/10.1007/978-3-031-26625-6_3



C. M. Magin (ed.), *Engineering Translational Models of Lung Homeostasis and Disease*, Advances in Experimental Medicine and Biology 1413,
https://doi.org/10.1007/978-3-031-26625-6_17

# Correction to: Multipotent Embryonic Lung Progenitors: Foundational Units of In Vitro and In Vivo Lung Organogenesis

**Laertis Ikonomou, Maria Yampolskaya, and Pankaj Mehta**

## Correction to:
## Chapter 4 in: C. M. Magin (ed.), *Engineering Translational Models of Lung Homeostasis and Disease*, Advances in Experimental Medicine and Biology 1413, https://doi.org/10.1007/978-3-031-26625-6_4

Chapter 4 "Multipotent Embryonic Lung Progenitors: Foundational Units of In Vitro and In Vivo Lung Organogenesis" was previously published as non-open access. It has now been converted to open access. The book has also been updated with these changes.

The updated original version of this chapter can be found at
https://doi.org/10.1007/978-3-031-26625-6_4



C. M. Magin (ed.), *Engineering Translational Models of Lung Homeostasis and Disease*, Advances in Experimental Medicine and Biology 1413,
https://doi.org/10.1007/978-3-031-26625-6_18

MIX
Papier aus verantwortungsvollen Quellen
Paper from responsible sources
FSC® C105338

If you have any concerns about our products,
you can contact us on
**ProductSafety@springernature.com**

In case Publisher is established outside the EU,
the EU authorized representative is:
**Springer Nature Customer Service Center GmbH**
**Europaplatz 3, 69115 Heidelberg, Germany**

Printed by Libri Plureos GmbH
in Hamburg, Germany

MIX
Papier aus verantwortungsvollen Quellen
Paper from responsible sources
FSC® C105338

If you have any concerns about our products,
you can contact us on
**ProductSafety@springernature.com**

In case Publisher is established outside the EU,
the EU authorized representative is:
**Springer Nature Customer Service Center GmbH**
**Europaplatz 3, 69115 Heidelberg, Germany**

Printed by Libri Plureos GmbH
in Hamburg, Germany